EYEWITNESS TRAVEL

THE GREEK ISLANDS

EYEWITNESS TRAVEL

THE GREEK
ISLANDS

Main Contributor **Marc Dubin**

DK

LONDON, NEW YORK,
MELBOURNE, MUNICH AND DELHI
www.dk.com

Project Editor Jane Simmonds
Art Editor Stephen Bere
Editors Isabel Carlisle, Michael Ellis, Simon Farbrother,
Claire Folkard, Marianne Petrou, Andrew Szudek
Designers Jo Doran, Paul Jackson, Elly King, Marisa Renzullo
Map Co-ordinators Emily Green, David Pugh
Visualizer Joy Fitzsimmons
Language Consultant Georgia Gotsi

Contributors and Consultants
Rosemary Barron, Marc Dubin, Stephanie Ferguson, Carole French, Mike Gerrard,
Andy Harris, Lynette Mitchell, Colin Nicholson, Robin Osborne, Barnaby Rogerson,
Paul Sterry, Tanya Tsikas

Maps
Gary Bowes, Fiona Casey, Christine Purcell (ERA-Maptec Ltd)

Photographers
Max Alexander, Joe Cornish, Paul Harris, Rupert Horrox,
Rob Reichenfeld, Linda Whitwam, Francesca Yorke

Illustrators
Stephen Conlin, Steve Gyapay, Maltings Partnership, Chris Orr & Associates,
Mel Pickering, Paul Weston, John Woodcock

Printed and bound in China

First published in Great Britain in 1997
by Dorling Kindersley Limited
80 Strand, London WC2R 0RL, UK

15 16 17 18 10 9 8 7 6 5 4 3 2 1

**Reprinted with revisions 1998, 1999, 2000, 2001,
2002, 2003, 2004, 2006, 2007, 2009, 2011, 2013, 2015**

Copyright 1997, 2015 © Dorling Kindersley Limited, London
A Penguin Random House Company

ISBN: 978-0-2411-8132-4

MIX
Paper from
responsible sources
FSC™ C018179

Front cover main image: The characteristic white rooftops of Oia in Santorini

◀ Préveli Beach, on the island of Crete

Mólyvos harbour, Lésvos

Contents

How to Use this Guide **6**

The Turkish Prince Cem arriving in Rhodes
(15th century)

Introducing the Greek Islands

Roman horse head in
Archaeological Museum

Ancient Greece

Almond biscuits eaten at
Christmas and Easter

The Greek Islands Area by Area

Basket of herbs and spices from a market
stall in Irákleio, Crete

Travellers' Needs

Kámpos beach on Ikaría in the Northeast
Aegean Islands

Survival Guide

Néa Moní on Chíos,
Northeast Aegean Islands

HOW TO USE THIS GUIDE

This guide helps you to get the most from your visit to the Greek Islands. *Introducing the Greek Islands* maps the country in its historical and cultural context, including a quick comparison chart with *Choosing Your Island*. *Ancient Greece* gives a background to the many remains and artifacts to be seen.

The seven regional chapters, plus *A Short Stay in Athens,* describe important sights, with maps and illustrations. Restaurant and hotel recommendations can be found in *Travellers' Needs*. The *Survival Guide* has tips on everything from the Greek telephone system to transport networks.

The Greek Islands Area by Area

The islands have been divided into six groups, each of which has a separate chapter. Crete has a chapter on its own. A map of these groups can be found inside the front cover of the book. Each island group is colour-coded for easy reference.

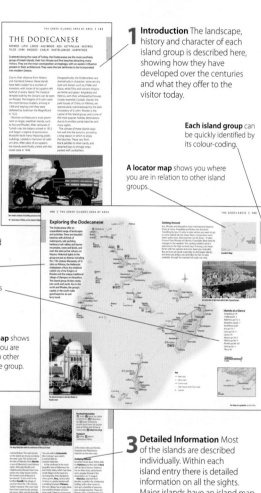

1 Introduction The landscape, history and character of each island group is described here, showing how they have developed over the centuries and what they offer to the visitor today.

Each island group can be quickly identified by its colour-coding.

A locator map shows you where you are in relation to other island groups.

2 Regional Map This shows all the islands covered in the chapter. Main ferry routes are marked and there are useful tips on getting around the islands.

Islands at a Glance lists the islands alphabetically. Each island has a cross-reference to its entry.

The main ferry routes, roads and transport points are marked on each map.

A locator map shows you where you are in relation to islands in the group.

3 Detailed Information Most of the islands are described individually. Within each island entry there is detailed information on all the sights. Major islands have an island map showing all the main towns, villages, sights and beaches.

Story boxes highlight special or unique aspects of a particular sight.

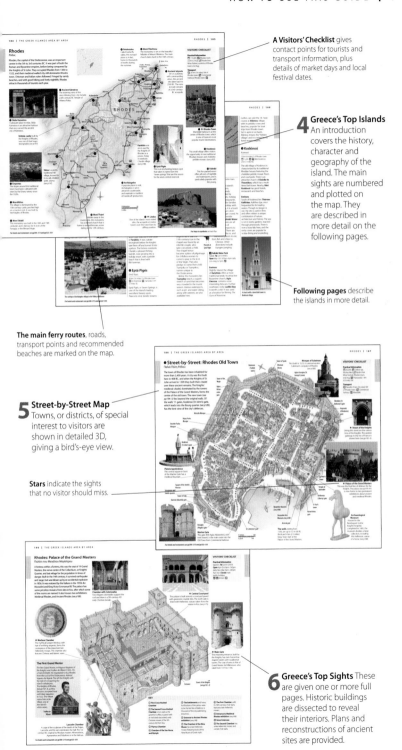

A Visitors' Checklist gives contact points for tourists and transport information, plus details of market days and local festival dates.

4 Greece's Top Islands An introduction covers the history, character and geography of the island. The main sights are numbered and plotted on the map. They are described in more detail on the following pages.

Following pages describe the islands in more detail.

The main ferry routes, roads, transport points and recommended beaches are marked on the map.

5 Street-by-Street Map Towns, or districts, of special interest to visitors are shown in detailed 3D, giving a bird's-eye view.

Stars indicate the sights that no visitor should miss.

6 Greece's Top Sights These are given one or more full pages. Historic buildings are dissected to reveal their interiors. Plans and reconstructions of ancient sites are provided.

INTRODUCING THE GREEK ISLANDS

DISCOVERING THE GREEK ISLANDS

The following tours have been designed to take in as many of the country's highlights as possible, while keeping long-distance travel to a minimum. Our first tour covers two days in Athens, with a trip to Salamína, one of the picturesque Argo-Saronic Islands. We then spend three day in Rhodes, followed by four days in Crete, Greece's southernmost island. These itineraries can be followed individually or combined to form a longer city break-style tour. Next come three island-hopping tours, covering the Ionian Islands; the Sporades, Evvoia and the Northeast Aegean Islands; and the Cyclades and Dodecanese Islands. Extra suggestions are provided for those who want to extend their stay. Pick, combine and follow your favourite tours, or simply dip in and out and be inspired.

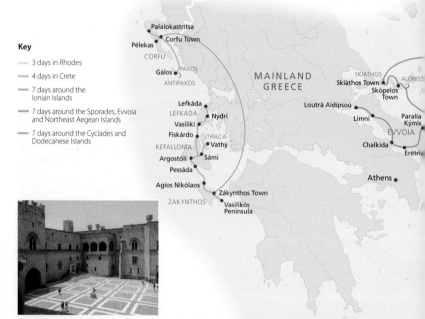

Key

— 3 days in Rhodes

— 4 days in Crete

— 7 days around the Ionian Islands

— 7 days around the Sporades, Evvoia and Northeast Aegean Islands

— 7 days around the Cyclades and Dodecanese Islands

The Palace of the Grand Masters in Rhodes Old Town

Island Hopping – 7 days around the Ionian Islands

- Take a boat trip from **Nydrí** on Lefkáda to the glorious island of **Meganísi**.

- Admire Kefalloniá's spectacular **Melissáni Cave-Lake** and **Drogkaráti Cave**.

- Explore the architecture and tiny lanes of **Argostóli**, Kefalloniá's capital.

- Relax on board ferries between **Lefkáda** and **Kefalloniá**, or Kefalloniá and **Zákynthos**.

- Head for atmospheric **Corfu Old Town** and dine on traditional Corfiot specialities such as *biánko* fish stew and *pastitsáda*.

- Don't miss **Gáïos** on Paxos, a lively, picturesque place with two harbours, Venetian houses and a welcoming community.

- Get a taste of rural life on the laid-back island of **Antípaxos**, which is home to just 60 people.

◀ Dolphin fresco in the Queen's Megaron at the Palace of Knosós

7 days around the Cyclades and Dodecanese Islands

- Admire the beauty of brilliant white-and-blue-washed **Santoríni** and the views from Firá and Oía.

- Visit the traditional Greek island of **Síkinos** – its natural beauty will leave a lasting impression.

- Admire the giant abandoned *koúroi* in **Náxos'** lush Mélanes Valley.

- Imagine how life was in ancient times at **Delos**, one of Greece's most important archaeological sites.

- Explore the picturesque port-resort of **Náousa** or the hill-village of **Léfkes**, on Páros.

- Wander around the **Palace of the Grand Masters** and the Street of the Knights in **Rhodes Old Town**.

- Look out for the Dodecanese islands of **Sými**, **Tílos** and **Nísyros** when crossing from Rhodes to Kos.

- Scramble around **Kos Town's** Castle of the Knights, or relax on one of the island's beautiful beaches.

- Marvel at the Holy Cave of the Apocalypse and the Monastery of St John on **Pátmos**.

| 0 km | 50 |
| 0 miles | 50 |

7 days around the Sporades, Evvoia and Northeast Aegean Islands

- Be captivated by the beauty of **Loutrá Aidipsoú**, a picturesque *belle époque* spa in northern Evvoia.

- Explore the remains of a once prosperous city at **Ancient Erétria**.

- Admire the cube-shaped homes on **Skýros**, in the Sporades, and enjoy the island's tranquillity as darkness falls.

- Stroll along the time-warped lanes of **Skópelos Town**.

- Marvel at the *kalývia* farmhouses set in rural **Skópelos**.

- Soak up the atmosphere of the medieval mastic villages of **Chíos**, and admire the nearby **Néa Moní** monastery, with its mosaics.

- Experience the spectacular sunset from western **Lésvos**.

- Relax on the ferry crossing from Linaria to Alónissos and skirting the **Sporades Marine Park**.

The characteristic blue and white rooftops of **Santorini**

2 days in Athens

- **Arriving** Elefthérios Venizélos International Airport lies around 27 km (17 miles) from Athens' city centre. A metro service connects the airport with Plateía Syntágmatos and Monastiráki; express bus line X95 also calls at Plateía Syntágmatos. Taxis operate from outside the terminal. Athens can also be reached by road and rail, and by boat via the port of Piraeus.

- **Transport** The main sights are all within easy walking distance; the city also has an efficient bus network.

- **Booking ahead** Many museums are closed on Mondays; the Benáki is closed on Tuesdays, too.

Day 1
Morning Start the day at the **Acropolis** (pp292–3). As you follow the winding path up from the entrance, stop to admire the Theatre of Herodes Atticus, built between AD 161 and 174. Ascend the steps to the Propylaia, taking a moment to admire the Temple of Athena Nike, before seeing the mighty Parthenon ahead. Later, make your way back down to the pedestrianized Dionysiou Areopagitou, turn right and follow it downhill for a selection of lunchtime restaurants along Adrianoú.

Afternoon Head for **Pláka** (p291). The historic heart of Athens is a labyrinth of tiny, picturesque lanes full of old buildings housing cafés and

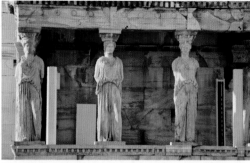
The Porch of the Caryatids at the Acropolis, Athens

souvenir shops. Look out for the beautiful Agios Nikólaos Ragavás, which dates from the 11th century. For dinner, head for one of the many tavernas in the buzzing district of **Psyrrí** (p290).

Day 2
Morning Make for **Monastiráki** (p290), famous for its flea market, the nearby ruins of the **Ancient Agora** (p291) and the Roman Forum. Among the ruins at the Forum is the octagonal Tower of the Winds. Continue east along Ermoú to Plateía Syntágmatos for metro transport to Piraeus and a short ferry ride to Paloúkia, on **Salamína** (p100).

Afternoon Head to Salamína Town, with its choice of tavernas, and on to the enchanting 17th-century monastery of Faneroménis. Trace your steps back to Paloúkia for your return to Athens.

> **To extend your trip…**
> Explore the other **Argo-Saronic Islands**: **Aígina** and its splendid Temple of Aphaia, laid-back **Póros**, **Ydra** and **Spétses**.

3 days in Rhodes

- **Arriving** Diagoras Airport is located in Paradísi, around 15 km (10 miles) from Rhodes Town. A municipal bus to Rhodes Town operates all day and into the night; taxis can be found outside the terminal.

- **Transport** The main sights of Rhodes Town are all within walking distance, while the island is best explored by car.

- **Booking ahead** Some remote archaeological sites are closed on Mondays.

Day 1
Morning Start your day in **Rhodes Old Town** (pp186–9). Visit the 14th-century **Palace of the Grand Masters** (pp190–91) and stroll down charming Odós Ippotón, also known as the **Street of the Knights** (pp192–3).

Afternoon After lunch, head to the **Archaeological Museum** (p188), housed in the Knights' Gothic-style former hospital. Next, relax with a little people-watching in **Plateía Ippokrátous** (p186), just inside the imposing Marine Gate, before dinner.

Day 2
Morning While away an hour or so at **Mandráki Harbour** (p194), in Rhodes New Town, before taking the coastal road to explore the western side of the island. The first stop is **Ancient Ialyssós** (p196), with its Temple of Athena Polias and Zeus Poliefs.

View across the bay towards Salamína Town on Salamína, in the Saronic Islands

For practical information on travelling around the Greek islands, see pp356–63

Adjacent is **Moní Filerímou** (*p196*), a monastery with an ancient mosaic floor in one corner. Take in the views from the top of the metal crucifix at the end of the nearby Via Crucis. Continue southwest, then slightly inland, to reach picturesque **Petaloúdes** (*p196*).

Afternoon See the nearby remains of **Ancient Kámeiros** (*p196*), then continue along the coast to the scenic village of **Skála Kameírou** (*p197*). Have a late lunch at a taverna, then explore the ruins of Kritinía castle, with superb views west to Alimiá and Chálki. As you head for the mountains, the road will begin to wind, passing through Siána and on to **Monólithos** (*p197*). Here, the dramatic coastline and castle perched atop a massive rock make for a breathtaking view. Head back to Rhodes Town.

Day 3
Morning Explore eastern Rhodes. Follow the coastal road through Kalithéa resort, after which you will see the turnoff to **Thérmes Kalithéas** (*p199*), a former Art Deco spa now housing a small museum. Continue on for another 16 km (10 miles) to **Eptá Pigés** (*pp198–9*), one of Rhodes' most magical sights and a great place to unwind. **Stegná** (*p198*) makes an ideal lunch stop; en route you will pass the turning for **Moní Tsampíkas** (*p199*), set high on a mountain and well worth the detour for the superb views.

Afternoon Explore the village of **Líndos** (*pp200–201*), with its white houses and ancient acropolis. After relaxing on Megálos Gialós beach, head to Mavrikos Restaurant (*p328*), one of the best on the island, for your final meal on Rhodes.

To extend your trip...
Rhodes' painted medieval chapels feature priceless frescos. The best are in **Askilipieío** and **Moní Thárri** (*p198*), as well as in **Panagía Katholikí**, on the road to Afántou (*p199*).

The Byzantine Moní Agias Triádas

4 days in Crete

- **Arriving** There is an airport in Irákleio and one in Chaniá. Buses connect them with their respective towns. There are also ferries from Piraeus. Taxis operate from outside both airports and seaports.

- **Transport** Crete is best explored by car.

- **Booking ahead** Some museums close on Monday.

Day 1
Explore the town of **Réthymno** (*pp262–3*), see its Fortétsa and visit its archaeological museum. Later, follow the coastal road to **Chaniá** (*pp256–7*), which has a picturesque old quarter and fine examples of 19th-century indigenous architecture. In the afternoon, drive to the **Akrotíri Peninsula** (*p255*) to see the monastery of Moní Agias Triádas.

Day 2
Start the day at the spectacular **Samariá Gorge** (*pp258–9*). Join a full-day walking tour of the gorge from Xylóskalo to Agía Rouméli, followed by a boat trip to **Sfakiá** (*p263*). There is also a shorter circular route at Xylóskalo, which can be followed by a drive through scenic countryside to Sfakiá. From here, take the road via **Frangokástello** (*p263*) to **Plakiás** (*p264*) or Réthymno.

Day 3
From either Plakiás or Réthymno, it's a short drive to **Agía Galíni** (*p267*), beyond which is the road junction for **Agía Triáda** (*p267*) and **Phaestos** (*pp270–71*). Explore these archaeological sites where the Minoan people once lived. Stop for lunch and a swim in **Mátala** (*p268*), near Phaestos, then head to Irákleio, calling into **Górtys** (*p268–9*), the island's Roman capital, on the way. Spend the evening in **Irákleio** (*pp272–3*), visiting its **Archaeological Museum** (*pp274–5*) or enjoying the nightlife.

Day 4
The **Palace of Knosós** (*pp276–9*), 5 km (3 miles) south of Irákleio, is one of the world's greatest archaeological sites, dating from around 1900 BC. After lunch, take the busy coastal highway to **Agios Nikólaos** (*p282*). This delightful town enjoys a superb setting overlooking Mirabéllou Bay and offers a selection of waterside *mezedopoleía* for your last evening on Crete.

To extend your trip...
Explore a smaller island by taking a boat trip from **Ierápetra** to Chrysí or from **Palaióchora** to Gávdos.

The fascinating Archaeological Museum in Rhodes Old Town

7 days around the Ionian Islands

- **Arriving** The Aktion National Airport in Préveza serves Lefkáda, while the islands of Corfu, Kefaloniá and Zákynthos all have international airports. Buses and taxis connect the airports with all areas of their respective islands.
- **Transport** The islands are best explored by car, scooter, bicycle or taxi.

Day 1
Arrive on **Lefkáda** *(p89)* by road from Préveza, pausing to admire the Sánta Mávra Fortress. Follow the coastal road to Nydrí, and take a short boat trip to pretty Meganísi. Back in Nydrí, drive through rugged countryside to the village of Kalamítsi, before turning south towards Vasilikí, where accommodation and dining options are plentiful.

Day 2
Take the morning ferry from Vasilikí to Fiskárdo in **Kefalloniá** *(pp92–3)*, a 2-hour crossing. Stroll around Fiskárdo's harbour, stop for lunch in unspoiled Asos and have a swim in Mýrtou Bay. Cross the island to the Melissáni Cave-Lake and Drogkaráti Cave, two wonders of nature. Finally, head to the capital, Argostóli, where you will spend the night.

Day 3
Head for Sámi and hop aboard a ferry to Pisaetós on **Ithaca** *(p90)*. The 45-minute journey allows just enough time to explore the island, charming Vathý and the bustling hill-village of Stavrós. Take a ferry back to Sámi and enjoy another night in Argostóli.

Day 4
Take the ferry from Pessáda to Agios Nikólaos on **Zákynthos** *(pp94–5)*, a 90-minute crossing. (Off-season, you'll have to sail from Argostóli to Zákynthos via Kyllíni, on the Peloponnese.) Zákynthos has superb beaches on the Vasilikós peninsula, charming westerly hill-villages, and the Blue Caves in the north. Spend the night near the airport.

Day 5
Take a flight from Zákynthos to **Corfu** *(pp76–87)* via Kefaloniá. Explore northern and western Corfu, especially Palaiokastrítsa's stunning topography, the beaches below Pélekas and the idiosyncratic Achílleion Palace, south of the airport. Spent the night in Corfu Old Town.

Day 6
Take a ferry from Corfu Town via Igoumenítsa to Gáïos on **Paxós** *(p88)*. Gáïos is a lively place with two harbours, Venetian houses and a welcoming community. Take a trip across the island, stopping at picturesque Longós, with its flanking beaches, and busy Lákka in the far northwest.

Day 7
Take a 20-minute boat ride from Gáïos to the laid-back island of **Antípaxos** *(p88)*. Unwind here for a few hours before the onward journey via Corfu Town.

7 days around the Sporades, Evvoia and Northeast Aegean Islands

- **Arriving** Chalkída, on Evvoia, is easily reached by motorway from Athens International Airport.
- **Transport** All of these islands are best reached by ferry or plane, and explored by road or on foot.

Day 1
Spend the morning on the atmospheric waterfront of **Chalkída** *(p124)*, on **Evvoia** *(pp122–7)*, before exploring Greece's second-largest island. Head north to picturesque **Límni** *(p127)* until you reach the fishing harbour of **Loutrá Aidipsoú** *(p127)*, where you can enjoy a *mezédes* lunch and a stroll. Spend the night at a spa hotel in Loutrá Aidipsoú.

Day 2
Return to Chalkída, then head south to Erétria and the site of **Ancient Erétria** *(p125)*, with its unmissable archaeological museum. After lunch in modern Erétria or nearby Amárynthos, make for Paralía Kýmis, via Lépoura, where you can take the evening ferry to Linariá on Skýros (about 90 minutes).

Day 3
Explore the island of **Skýros** *(pp120–21)*, including its scenic capital Skýros Town, with its unusual cube-shaped homes and two museums. There are good beaches near the town, just to the north. Spend another night on Skýros.

Day 4
In summer, there's an early-morning ferry (three times a week) from Linariá to Paralía Kýmis, which then continues to **Alónnisos** *(p118)* and **Skópelos** *(pp116–17)*. It's a lengthy crossing (about 6.5 hours to Skópelos), but it gives you a full afternoon to explore this green, unspoiled island where you will overnight. Spend the remaining daylight at one of the excellent

The stunning Blue Caves on the north of Zákynthos island

For practical information on travelling around the Greek islands, see pp356–63

beaches, then pass the evening wandering around its charming capital, Skópelos Town.

Day 5
Take a ferry to **Skiáthos** *(pp112–13)* and stroll around its Old Town, dominated by the Trión Ierarchón and Panagía Limniá churches. Later, fly to Athens and onward to the Northeast Aegean island of **Chíos** *(pp150–57)*, your overnight stop.

A traditional Greek village scene

Day 6
Wake up early to enjoy sunrise on Chíos Town, then hire a car or taxi to tour the medieval **mastic villages** *(pp152–3)* and Byzantine **Néa Móni** *(pp154–5)*, with its superb mosaics. Spend the afternoon at the beach and the evening at a town taverna, before an early bedtime.

Day 7
Rise very early for a dawn ferry to **Lésvos** *(pp140–49)*. Arrive in time for breakfast in **Mytilíni** port town *(p142)*, then head to the town's archaeological museum, with its Roman mosaics. Next, pick up a hire car to explore this big island. Highlights include **Agiásos** *(p144)*, **Mólyvos** *(p145)* and **Sígri** *(p149)*. Spend your final night at one of these last two resorts. From either, there are glorious sunsets to watch.

To extend your trip...
Spend some time on sandy, volcanic **Límnos** *(pp138–9)*, which is easily reached from Lésvos by ferry or plane.

---

7 days around the Cyclades and Dodecanese Islands

- **Arriving** International and domestic flights arrive into Mýkonos and Santoríni airports. Taxis operate outside the terminal. Rhodes and Kos airports, each with bus connections to the islands' capitals, are the main gateways to the Dodecanese.

- **Transport** The Cyclades and Dodecanese islands are linked by frequent ferries; the islands are best explored by hire car, taxi or scooter.

Day 1
Mýkonos *(pp218–19)* is an alluring introduction to the Cycladic Islands. After exploring the main town, spend a leisurely afternoon on one of the famous beaches on the south coast. Overnight on the island.

Day 2
Take the first excursion boat to **Delos** *(pp222–3)*, one of the most important sites in Greece. Return in time to catch the afternoon catamaran to **Náxos Town** *(p234)*, which offers an unusual Greek Catholic cathedral and a good archaeological museum. If you have any daylight left, explore **Náxos island** *(pp234–5)*, including the Mélanes Valley, with its two *koúroi*, and the hill-village of **Apeíranthos** *(p236)*. Spend the night in Náxos Town.

Day 3
Take a morning ferry to **Páros** *(pp230–33)*, a 90-minute crossing. Hire a scooter or a car, and you'll have time to visit pretty **Náousa** *(p232)* and **Léfkes** *(p232)*, inland. Next, head for **Paroikiá** *(p230)*, with its imposing Byzantine cathedral of Ekantontapyliani.

Day 4
Take a ferry from Paroikiá to **Santoríni** *(pp242–5)*. Its capital **Firá** *(p242–3)*, which looks out over the caldera, boasts the excellent Prehistoric Museum. At **Oía** *(p244)* you can enjoy spectacular views, particularly at sunset. Overnight in Firá or Oía.

Day 5
Take a pre-dawn ferry (typically 3 days weekly in season) from Santoríni to **Rhodes** *(pp184–201)*, arriving at around noon. Explore **Rhodes Old Town** *(pp186–9)*, making a beeline for the **Palace of the Grand Masters** *(pp190–91)* and the delightful **Street of the Knights** *(pp192–3)*. With any daylight remaining, head down the east coast to **Líndos** *(pp200–201)*.

Day 6
Catch the catamaran from Rhodes Town to Kos Town, on **Kos** *(pp174–7)*, a 2.5-hour crossing. Once ashore, take in the Castle of the Knights, the Ancient Agora and the archaeological museum (due to reopen in 2015–16). Later, relax on one of Kos' beaches. Watch the sunset from Ziá, one of the **Asfendioú villages** *(p176)*, before heading back Kos Town.

Day 7
Hop aboard the catamaran to **Skála** *(p166)*, on **Pátmos** *(pp166–9)*. After lunch in Skála, explore the island, with its Holy Cave of the Apocalypse and the **Monastery of St John** *(pp168–9)*. Spend the afternoon on one of Pátmos' excellent beaches, then return to Skála for the night.

To extend your trip...
The Cycladic islands of Tínos, Andros and Sýros can be reached from Mýkonos; Santoríni is a good base for trips to Folígandros, Sífnos and Mílos. Kárpathos, Chálki, Sými, Nísyros and Tílos, in the Dodecanese, are easily accessed from Rhodes or Kos.

Venetía, or Little Venice, in Mýkonos Town

Choosing Your Island

One great appeal of the Greek Islands is the sheer variety of attractions and activities on offer. Choosing the right island for the type of holiday you want – whether it be action-packed, historical or lazy (or a combination) – can be a bewildering decision, however. This chart gives a quick reference point to the strengths, charms and facilities of each island covered in this guide.

Key

- 🟦 The Ionian Islands
- 🟦 The Argo-Saronic Islands
- 🟦 The Sporades and Evvoia
- 🟦 The Northeast Aegean Islands
- 🟦 The Dodecanese
- 🟦 The Cyclades
- 🟦 Crete

Kayaks for hire in Skiáthos, the Sporades and Evvoia

Key

★ Excellent

• Available

	Diving	Snorkelling	Day Trips	Watersports
The Ionian Islands				
Corfu (see pp76–87)	•	•	★	★
Paxós (see p88)				
Lefkáda (see p89)	•	•	★	★
Ithaca (see pp90–91)				
Kefalloniá (see pp92–93)	•	•	•	•
Zákynthos (see pp94–5)	•	•		
The Argo-Saronic Islands				
Salamína (see p100)			•	
Aígina (see pp100–103)	•			
Póros (see p104)	•			
Ýdra (see pp104–5)	•			
Spétses (see p105)	•			•
Kýthira (see pp106–7)	•			
The Sporades and Evvoia				
Skiáthos (see pp112–13)	•	•		•
Skópelos (see pp116–17)	•	•		•
Alónnisos (see p118)				
Skýros (see pp120–21)				
Evvoia (see pp122–27)	•	•	•	•
The Northeast Aegean Islands				
Thásos (see pp132–35)	•	•		
Samothráki (see pp136–7)	•			
Límnos (see pp138–95)	•			
Lésvos (see pp140–9)	•			•
Chíos (see pp150–57)	•			•
Ikaría (see p157)				
Sámos (see pp158–61)	•		•	•
The Dodecanese				
Pátmos (see pp166–9)	•	•	•	•
Lipsí (see p170)				
Léros (see pp170–71)	•			
Kálymnos (see pp172–3)	•	•	•	•
Kos (see pp174–7)	•		•	•
Astypálaia (see p178)				
Nísyros (see pp178–80)				
Tílos (see p181)	•			
Sými (see pp182–3)		•	•	
Rhodes (see pp184–201)	•	•	★	•
Chálki (see pp202–3)				
Kastellórizo (see p203)		★		
Kárpathos (see pp206–7)	•			•
The Cyclades				
Andros (see pp212–15)	•			
Tínos (see pp216–17)	•			
Mýkonos (see pp218–19)	•	•		•
Delos (see pp222–3)				
Syros (see pp224–7)	•			
Kéa (see p227)	•			
Kythnos (see p228)	•			
Sérifos (see pp228–9)	•			•
Sífnos (see p229)		•		
Páros (see pp230–33)	★	★	•	•
Náxos (see pp234–7)	•			
Amorgós (see p237)	•	•		
Íos (see p238)	•			
Síkinos (see pp238–9)				
Folégandros (see p239)				
Mílos (see pp240–41)	•			
Santoríni (see pp242–5)	•	•	•	•
Crete (see pp248–85)	★	★	★	•

The following geographic labels appear on the map:

Bulgaria

Republic of Macedonia

Albania

Mainland Greece

AEGEAN SEA

Turkey

Camping	Horseriding	Beach Sports	Swimming	Hiking	Fishing	Birdwatching	Sailing/Boat Hire	Marine Life	Cycling	Kayaking	Golf	Archaeology	Family Beaches	Nudist Beaches	Nightlife	Unspoiled Islands	Gay Scene	Restaurants	Wine	Shopping	Scenic	Spa/Luxury Hotel	Art Scene	Museums	Major Airport
★	•	•	•	★	•	•	•		•	•	★	•	★	•	★			★	•	•	•	•	•	•	•
			•							•			•	•											
	•		•	•	★		•		•	•		•		•			•								
•	•	•	•	•			•		•	•			•	•	•				•		•				
•		•	•				•	•					•	•	•										
		•																							
•											•							•							
									•				•		•	•	•		•				★		
		•	•	•			•		•			•									•				
•	•	•	•	•			•		•			•	•	•		•	•		•						
	•	•	•				•		•			•	•												
		•	•		•			•	•				•			•									
•		•	•	•	•	•	•		•			•	•			•		•			•				
•	•	•	•	•			•		•			•	•	•							•				
		•	•				•		•		•		•	•		•			•		•				
•		•	•	•	•		•		•	•		•	•		★			•		•			•	•	•
•		•	•	•			•		•			•	•						•			★		★	•
•		•	•	•			•		•			•	•		•						•			•	•
		•	•	•			•		•			•	•					•		•			•	•	•
•		•	•	•	•		•		•			•	•						•				•		
•			•	•									•												
•			•	•									•												
		•	•	•			•					•	•	•	•										
•		•	•	•			•					•	•	•	•	•									•
•			•											•											
		•	•				•		•					•											
		•	•	•			•		•			•	•	•				•							
		•	•	•	•								•				•								
•	•	•	•	•			•		•	•	•	•	•	•	★		•	•	•	★		•		•	•
		•											•		•											
•		•	•	•	•		•	•	•	•			•		•					★					•
•			•	★					★	•			•	★											
•			•	•									•		•									•	
•		•	•				•					•	★	★		★	•	★	•		★		★	•	•
•											★			★											•
•			•	•			•		•				•		•										•
•			•	•			•			•			•		•										
•			•	•									•		•										
•		•	•	★			•						•		•										
•			•										•				•								
•	•	•	•	•	•		•		•	•		•	•	★		•	•								•
•		•		•					•			•	•	•						•					•
•			•	★								•	★						★						
•		•	•				•					•													
•			•											★											
•		•	•						•	★		•						•			★				•
•		•					•				★		★			★	•	•	★	★	★	•	★	•
★	★	•	•	★	•	★	•	•	★	•	•	★	★	★	★		•	★	★	•	★	★	★	★	•

Putting Greece on the Map

Occupying the southernmost tip of the Balkan peninsula, Greece divides into over 2,000 islands stretching from the Ionian Sea in the west to the Aegean Sea in the east. The mainland has borders with Albania, Bulgaria, Turkey and the Former Yugoslav Republic of Macedonia and is home to most of Greece's 10.9 million people, with a third of these in Athens.

Key

— Motorway

— Major road

— Railway line

---- Ferry route

— National boundary

For keys to symbols *see back flap*

A PORTRAIT OF THE GREEK ISLANDS

Greece is one of the most visited European countries, but also one of the least known. At a geographical crossroads, the modern Greek state dates only from 1830, and combines elements of the Balkans, Middle East and Mediterranean.

Of the thousands of Greek islands, large and small, only about 100 are permanently inhabited today. Around 10 per cent of the country's population of just over 11 million lives on the islands, and for centuries a large number of Greek islanders have lived abroad: currently there are over half as many Greeks outside the country as in. The proportion of their income sent back to relatives significantly bolsters island economies. Recently there has been a trend for reverse immigration, with expatriate Greeks returning home to influence the architecture and cuisine on many islands.

Islands lying within sight of each other can have vastly different histories. Most of the archipelagos along sea lanes to the Levant played a crucial role between the decline of Byzantium and the rise of modern Greece. Crete, the Ionian group and the Cyclades were occupied by the Venetians and exposed to the influence of Italian culture. The Northeast Aegean and Dodecanese islands were ruled by Genoese and Crusader overlords in medieval times, while the Argo-Saronic isles were completely resettled by Albanian Christians.

Island and urban life in contemporary Greece were transformed in the 20th century despite years of occupation and war, including a civil war, which only ended after the 1967–74 colonels' Junta. Recently, based on the revenues from tourism and the EU, there has been

Fishermen mending their nets on Páros in the Cyclades

◀ A whitewashed Cycladic-style house, with bougainvillea

A village café on Crete's Lasíthi Plateau

a rapid transformation of many of the islands from backwater status to prosperity. Until the 1960s most of the Aegean Islands and many of those in the Cyclades and Dodecanese, for example, lacked paved roads and basic utilities. Even larger islands boasted just a single bus and only a few taxis as transport and emigration, either to Athens or overseas, increased.

Religion, Language and Culture

During the centuries of domination by Venetians and Ottomans (see pp44–5) the Greek Orthodox church preserved the Greek language, and with it Greek identity, through its liturgy and schools. The query *Eísai Orthódoxos* (Are you Orthodox?) is virtually synonymous with *Ellinas eísai* (Are you Greek?). Today, the Orthodox Church is still a powerful force, despite the secularizing reforms of the first democratically elected PASOK government of 1981–5. While no self-respecting couple would dispense with church baptisms for their children, civil marriages are now as valid in law as the religious service. Sunday Mass is popular, particularly with women, who often socialize there as men do at *kafeneía* (cafés).

Many parish priests, recognizable by their tall stovepipe hats and long beards, marry and have a second trade (a custom that helps keep up the numbers of entrants to the church). However, there has also been a renaissance in celibate monastic life, perhaps as a reaction to postwar materialism.

Frescoed saint from monastery of St John, Pátmos

The beautiful and subtle Greek language, that other hallmark of national identity, was for a long time a field of conflict between the written *katharévousa*, an artificial form hastily devised around the time of Independence, and the

Traditional houses by the sea on Kefalloniá, the Ionian Islands

Stepped streets and blue rooftops at Oía on Santoríni in the Cyclades

imports, since Greece is one of the very few European countries not to manufacture any of its own.

Greece still bears the hallmarks of a developing economy, with profits from the service sector and agriculture accounting for two-thirds of its GNP. With EU membership since 1981, and an economy that is more capitalist than not, Greece lost its economic similarity to Eastern Europe before the fall of the Iron Curtain. After the turn of the millennium, Greece saw an initial growth in its economy above the EU average and became a member of the EU monetary union, with the euro as its sole currency, in 2002. However, due to uncontrolled government spending resulting in huge debts and the world financial crisis that began in the late 2000s, the Greek economy entered a severe crisis in 2010, resulting in a financial bailout by the International Monetary Fund and EU countries. A second bailout followed. In total, Greece received around €240 billion. There has been social unrest as a result of austerity measures, though riots have been limited to small areas in a few large cities. Although Greece has

slowly evolved everyday speech, or *dimotikí* (demotic Greek).

Today's prevalence of the more supple *dimotikí* was perhaps a foregone conclusion in an oral culture. Storytelling is still as prized in Greece as in Homer's time, with conversation pursued for its own sake in *kafeneía*. The bardic tradition is alive with poet-lyricists such as Mános Eleftheríou, Níkos Gátsos and Apóstolos Kaldáras. Collaborations such as theirs have produced accessible works which have played an important role keeping *dimotikí* alive from the 19th century until today. During times of censorship under past dictatorship or foreign rule, writers and singers have been a vital source of news and information.

A couple riding a motorbike in Kéfalos, on Kos

Development and Diplomacy

Greece's persistent negative trade deficit is aggravated by the large number of luxury goods imported on the basis of *xenomanía* – the belief that goods from abroad are of a superior quality to those made at home. Cars are the most conspicuous of these

A beach at Plakiás on Crete, blessed with crystal-clear waters and a beautiful blue sky

Windmills at Olympos on the island of Kárpathos, in the Dodecanese

been in recession since 2008 and, as a result, has seen its industry sector shrink and high unemployment, in 2014 its government announced that its austerity measures and EU bailouts have put the country back on course for a return to economic growth.

Tourism ranks as the largest hard currency earner, compensating for the depression in world shipping and the fact that Mediterranean agricultural products are duplicated within the EU.

Children dressed for a festival in Koskinoú village, Rhodes

Now the lifeblood of many islands, tourism has only been crucial since the late 1960s. While some of the islands' tourist facilities owe much to a megadevelopment ethos and permit-granting policy formulated under the Junta, subsequent developments have an appearance that is more in harmony with their natural surroundings. In order to attract higher spenders, many farmhouses, cottages and town houses have been renovated since the early 2000s to form luxury and boutique hotels. Yacht marinas and spas have been developed and, increasingly, special-interest tourism has been catered for. The fact that the Greek state is less than 200 years old and in the years since 1922 has been politically unstable means that Greeks have very little faith in government institutions. Everyday life operates on networks of personal friendships and official contacts. The classic political designations of Right and Left only acquired

Threshing with donkeys in the Cyclades

Festival bread from Chaniá's covered market on Crete

their conventional meanings in Greece as late as the 1930s. Among politicians, the dominant figure of the early 20th century was the anti-royalist Liberal Elefthérios Venizélos, who came from Crete. The years since World War II have been overshadowed by two politicians: the late Andréas Papandréou, three times premier as head of the Panhellenic Socialist Movement (PASOK), and the late conservative premier Konstantínos Karamanlís, who died in 1998. The current president of Greece is veteran politician Karolos Papoulias. He has been in office since 2004. The Prime Minister is Alexis Tsipras, leader of the anti-austerity party Syriza, who was elected in January 2015.

Since the end of the Cold War, Greece has been asserting its underlying Balkan identity. Relations with its nearest neighbours, and particularly with Albania, have improved considerably since the fall of the Communist regime there in 1990. Greece is a major investor in neighbouring Bulgaria; and after

Thriving Pythagóreio harbour on the island of Sámos

a rapprochement with Skopje (formerly Yugoslavian Macedonia) in the 1990s, Greece is now a significant regional power.

Home Life

The family is still the basic Greek social unit. Under traditional island land distribution and agricultural practices, one family could sow, plough and reap its own fields, without the help of cooperative work parties. Today's family-run businesses are still the norm, especially among rural communities and in the many port towns. Arranged marriages and granting of dowries, though not very common, persist; most single young people live with their parents or another relative until marriage; and outside the largest university towns, such as Rhodes Town, Irákleio or Mytilíni, few couples dare to cohabit "in sin". Children from the smaller islets board with a relative while attending secondary school on the larger islands. Despite the renowned Greek love of children, Greece has a very low birth rate. Currently, the Greek birth rate stands at less than half of pre-World War II levels.

Fish at Crete's Réthymno market

Macho attitudes persist on the islands and women often forgo any hope of a career in order to look after the house and children. Urban Greek women are seeing a rise in status as imported attitudes have started to creep in. Many now attend university on the mainland or overseas, to train as a teacher or lawyer or for a business career. However, no amount of outside influence is likely to jeopardize the essentially Greek way of life, which remains vehemently traditional.

A man with his donkey in Mýkonos town in the Cyclades

Vernacular Architecture on the Greek Islands

Greek island architecture varies greatly, even between neighbouring islands. Yet despite the fact that the generic island house does not exist, there are shared characteristics within and between island groups. The Venetians in Crete, the Cyclades, Ionian Islands and Dodecanese, and the Ottomans in the Northeast Aegean, strongly influenced the indigenous building styles developed by vernacular builders.

The town of Chóra on Astypálaia in the Dodecanese, with the kástro above

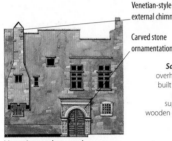

Venetian-style external chimney

Carved stone ornamentation

Venetian-style town houses on Crete date from Venice's 15th- to 17th-century occupation. Often built around a courtyard, the ground floor was used for storage.

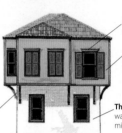

Sash windows with shutters

The top floor was for receiving guests and sleeping.

Sachnísia, or overhangs, were built of lath and plaster and supported by wooden cantilevers.

The kitchen was on the middle storey.

Arcade on ground floor supporting veranda

The stone ground floor housed animals and tools.

Lesvian *pýrgoi* are fortified tower-dwellings at the centre of a farming estate. First built in the 18th century, most surviving examples are 19th century and found near Mytilíni town.

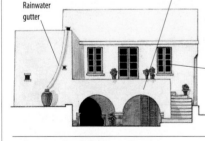

Rainwater gutter

Double "French" windows of the parlour

Sífnos archontiká or town houses are found typically in Kástro, Artemónas and Katavatí. They are two-storeyed, as opposed to the one-storey rural cottage.

Kástro Architecture

The kástro or fortress dwelling of Antíparos dates from the 15th century. It is the purest form of a Venetian pirate-safe town plan in the Cyclades.

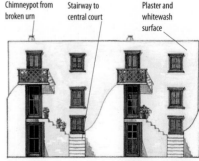

Chimneypot from broken urn

Stairway to central court

Plaster and whitewash surface

Kástro housefronts, with their right-angled staircases, face either on to a central courtyard or a grid of narrow lanes with limited access from outside. The seaward walls have tiny windows. Kástra are found on Síkinos, Kímolos, Sífnos, Antíparos and Folégandros.

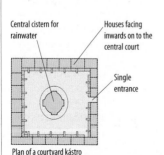

Central cistern for rainwater

Houses facing inwards on to the central court

Single entrance

Plan of a courtyard kástro

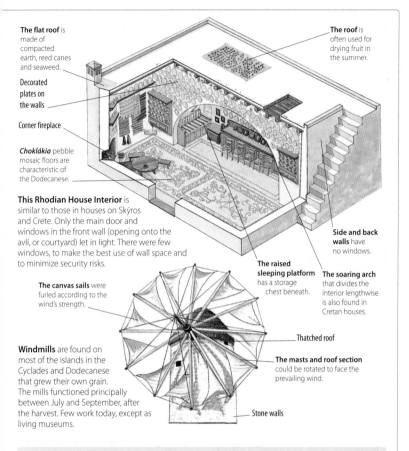

The flat roof is made of compacted earth, reed canes and seaweed.

Decorated plates on the walls

Corner fireplace

Choklákia pebble mosaic floors are characteristic of the Dodecanese.

This Rhodian House Interior is similar to those in houses on Skýros and Crete. Only the main door and windows in the front wall (opening onto the avlí, or courtyard) let in light. There were few windows, to make the best use of wall space and to minimize security risks.

The roof is often used for drying fruit in the summer.

Side and back walls have no windows.

The raised sleeping platform has a storage chest beneath.

The soaring arch that divides the interior lengthwise is also found in Cretan houses.

The canvas sails were furled according to the wind's strength.

Windmills are found on most of the islands in the Cyclades and Dodecanese that grew their own grain. The mills functioned principally between July and September, after the harvest. Few work today, except as living museums.

Thatched roof

The masts and roof section could be rotated to face the prevailing wind.

Stone walls

Local Building Methods and Materials

Lava masonry is found on the volcanic islands of Lésvos, Límnos, Nísyros and Mílos. The versatile and easily split schist is used in the Cyclades, while light-weight lath and plaster indicates Ottoman influence and is prevalent on Sámos, Lésvos, the Sporades and other northern islands. Mud-and-rubble construction is common on all the islands for modest dwellings, as is the *dóma* or flat roof of tree trunks supporting packed reed canes overlaid with seaweed and earth. Buttresses are often used to help strengthen buildings.

Unmortared wall of schist slabs

Masoned volcanic boulders

Slate (or "fish-scale") roof

Pantiled roof, found in the Dodecanese

Flat earthen roof or *dóma*

Arched buttresses for earthquake protection

Marine Life

By oceanic standards, the Mediterranean and Aegean are small, virtually landlocked seas with a narrow tidal range. This means that relatively little marine life is exposed at low tide, although coastal plants and shoreline birds are often abundant. However, if you snorkel close to the shore or dive below the surface of the azure coastal waters, a wealth of plant and animal life can be found. The creatures range in size from myriad shoals of tiny fish and dainty sea slugs to giant marine turtles, huge fish and imposing spider crabs.

The great pipefish's elongated body is easily mistaken for a piece of drifting seaweed. It lives among rocks, pebbles and weed, often in rather shallow water, and can be spotted when snorkelling.

Mediterranean gull   Masked crab

Sea spurge

Tamarisk

Yellow-horned poppy

The spiny spider crab is ungainly when removed from water but agile and surprisingly fast-moving in its element. The long legs allow it to negotiate broken, stony ground easily.

Neptune grass (*Posidonia*)

Fan mussels

Red mullet

Codium bursa

Sea slug

Murex

Top Snorkelling Areas

Snorkelling can be enjoyed almost anywhere around the Greek coast, although remoter areas are generally more rewarding.
- Kefalloniá and Zákynthos: you may find a rare loggerhead turtle (*see p95*) off the east coast.
- Rhodes: wide variety of fish near Líndos on the sheltered east coast.
- Evvoia: the sheltered waters of the west coast harbour sponges.
- Santoríni: the volcanic rock of the caldera has sharp drop-offs to explore.

The octopus catches its prey of crabs and small fish with the rows of powerful suckers along each of its eight legs. It can also change its colour and squeeze through the tiniest of crevices.

The sea turtle, or loggerhead, needs sandy beaches to lay its eggs and has been badly affected by the intrusion of tourists. The few remaining nesting beaches are now given a degree of protection from disturbance.

This jellyfish, called a "by-the-wind-sailor", uses a buoyant float to catch the wind and skim across the sea. Storms will often wash them up on to the beach. Swimmers beware: even the detached threadlike tentacles of some species can inflict painful stings.

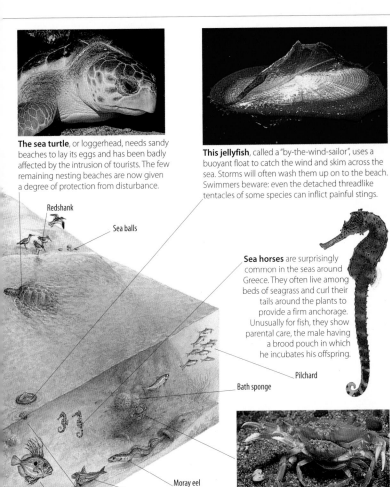

Redshank

Sea balls

Sea horses are surprisingly common in the seas around Greece. They often live among beds of seagrass and curl their tails around the plants to provide a firm anchorage. Unusually for fish, they show parental care, the male having a brood pouch in which he incubates his offspring.

Pilchard

Bath sponge

Moray eel

Red gurnard

Violet sea snail

Shore crab

The swimming crab is one of the most aggressive of all crabs and can inflict a painful nip. It can swim using the flattened, paddlelike tips of its back legs.

A John Dory is a majestic sight as it patrols among offshore rocks. It has a flattened, oval-shaped body and long rays on its dorsal fin. Where the species is not persecuted or exploited, some individuals can become remarkably confident and even inquisitive.

Safety Tips for Snorkelling

- Mediterranean storms can arrive out of nowhere so seek local advice about weather and swimming conditions before you go snorkelling.
- Do not go snorkelling if jellyfish are in the area.
- Take your own snorkel and mask with you to ensure you use one that fits properly.
- Never snorkel unaccompanied.
- Wear a T-shirt or wet suit to avoid sunburn.
- Avoid swimming near river mouths and harbours. The waters will be cloudy and there may be risks from boats and pollution.
- Always stick close to the shore and check your position from time to time.

THE HISTORY OF GREECE

The history of Greece is that of a nation, not of a land: the Greek idea of nationality is governed by language, religion, descent and customs, not so much by location. Early Greek history is the story of internal struggles, from the Mycenaean and Minoan cultures of the Bronze Age to the competing city-states that emerged in the 1st millennium BC.

After the defeat of the Greek army by Philip II of Macedon at Chaironeia in 338 BC, Greece became absorbed into Alexander the Great's empire. With the defeat of the Macedonians by the Romans in 168 BC, Greece became a province of Rome. As part of the Eastern Empire she was ruled from Constantinople and became a powerful element within the new Byzantine world.

In 1453, when Constantinople fell to the Ottomans, Greece disappeared as a political entity. The Venetian republic quickly established fortresses on the coast and islands in order to compete with the Ottomans for control of the important trade routes in the Ionian and Aegean seas. Eventually the realization that it was the democracy of Classical Athens that had inspired so many revolutions abroad gave the Greeks themselves the courage to rebel and, in 1821, to fight the Greek War of Independence. In 1830 the Great Powers that dominated Europe established a protectorate over Greece, marking the end of Ottoman rule.

After almost a century of border disputes, Turkey defeated Greece in 1922. This was followed by the dictatorship of Metaxás, and then by the war years of 1940–4, during which half a million people were killed. The present boundaries of the Greek state have only existed since 1948, when Italy returned the Dodecanese. Now an established democracy and member of the European Union, Greece's fortunes seem to have come full circle after 2,000 years of foreign rule.

A map of Greece from the 1595 Atlas of Abraham Ortelius called *Theatrum Orbis Terrarum*

◄ The Knights of the Order of St John from a 15th-century history of the siege of Rhodes

Prehistoric Greece

During the Bronze Age three separate civilizations flourished in Greece: the Cycladic, during the 3rd millennium; the Minoan, based on Crete but with an influence that spread throughout the Aegean Islands; and the Mycenaean, which was based on the mainland but spread to Crete in about 1450 BC, when the Minoans went into decline. Both the Minoan and Mycenaean cultures found their peak in the Palace periods of the 2nd millennium, when they were dominated by a centralized religion and bureaucracy.

Prehistoric Greece
Areas settled in the Bronze Age

Neolithic Head (3000 BC)
This figure was found on Alónnisos in the Sporades. It probably represents a fertility goddess who was worshipped by farmers to ensure a good harvest. These figures indicate a certain stability in early communities.

The town is unwalled, showing that inhabitants did not fear attack.

Cycladic Figurine
Marble statues such as this, produced in the Bronze Age from about 2800 to 2300 BC, have been found in a number of tombs in the Cyclades.

Multistorey houses

Minoan Bathtub Sarcophagus
This type of coffin, dating to 1400 BC, is found only in Minoan art. It was probably used for a high-status burial.

	7000 Neolithic farmers in northern Greece	**3200** Beginnings of Bronze Age cultures in Cyclades and Crete	**2000** Arrival of first Greek-speakers on mainland Greece
200,000 BC	**5000 BC**	**4000 BC**	**3000 BC**	200
200,000 Evidence of Palaeolithic civilization in northern Greece and Thessaly		"Frying Pan" vessel from Sýros (2500–2000 BC)	**2800–2300** Kéros-Sýros culture flourishes in Cyclades **2000** Building of palaces begins in Crete, initiating First Palace period

Mycenaean Death Mask
Large amounts of worked gold were discovered in the Peloponnese at Mycenae, the ancient city of Agamemnon. Masks like this were laid over the faces of the dead.

Forested hills

The inhabitants are on friendly terms with the visitors.

Where to See Prehistoric Greece

The Museum of Cycladic Art in Athens (p295) has the leading collection of Cycladic figurines in Greece. In the National Archaeological Museum (p290) Mycenaean gold and other prehistoric artifacts are on display. It houses one of the world's finest collections of ancient ceramics. Akrotíri (p245) on Santoríni in the Cyclades has Minoan buildings surviving up to the third storey. The city of Phylakopi on Mílos (p241) also has Mycenaean walls dating to 1500 BC. Crete, the centre of Minoan civilization, has the palaces of Knosós (pp276–9), Phaestos (pp270–71) and Agía Triáda (p267).

Cyclopean Walls
Mycenaean citadels, as this one at Tiryns in the Peloponnese, were encircled by walls of stone so large that later civilizations believed they had been built by giants. It is unclear whether the walls were used for defence or just to impress.

Oared sailing ships

Mycenaean Octopus Jar
This 14th-century-BC vase's decoration follows the shape of the pot. Restrained and symmetrical, it contrasts with relaxed Minoan prototypes.

Minoan Sea Scene

The wall paintings on Santoríni (see pp242–5) were preserved by the volcanic eruption at the end of the 16th century BC. This section shows ships departing from a coastal town. In contrast to the warlike Mycenaeans, Minoan art reflects a more stable community which dominated the Aegean through trade, not conquest.

1750–1700
Start of Second Palace Period and golden age of Minoan culture in Crete

1525 Volcanic eruption on Santoríni devastates the region

1250–1200 Probable destruction of Troy, after abduction of Helen (see p58)

1450 Mycenaeans take over Knosós; use of Linear B script

Helen of Troy

1800 BC	1600 BC	1400 BC	1200 BC

1730 Destruction of Minoan palaces; end of First Palace period

1600 Beginning of high period of Mycenaean prosperity and dominance

Minoan figurine of a snake goddess, 1500 BC

1200 Collapse of Mycenaean culture

1370–1350 Palace of Knosós on Crete destroyed for second time

The Dark Ages and Archaic Period

In about 1200 BC, Greece entered a period of darkness. There was widespread poverty, the population decreased and many skills were lost. A cultural revival in about 800 BC accompanied the emergence of the city-states across Greece and inspired new styles of warfare, art and politics. Greek colonies were established as far away as the Black Sea, present-day Syria, North Africa and the western Mediterranean. Greece was defined by where Greeks lived.

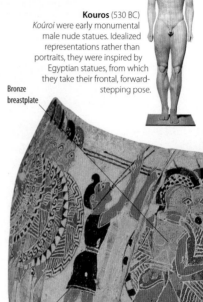

Kouros (530 BC)
Koúroi were early monumental male nude statues. Idealized representations rather than portraits, they were inspired by Egyptian statues, from which they take their frontal, forward-stepping pose.

Bronze breastplate

Mediterranean Area, 479 BC

■ Areas of Greek influence

The double flute player kept the men marching in time.

Bronze greaves protected the legs.

Solon (640–558 BC)
Solon was appointed to the highest magisterial position in Athens. His legal, economic and political reforms heralded democracy.

Hoplite Warriors

The "Chigi" vase from Corinth, dating to about 750 BC, is one of the earliest clear depictions of the new style of warfare that evolved at that period. This required rigorously trained and heavily armed infantrymen called hoplites to fight in a massed formation or phalanx. The rise of the city-state may be linked to the spirit of equality felt by citizen hoplites fighting for their own community.

Vase fragment showing bands of distinctive geometric line patterns

900
Appearance of first Geometric pottery

1100 BC	1000 BC	900 BC

1100 Migrations of different peoples throughout the Greek world

1000–850 Formation of the Homeric kingdoms

6th-Century Vase
This bowl *(krater)* for mixing wine and water at elegant feasts is an early example of the art of vase painting. It depicts mythological and heroic scenes.

Bronze helmets for protection

Spears were used for thrusting.

The phalanxes shoved and pushed, aiming to maintain an unbroken shield wall, a successful new technique.

Gorgon's head decoration

Characteristic round shields

Hunter Returning Home (500 BC)
Hunting for hares, deer, or wild boar was an aristocratic sport pursued by Greek nobles on foot with dogs, as depicted on this cup.

Darius I (ruled 521–486 BC)
This relief from Persepolis shows the Persian king who tried to conquer the Greek mainland, but was defeated at the Battle of Marathon in 490.

Where to See Archaic Greece

Examples of *koúroi* can be found in the National Archaeological Museum *(see p290)* and in the Acropolis Museum *(p294)*, both in Athens. The National Archaeological Museum also houses the national collection of Greek Geometric, red-figure and black-figure vases. Old *koúroi* lie in the old marble quarry on Náxos *(pp234–7)*. Sámos boasts the impressive Efpalineio tunnel *(p159)* and a collection of *koúroi* *(p158)*. Delos has a terrace of Archaic lions *(pp222–3)* and the Doric temple of Aphaia on Aígina is well preserved *(pp102–3)*. Palaiókastro on Nísyros has huge fortifications *(p179)*.

776 Traditional date for the first Olympic Games

630 Poet Sappho writing in Lésvos

600 First Doric columns built at Temple of Hera, Olympia

Doric capital

490 Athenians defeat Persians at Marathon

675 Lykourgos initiates austere reforms in Sparta

800 BC

700 BC

600 BC

500 BC

770 Greeks start founding colonies in Italy, Egypt and elsewhere

750–700 Homer records epic tales of the *Iliad* and *Odyssey*

Spartan votive figurine

546 Persians gain control over Ionian Greeks; Athens flourishes under the tyrant Peisistratos and his sons

480 Athens destroyed by Persians who defeat Spartans at Thermopylae; Greek victory at Salamis

479 Persians annihilated at Plataiai by Athenians, Spartans and allies

Classical Greece

The Classical period has always been considered the high point of Greek civilization. Around 150 years of exceptional creativity in thinking, writing, theatre and the arts produced the great tragedians Aeschylus, Sophocles and Euripides as well as the great philosophical thinkers Socrates, Plato and Aristotle. This was also a time of warfare and bloodshed, however. The Peloponnesian War, which pitted the city-state of Athens and her allies against the city-state of Sparta and her allies, dominated the 5th century BC. In the 4th century Sparta, Athens and Thebes struggled for power only to be ultimately defeated by Philip II of Macedon in 338 BC.

Classical Greece, 440 BC
Athens and her allies
Sparta and her allies

Fish Shop
This 4th-century-BC Greek painted vase comes from Cefalù in Sicily. Large parts of the island were inhabited by Greeks who were bound by a common culture, religion and language.

Theatre used in Pythian Games

Temple of Apollo

Siphnian Treasury

The Sanctuary of Delphi
The sanctuary in central Greece, shown in this 1894 reconstruction, reached the peak of its political influence in the 5th and 4th centuries BC. Of central importance was the Oracle of Apollo, whose utterances influenced the decisions of city-states such as Athens and Sparta. Rich gifts dedicated to the god were placed by the states in treasuries that lined the Sacred Way.

Perikles
This great democratic leader built up the Greek navy and masterminded the extensive building programme in Athens between the 440s and 420s, including the Acropolis temples.

Detail of the Parthenon frieze

462 Ephialtes's reforms pave the way for radical democracy in Athens

431–404 Peloponnesian War, ending with the fall of Athens and start of 33-year period of Spartan dominance

c.424 Death of Herodotus, historian of the Persian Wars

475 BC

450 BC

425 BC

478 With the formation of the Delian League, Athens takes over leadership of Greek cities

451–429 Perikles rises to prominence in Athens and launches a lavish building programme

447 Construction of the Parthenon begins

Bust of Herodotus, probably of Hellenistic origin

Gold Oak Wreath from Vergína
By the mid-4th century BC, Philip II of Macedon dominated the Greek world through diplomacy and warfare. This wreath comes from his tomb.

Where to See Classical Greece

Athens is dominated by the Acropolis and its religious buildings, including the Parthenon, erected as part of Perikles's mid-5th-century-BC building programme *(see pp292–4)*. The island of Delos, the mythological birthplace of Artemis and Apollo, was the centre for the Delian League, the first Athenian naval league. The site contains examples of 5th-century-BC sculpture *(pp222–3)*. On Rhodes, the 4th-century Temple of Athena at Líndos *(pp200–201)* is well preserved.

Votive of the Rhodians

Stoa of the Athenians

Sacred Way

Athenian Treasury

Athena Lemnia
This Roman copy of a statue by Pheidias (c.490–c.430 BC), the sculptor-in-charge at the Acropolis, depicts the goddess protector of Athens in an ideal rather than realistic way, typical of the Classical style in art.

Slave Boy (400 BC)
Slaves were fundamental to the Greek economy and used for all types of work. Many slaves were foreign; this boot boy came from as far as Africa.

Sculpture of Plato

387 Plato founds Academy in Athens

359 Philip II becomes King of Macedon

337 Foundation of the League of Corinth legitimizes Philip II's control over the Greek city-states

400 BC

375 BC

350 BC

399 Trial and execution of Socrates

371 Sparta defeated by Thebes at Battle of Leuktra, heralding a decade of Theban dominance in the area

338 Greeks defeated by Philip II of Macedon at Battle of Chaironeia

336 Philip II is assassinated at Aigai and is succeeded by his son, Alexander

Hellenistic Greece

Alexander the Great of Macedon fulfilled his father Philip's plans for the conquest of the Persians. He went on to create a vast empire that extended to India in the east and Egypt in the south. The Hellenistic period was extraordinary for the dispersal of Greek language, religion and culture throughout the territories conquered by Alexander. It lasted from after Alexander's death in 323 BC until the Romans began to dismantle his empire in the mid-2nd century BC. For Greece, Macedonian domination was replaced by that of Rome in AD 168.

Relief of Hero-Worship (c.200 BC)
Hero-worship was part of Greek religion. Alexander, however, was worshipped as a god in his lifetime.

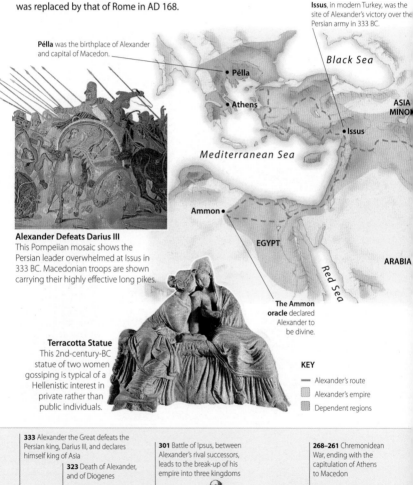

Pélla was the birthplace of Alexander and capital of Macedon.

Issus, in modern Turkey, was the site of Alexander's victory over the Persian army in 333 BC.

Black Sea

Pélla

Athens

ASIA MINO[R]

Issus

Mediterranean Sea

Ammon

EGYPT

ARABIA

Red Sea

The Ammon oracle declared Alexander to be divine.

Alexander Defeats Darius III
This Pompeiian mosaic shows the Persian leader overwhelmed at Issus in 333 BC. Macedonian troops are shown carrying their highly effective long pikes.

Terracotta Statue
This 2nd-century-BC statue of two women gossiping is typical of a Hellenistic interest in private rather than public individuals.

KEY

— Alexander's route

▢ Alexander's empire

▢ Dependent regions

333 Alexander the Great defeats the Persian king, Darius III, and declares himself king of Asia

323 Death of Alexander, and of Diogenes

301 Battle of Ipsus, between Alexander's rival successors, leads to the break-up of his empire into three kingdoms

268–261 Chremonidean War, ending with the capitulation of Athens to Macedon

325 BC — **300 BC** — **275 BC** — **250 BC**

322 Death of Aristotle

287–275 "Pyrrhic victory" of King Pyrros of Epirus who defeated the Romans in Italy but suffered heavy losses

Diogenes, the Hellenistic philosopher

331 Alexander founds Alexandria after conquering Egypt

Fusing Eastern and Western Religion
This plaque from Afghanistan shows the Greek goddess Nike, and the Asian goddess Cybele, in a chariot pulled by lions.

Where to See Hellenistic Greece

The Aegean was ruled by the Ptolemies in the 3rd and 2nd centuries BC from ancient Thíra on Santoríni, where there are Hellenistic remains: the Sanctuary of Artemídoros of Perge, the Royal Portico, and the Temple of Ptolemy III (see p244). In Rhodes Town, the Hospital of the Knights, now the Archaeological Museum (p188), houses a collection of Hellenistic sculpture. The Asklepieíon on Kos (p176) was the seat of an order of medical priests. The Tower of the Winds (p291), in Athens, was built by the Macedonian astronomer Andronikos Kyrrestes.

Susa, capital of the Persian Empire, was captured in 331 BC. A mass wedding of Alexander's captains to Asian brides was held in 324 BC.

Caspian Sea

SOGDIANI

Alexandropolis

Taxil

BACTRIA

PERSIA

Susa

Persopolis

Beas

INDIA

GEDROSIA

Persian Gulf

Arabian Sea

Alexander's army turned back at the River Beas.

The Persian religious centre of Persepolis, in modern Iran, fell to Alexander in 330 BC.

Alexander's army suffered heavy losses in the Gedrosia desert.

Alexander the Great's Empire
In forming his empire Alexander covered huge distances. After defeating the Persians in Asia he moved to Egypt, then returned to Asia to pursue Darius, and then his murderers, into Bactria. In 326 his troops revolted in India and refused to go on. Alexander died in 323 in Babylon.

The Death of Archimedes
Archimedes was the leading Hellenistic scientist and mathematician. This mosaic from Renaissance Italy shows his murder in 212 BC by a Roman.

227 Colossus of Rhodes destroyed by earthquake

Colossus of Rhodes

197 Romans defeat Philip V of Macedon and declare Greece liberated

146 Romans sack Corinth and Greece becomes a province of Rome

225 BC

200 BC

175 BC

150 BC

222 Macedon crushes Sparta

217 Peace of Náfpaktos: a call for the Greeks to settle their differences before "the cloud in the west" (Rome) settles over them

Roman coin (196 BC) commemorating Roman victory over the Macedonians

168 Macedonians defeated by Romans at Pydna

Roman Greece

After the Romans gained control of Greece with the sack of Corinth in 146 BC, Greece became the cultural centre of the Roman Empire. The Roman nobility sent their sons to be educated in the schools of philosophy in Athens. The end of the Roman civil wars between leading Roman statesmen was played out on Greek soil, finishing in the Battle of Actium in Thessaly in 31 BC. In AD 323 the Emperor Constantine founded the new eastern capital of Constantinople; the empire was later divided into the Greek-speaking East and the Latin-speaking West.

Roman Provinces, AD 211

Mithridates
In a bid to extend his territory, this ruler of Pontus, on the Black Sea, led the resistance to Roman rule in 88 BC. He was forced to make peace three years later.

Bema, or raised platform, where St Paul spoke

Roman basilica

Bouleuterion

Notitia Dignitatum (AD 395)
As part of the Roman Empire, Greece was split into several provinces. The proconsul of the province of Achaia used this insignia.

Springs of Peirene, the source of water

Reconstruction of Roman Corinth

Corinth, in the Peloponnese, was refounded and largely rebuilt by Julius Caesar in 46 BC, becoming the capital of the Roman province of Achaia. The Romans built the forum, covered theatre and basilicas. St Paul visited the city in AD 50–51, working as a tent maker.

Baths of Eurycles

A coin of Cleopatra, Queen of Egypt

49–31 BC Rome's civil wars end with the defeat of Mark Antony and Cleopatra at Actium, in Greece

AD 49–54 St Paul preaches Christianity in Greece

AD 124–131 Emperor Hadrian oversees huge building programme in Athens

| 100 BC | AD 1 | AD |

86 BC Roman commander, Sulla, captures Athens

46 BC Corinth refounded as Roman colony

St Paul preaching

AD 66–7 Emperor Nero tours Greece

Mosaic (AD 180) This highly sophisticated Roman mosaic of Dionysos riding on a leopard comes from the House of Masks, on Delos.

Temple of Octavia

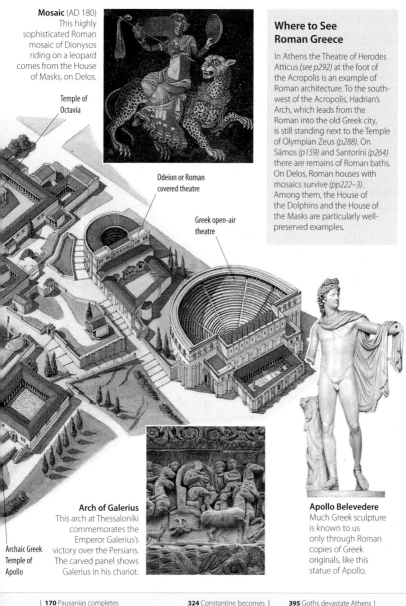

Odeion or Roman covered theatre

Greek open-air theatre

Where to See Roman Greece

In Athens the Theatre of Herodes Atticus (see p292) at the foot of the Acropolis is an example of Roman architecture. To the south-west of the Acropolis, Hadrian's Arch, which leads from the Roman into the old Greek city, is still standing next to the Temple of Olympian Zeus (p288). On Sámos (p159) and Santoríni (p264) there are remains of Roman baths. On Delos, Roman houses with mosaics survive (pp222–3). Among them, the House of the Dolphins and the House of the Masks are particularly well-preserved examples.

Archaic Greek Temple of Apollo

Arch of Galerius This arch at Thessaloníki commemorates the Emperor Galerius's victory over the Persians. The carved panel shows Galerius in his chariot.

Apollo Belevedere Much Greek sculpture is known to us only through Roman copies of Greek originals, like this statue of Apollo.

170 Pausanias completes Guide to Greece for Roman travellers

267 Goths pillage Athens

324 Constantine becomes sole emperor of Roman Empire and establishes his capital in Constantinople

395 Goths devastate Athens and Peloponnese

381 Emperor Theodosius I makes Christianity state religion

AD 200

AD 300

Coin of the Roman Emperor Galerius

293 Under Emperor Galerius, Thessaloníki becomes second city to Constantinople

393 Olympic games banned

395 Death of Theodosius I; formal division of Roman Empire into Latin West and Byzantine East

Byzantine and Crusader Greece

Under the Byzantine Empire, which at the end of the 4th century succeeded the old Eastern Roman Empire, Greece became Orthodox in religion and was split into administrative *themes*. When the capital, Constantinople, fell to the Crusaders in 1204 Greece was again divided, mostly between the Venetians and the Franks. Constantinople and Mystrás were recovered by the Byzantine Greeks in 1261, but the Turks' capture of Constantinople in 1453 marked the final demise of the Byzantine Empire. It left a legacy of hundreds of churches and a wealth of religious art.

Byzantine Greece in the 10th Century

Chapel

Watchtower of Tsimiskís

Refectory

Two-Headed Eagle
In the Byzantine world, the emperor was also patriarch of the church, a dual role represented in this pendant of a two-headed eagle.

Great Lavra

This monastery is the earliest (AD 963) and largest of the religious complexes on Mount Athos in northern Greece. Many parts have been rebuilt, but its appearance remains essentially Byzantine. The monasteries became important centres of learning and religious art.

Defence of Thessaloníki
The fall of Thessaloníki to the Saracens in AD 904 was a blow to the Byzantine Empire. Many towns in Greece were heavily fortified against attack from this time.

578–86 Avars and Slavs invade Greece

Gold solidus of the Byzantine Empress Irene, who ruled AD 797–802

400	600	800
529 Aristotle's and Plato's schools of philosophy close as Christian culture supplants Classical thought	**680** Bulgars cross Danube and establish empire in northern Greece	**726** Iconoclasm introduced by Pope Leo III (abandoned in 843) **841** Parthenon becomes a cathedral

Constantine the Great
The first eastern emperor to recognize Christianity, Constantine founded the city of Constantinople in AD 324. Here he is shown with his mother, Helen.

Where to See Byzantine and Crusader Greece

In Athens, the Benáki Museum *(see p295)* contains icons, metalwork, sculpture and textiles. On Pátmos, the treasury of the Monastery of St John, founded in 1088 *(pp168–9)*, is the richest outside Mount Athos. The 11th-century convent of Néa Moní on Chíos *(pp154–5)* has magnificent gold-ground mosaics. The medieval architecture of the Palace of the Grand Masters and the Street of the Knights on Rhodes *(pp190–93)* is particularly fine. Buildings by the Knights on Kos *(pp174–7)* are also worth seeing. The Venetian castle on Páros *(p231)* dates from 1260.

Cypress tree of Agios Athanásios

Christ Pantokrátor
This 14th-century fresco of Christ as ruler of the world is in the Byzantine city and monastic centre of Mystrás.

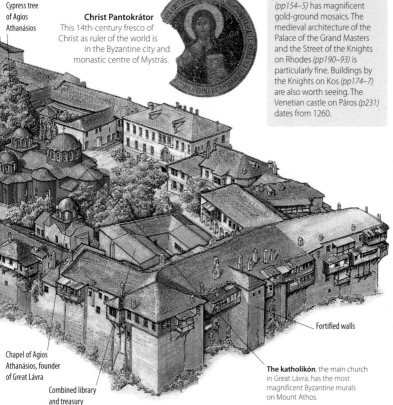

Fortified walls

Chapel of Agios Athanásios, founder of Great Lávra

Combined library and treasury

The katholikón, the main church in Great Lávra, has the most magnificent Byzantine murals on Mount Athos.

1054 Patriarch of Constantinople and Pope Leo IX excommunicate each other

Frankish Chlemoútsi Castle

1081–1149 Normans invade Greek islands and mainland

1354 Ottoman Turks enter Europe, via southern Italy and Greece

1390–1450 Turks gain power over much of mainland Greece

1000 | **1200** | **1400**

Basil the Bulgar Slayer, Byzantine emperor (lived 956–1025)

1204 Crusaders sack Constantinople. Break-up of Byzantine Empire as result of occupation by Franks and Venetians

1210 Venetians win control over Crete

1261 Start of intellectual and artistic flowering of Mystrás

1389 Venetians in control of much of Greece and the islands

Venetian and Ottoman Greece

Following the Ottomans' momentous capture of
Constantinople in 1453, and their conquest of almost all the
remaining Greek territory by 1460, the Greek state effectively
ceased to exist for the next 350 years. Although the city
became the capital of the vast Ottoman Empire, it remained
the principal centre of Greek population and the focus of Greek
dreams of resurgence. The small Greek population of what
today is modern Greece languished in an impoverished and
underpopulated backwater, but even there rebellious bands of
brigands and private militias were formed. The Ionian Islands,
Crete and a few coastal enclaves were seized for long periods
by the Venetians – an experience more intrusive than the
inefficient tolerance of the Ottomans, but one which left a rich
cultural and architectural legacy.

Greece in 1493

   Areas occupied by Venetians

   Areas occupied by
Ottomans

Battle of Lepanto (1571)
The Christian fleet, under Don
John of Austria, decisively defeated
the Ottomans off Náfpaktos,
halting their advance westwards.

Cretan Painting
This 15th-century icon is typical of
the style developed by Greek artists
in the School of Crete, active until
the Ottomans took Crete in 1669.

Arrival of Turkish Prince
Cem on Rhodes
*Prince Cem, Ottoman rebel and son of
Mehmet II, fled to Rhodes in 1481 and was
welcomed by the Christian Knights of St
John (see pp192–3). In 1522, however,
Rhodes fell to the Ottomans after a siege.*

1453 Mehmet II captures
Constantinople which is
renamed Istanbul and made
capital of the Ottoman Empire

1503 Ottoman Turks win
control of the Peloponnese
apart from Monemvasía

1571 Venetian and
Spanish fleet defeats
Ottoman Turks at the
Battle of Lepanto

1500

1550

1600

1460 Turks
capture Mystrás

*Cretan chain-mail armour
from the 16th century*

1456 Ottoman Turks
occupy Athens

1522 The Knights of St John
forced to cede Rhodes to
the Ottomans

Shipping

Greek merchants traded throughout the Ottoman Empire. By 1800 there were merchant colonies in Constantinople and as far afield as London and Odessa. This 19th-century embroidery shows the Turkish influence on Greek decorative arts.

Where to See Venetian and Ottoman Architecture

The Ionian Islands are particularly rich in buildings dating from the Venetian occupation. The old town of Corfu (see pp78–83) is dominated by its two Venetian fortresses. The citadel in Zákynthos (p94) is also Venetian. Crete has a number of Venetian buildings: the old port of Irákleio (pp272–3) and some of the backstreets of Chaniá (pp256–7) convey an overwhelming feeling of Venice. Irákleio's fort withstood the Great Siege of 1648–69. Some Ottoman-era houses survive on Thásos (p135). Several mosques and other Ottoman buildings, including a library and hammam (baths), can be seen in Rhodes Old Town (pp186–95).

The Knights of St John defied the Turks until 1522.

The massive fortifications eventually succumbed to Turkish artillery.

The Knights supported Turkish rebel, Prince Cem.

Dinner at a Greek House in 1801

Nearly four centuries of Ottoman rule profoundly affected Greek culture, ethnic composition and patterns of everyday life. Greek cuisine incorporates Turkish dishes still found thoughout the old Ottoman Empire.

1687 Parthenon seriously damaged during Venetian artillery attack on Turkish magazine

1715 Turks reconquer the Peloponnese

Ali Pasha (1741–1822), a governor of the Ottoman Empire

1814 Britain gains possession of Ionian Islands

| 1650 | 1700 | 1750 | 1800 |

1684 Venetians reconquer the Peloponnese

Parthenon blown up

1778 Ali Pasha becomes Vizier of Ioánnina and establishes powerful state in Albania and northern Greece

1801 Frieze on Parthenon removed by Lord Elgin

1814 Foundation of Filikí Etaireía, Greek liberation movement

The Making of Modern Greece

The Greek War of Independence marked the overthrow of the Ottomans and the start of the "Great Idea", an ambitious project to bring all Greek people under one flag *(Enosis)*. The plans for expansion were initially successful, and during the 19th century the Greeks succeeded in doubling their national territory and reasserting Greek sovereignty over many of the islands. However, an attempt to take the city of Constantinople by force after World War I ended in disaster: in 1922, millions of Greeks were expelled from Smyrna in Turkish Anatolia, ending thousands of years of Greek presence in Asia Minor.

The Emerging Greek State
🔲 Greece in 1832
🔲 Areas gained 1832–1923

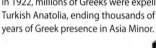

Klephts (mountain brigands) were the basis of the Independence movement.

Massacre at Chíos
This detail of Delacroix's shocking painting *Scènes de Massacres de Scio* shows the events of 1822, when Turks took savage revenge for an earlier killing of Muslims.

Weapons were family heirlooms or donated by philhellenes.

Declaration of the Constitution in Athens
Greece's Neo-Classical parliament building in Athens was the site of the Declaration of the Constitution in 1843. It was built as the Royal Palace for Greece's first monarch, King Otto, in the 1830s.

1824 The poet Lord Byron dies of a fever at Mesolóngi

1831 President Kapodístrias assassinated

1832 Great Powers establish protectorate over Greece and appoint Otto, Bavarian prince, as king

1834 Athens replaces Náfplio as capital

German archaeologist Heinrich Schliemann

| 1830 | 1840 | 1850 | 1860 | 187 |

1827 Battle of Navaríno

1828 Ioánnis Kapodístrias becomes first President of Greece

King Otto (ruled 1832–62)

1862 Revolution drives King Otto from Greece

1874 Heinrich Schliemann begin excavation of Mycenae

1821 Greek flag of independence raised on 25 March; Greeks massacre Turks at Tripolitsá in Morea

1864 New constitution makes Greece a "crowned democracy"; Greek Orthodoxy made the state religion

Life in Athens

By 1836 urban Greeks still wore a mixture of Greek traditional and Western dress. The Ottoman legacy had not totally disappeared and is visible in the fez worn by men.

Where to See 19th-century Greece

In Crete, Moní Arkadíou *(see p264)* is the site of mass suicide by freedom fighters in 1866; the tomb of Venizélos is at Akrotíri *(p255)*. The harbour and surrounding buildings at Sýros *(p224)* are evidence of the importance of Greek sea-power in the 19th century.

Flag Raising of 1821 Revolution

In 1821, the Greek secret society Filikí Etaireía was behind a revolt by Greek officers which led to anti-Turk uprisings throughout the Peloponnese. Tradition credits Archbishop Germanós of Pátra with raising the rebel flag near Kalávryta in the Peloponnese on 25 March. The struggle for independence had begun.

Corinth Canal
This spectacular link between the Aegean and Ionian seas opened in 1893.

Elefthérios Venizélos
This great Cretan politician and advocate of liberal democracy doubled Greek territory during the Balkan Wars (1912–13) and joined the Allies in World War I.

	1893 Opening of Corinth Canal	**1896** First Olympics of modern era, held in Athens	**1921** Greece launches offensive in Asia Minor	**1922** Turkish burning of Smyrna signals end of the "Great Idea"
			1908 Crete united with Greece	**1917** King Constantine is deposed; Greece joins World War I	
1880	**1890**	**1900**	**1910**	**1920**	
Spyrídon Louis, Marathon winner at the first modern Olympics	**1899** Arthur Evans begins excavations at Knosós	**1912–13** Greece extends its borders during the Balkan Wars	**1920** Treaty of Sèvres gives Greece huge gains in territory		
		1923 Population exchange agreed between Greece and Turkey at Treaty of Lausanne. Greece loses previous gains			

Twentieth-Century Greece

The years after the 1922 defeat by Turkey were terrible ones for Greek people. The influx of refugees contributed to the political instability of the interwar years. The dictatorship of Metaxás was followed by invasion in 1940, then Italian, German and Bulgarian occupation and, finally, the Civil War between 1946 and 1949, with its legacy of division. After experiencing the Cyprus problem of the 1950s and the military dictatorship of 1967 to 1974, Greece is now an established democracy and became a member of the European Economic and Monetary Union in 2000.

1947 Internationally acclaimed Greek artist Giánnis Tsaroúchis holds his first exhibition of set designs, in the Romvos Gallery, Athens

1938 Death of sculptor Giannoúlis Chalepás, best known for his *Sleeping Girl* funerary statue

1946 Government institutes "White Terror" against Communists

1958 USSR threatens Greece with economic sanctions if NATO missiles installed

1945 Níkos Kazantzákis publishes *Zorba the Greek*, later made into a film

1967 Right-wing colonels form Junta, forcing King Constantine into exile

1933 Death of Greek poet, Constantine (C P) Cavafy

1957 Mosaics found by chance at Philip II's 300-BC palace at Pélla

| 1925 | 1935 | 1945 | 1955 | 1965 |

| 1925 | 1935 | 1945 | 1955 | 1965 |

1951 Greece enters NATO

1955 Greek Cypriots start campaign of violence in Cyprus against British rule

1939 Greece declares neutrality at start of World War II

1948 Dodecanese becomes part of Greece

1932 Aristotle Onassis purchases six freight ships, the start of his shipping empire

1925 Mános Chatzidákis, who wrote music for the film *Never on Sunday*, is born

1946–9 Civil War between Greek government and the Communists, who take to the mountains

1960 Cyprus declared independent

ΟΙ ΗΡΩΙΔΕΣ ΤΟΥ 1940

1940 Italy invades Greece. Greek soldiers defend northern Greece. Greece enters World War II

1944 Churchill visits Athens to show his support for Greek government against Communist Resistance

1963 Geórgios Papandréou's centre-left government voted into power

1973 University students in Athens rebel against dictatorship and are crushed by military forces. Start of decline in power of dictatorship

1988 Eight million visitors to Greece; tourism continues to expand

1993 Andréas Papandréou wins Greek general election for the third time

2012 A coalition government is formed of three political parties under Prime Minister Antonis Samoras, with promises to reduce Greece's budget deficit

ΕΛΛΗΝΙΚΗ ΔΗΜΟΚΡΑΤΙΑ HELLAS 60

1981 Melína Merkoúri appointed Minister of Culture. Start of campaign to restore Elgin Marbles to Greece

2002 Drachma replaced by the euro at the beginning of March

2004 The Olympic Games take place in Athens

2010 The economic crisis obliges an application for support to the International Monetary Fund and the Eurozone

2013 Greece records the highest unemployment figure in the EU at 26.8 per cent. Continuing austerity measures see the closure of state broadcaster ERT, but mass protests result in the launch of new station EDT

1974 Cyprus is partitioned after Turkish invasion

1994 Because of the choking smog (néfos), central Athens introduces traffic restrictions

1975	1985	1995	2005	2015
1975	1985	1995	2005	2015

1975 Death of Aristotle Onassis

1974 Fall of Junta; Konstantínos Karamanlís elected Prime Minister

1990 New Democracy voted into power; Konstantínos Karamanlís becomes President

2004 Greece win Euro 2004 Football Championship

2015 Alexis Tsipras, the leader of the anti-austerity party Syriza, becomes Prime Minister following parliamentary elections in January.

2014 Greece sells government bonds to help its economic recovery

1981 Andréas Papandréou's left-wing PASOK party forms first Greek Socialist government

1998 Karamanlís dies. Kostis Stefanopoulos succeeds him

1973 Greek bishops give their blessing to the short-lived presidency of Colonel Papadópoulos

1997 Athens is awarded the 2004 Olympics

2011 Indignant Citizens Movement (Kinima Aganaktismenon Politon) demonstrates against austerity measures in major cities across Greece

1994 European leaders meet in Corfu under Greek presidency of the EU

2009 Left-wing PASOK party voted into power; Géorgios Papandréou becomes Prime Minister

1996 Andréas Papandréou dies; Kóstas Simítis succeeds him

THE GREEK ISLANDS THROUGH THE YEAR

Greek island life revolves around the seasons, and is punctuated by saints' days and colourful religious festivals, or *panigýria*. Easter is the most important Orthodox festival of the year, but there are lively pre-Lenten carnivals on some islands as well. The Greeks mix piety and pleasure, with a great enthusiasm for their celebrations, from the most important to the smallest village fair. Folkloric music and dance are key elements of almost every festival. There are also festivals that have ancient roots in pagan revels. Other festivals celebrate harvests of local produce, such as grapes, olives and corn, or re-enact various victories for Greece in its struggle for Independence.

Spring

The Greek word for "spring" is *ánoixi* (the opening), and it heralds the beginning of the tourist season on the islands. After wintering in Athens or Rhodes, hoteliers and shopkeepers head for the smaller islands to open up. The islands in spring are at their most beautiful, carpeted with red poppies, camomile and wild cyclamen. Fruit trees are in blossom, fishing boats and houses are freshly painted and people are at their most welcoming. Orthodox Easter is the main spring event, preceded in late February or March with pre-Lenten carnivals. While northern island groups can be showery, by late April, Crete, the Dodecanese and east Aegean islands are usually warm and sunny.

Children in national dress, 25 March

March

Apókries, or Carnival Sunday *(first Sun before Lent)*. There are carnivals on many islands for three weeks leading up to this date, the culmination of pre-Lenten festivities. Celebrations are exuberant at Agiásos on Lésvos and on Kárpathos, while a goat dance is performed on Skýros.

Katharí Deftéra, or Clean Monday *(seven Sundays before Easter)*. This marks the start of Lent. Houses are spring-cleaned and the unleavened bread *lagána* is baked. Clean Monday is also the day for a huge kite-flying contest

Celebrating Easter in Greece

Greek Orthodox Easter can fall up to three weeks either side of Western Easter. It is the most important religious festival in Greece, and Holy Week is a time for Greek families to reunite. It is also a good time to visit Greece, to see the processions and church services and to sample the Easter food. The ceremony and symbolism is a direct link with Greece's Byzantine past, as well as with earlier more primitive beliefs.

The festivities reach a climax at midnight on Easter Saturday when, as priests intone "Christ is risen", fireworks explode to usher in a Sunday of feasting, music and dancing. The Sunday feasting on roast meat marks the end of the Lenten fast, and a belief in the renewal of life in spring. Particularly worthwhile visiting for the Holy Week processions and the Friday and Saturday night services are Olympos on Kárpathos, Ýdra, Pátmos and just about any village on Crete.

Priests in robes at the Easter parade of icons

Christ's bier, decorated with flowers and containing His effigy, is carried in solemn procession through the streets at dusk on Good Friday.

Candle lighting takes place at the end of the Easter Saturday Mass. In pitch darkness, a single flame is used to light the candles held by worshippers.

Participants at a workers' rally in Athens on Labour Day, 1 May

that takes place in Chalkída on Evvoia.

Independence Day and **Evangelismós** *(25 Mar)*. A national holiday, with parades and dances nationwide to celebrate the 1821 revolt against the Ottoman Empire. The religious festival, one of the Orthodox church's most important, marks the Archangel Gabriel's announcement to the Virgin Mary that she was to become the Holy Mother. Name day for Evángelos and Evangelía.

April
Megáli Evdomáda, Holy Week *(Apr or May)*, including

Kite-flying competition in Chalkída, Evvoia

Kyriakí ton Vaïón (Palm Sunday), *Megáli Pémpti* (Maundy Thursday), *Megáli Paraskeví* (Good Friday), *Megálo Sávvato* (Easter Saturday) and the most important date in the Orthodox calendar, *Páscha* (Easter Sunday).

Agios Geórgios, St George's Day *(23 Apr)*. This is a day for celebrating the patron saint of shepherds. The date traditionally marks the beginning of the grazing season in Greece.

May
Protomagiá, May Day or Labour Day *(1 May)*. Traditionally, wreaths made with wild flowers and garlic are hung up to ward off evil. In major towns and cities, the day is marked by workers' demonstrations and rallies.

Agios Konstantínos kai Agía Eléni *(21 May)*. A nationwide celebration for the saint and his mother, who were the first Orthodox Byzantine rulers.

Análipsi, Ascension *(40 days after Easter, usually in May)*. This is an important Orthodox feast day that is celebrated all across the nation.

Easter dancing, for young and old alike, continues the outdoor festivities after the midday meal on Sunday.

Easter biscuits celebrate the end of Lent. Another Easter dish, *mayerítsa* soup, is made of lamb's innards and is eaten in the early hours of Easter Sunday.

Egg loaves *(tsouréki)*, made of sweet plaited dough, contain eggs with shells dyed red to symbolize the blood of Christ. Red eggs are also traditionally given as presents on Easter Sunday.

Lamb roasting is traditionally done in the open air on giant spits over charcoal, for lunch on Easter Sunday. The first retsina wine from last year's harvest is opened, and for dessert there are sweet cinnamon-flavoured pastries.

Harvesting barley in July, on the island of Folégandros

Summer

With islands parched and sizzling, the tourist season is now in full swing. Villagers with rooms to let meet backpackers from the ferries, and prices go up. The islands are sometimes cooled by the strong, blustery *meltémi*, a northerly wind from the Aegean, which can blow up at any time to disrupt ferry schedules and delight windsurfers.

In June, the corn is harvested and cherries, apricots and peaches are at their best. In July herbs are gathered and dried, and figs begin to ripen. August sees the mass exodus from Athens to the islands, especially for the festival of the Assumption on 15 August. By late summer the first of the grapes have ripened, while temperatures soar.

Consecrated bread for religious festivals

June

Pentikostí, Pentecost, or Whit Sunday. *(seven weeks after Orthodox Easter)*. An important Orthodox feast day, celebrated throughout Greece.

Agíou Pnévmatos, Feast of the Holy Spirit, or Whit Monday *(the following day)*. A national holiday.

Athens Festival *(mid-Jun to mid-Sep)*, Athens. A cultural festival with modern and ancient theatre and music.

Klídonas *(24 Jun)* Chaniá, Crete *(see pp256–7)*. A festival celebrating the custom of water-divining for a husband. An amusing song is sung while locals dance.

Agios Ioánnis, St John's Day *(24 Jun)*. On some islands bonfires are lit on the evening before. May wreaths are consigned to the flames and youngsters jump over the fires.

Agioi Apóstoloi Pétros kai Pávlos, Apostles Peter and Paul *(29 Jun)*. There are festivals at dedicated churches, such as St Paul's Bay, Líndos, Rhodes *(see p201)*.

Agioi Apóstoloi, Holy Apostles *(30 Jun)*. This time the celebrations are for anyone named after one of the 12 Apostles.

July

Agios Nikódimos *(14 Jul)*, Náxos town. A small folk festival and procession for the town's patron saint.

Agía Marína *(17 Jul)*. This day is widely celebrated in rural

Festivities on Tínos for Koímisis tis Theotókou, 15 August

areas, with feasts to honour this saint. She is revered as an important protector of crops and healer of snakebites. There are festivals throughout Crete and at the town of Agía Marína, Léros.

Profítis Ilías, the Prophet Elijah *(18–20 Jul)*. There are high-altitude celebrations in the Cyclades, Rhodes and on Evvoia at the mountaintop chapels dedicated to him. The chapels were built on former sites of Apollo temples.

Agíou Panteleïmonos Festival *(25–28 Jul)*, Tílos *(see p181)*. Three days of song and dance at Moní Agíou Panteleïmonos, culminating in "Dance of the Koupa", or Cup, at Taxiárchis, Megálo Chorió. There are also celebrations at Moní Pana-chrántou, Andros *(see p213)*.

Simonídeia Festival *(1–19 Aug)*, Kéa. A celebration of the work of the island's famous lyric poet, Simonides (556–468 BC), with drama, exhibitions and dance.

Réthymno Festival *(Jul and Aug)*, Réthymno, Crete. The event includes a wine festival and Renaissance fair.

August

Ippokráteia, Hippocrates Cultural Festival *(throughout Aug)*, Kos *(see pp174–5)*. Art exhibitions are combined with concerts and films, plus the ceremony of the Hippocratic Oath at the Asklepieion.

Dionysía Festival *(first week of Aug)*, Náxos town. A festival of

One of the many local church celebrations during summer, Pátmos

folk dancing in traditional costume, with free food and plenty of wine.

Metamórfosi, Transfiguration of Christ *(6 Aug)*. An important day in the Orthodox calendar, celebrated throughout Greece. It is a fun day in the Dodecanese, and particularly on the island of Chálki, where you may get pelted with eggs, flour, yogurt and squid ink.

Koímisis tis Theotókou, Assumption of the Virgin Mary *(15 Aug)*. A national holiday, and the most important festival in the Orthodox calendar after Easter, and the name day for Maria, Despina, Panayiota (female) and Panayiotis (male). Following the long liturgy on the night of the 14th, the icon of the Madonna is paraded and kissed. Then the celebrations proceed, and continue for days, providing an excellent opportunity to experience traditional music and spontaneous dance. There are celebrations at Olympos on Kárpathos *(see p207)* and at Panagía Evangelístria on Tínos *(see pp216–17)*.

Women in ceremonial costume, Kárpathos

The year's first wine

Autumn

The wine-making months of September and October are still very warm in the Dodecanese, Crete and the Cyclades, although they can be showery further north, and the sea can be rough. October sees the "little summer of St Dimitrios", a pleasant heatwave when the first wine is ready to drink. The shooting season

begins and hunters take to the hills in search of pigeon, partridge and other game. The main fishing season begins, with fish such as bream and red mullet appearing on restaurant menus. By the end of October many islanders are heading for Athens, where they will live during the winter, packing the ferries and wishing each other *Kaló Chimóna* (good winter). But traditional island life goes on: olives are harvested and strings of garlic, onions and tomatoes are hung up to dry for the winter; flocks of sheep are brought down from the mountains; and fishing nets are mended.

September

Génnisis tis Theotókou, birth of the Virgin Mary *(8 Sep)*. An important feast day in the Orthodox church calendar. Also on this day, there is a re-enactment of the Battle of Spétses (1822) in the town's harbour *(see p105)*, followed by a fireworks display and feast.

Ypsosis tou Timíou Stavroú, Exaltation of the True Cross *(14 Sep)*. Though in autumn, this is regarded as the last of Greece's summer festivals. It is celebrated with fervour on Chálki, where locals gather at the church of Stavros *(see pp202–3)*.

October

Agios Dimítrios *(26 Oct)*. A popular and widely celebrated name day. It is also traditionally the day when the first wine of the year is ready to drink.

Ochi Day *(28 Oct)*. A national holiday, with patriotic parades in the cities, and plenty of dancing. The day commemorates the famous reply by Greece's prime minister of the time, Metaxás, to Mussolini's 1940 call for Greek surrender: an emphatic no *(óchi)*.

Greek veterans on Ochi Day

November

Ton Taxiarchón Michaíl kai Gavriíl, *(8 Nov)*. Ceremonies at many monasteries named after Archangels Gabriel and Michael, such as at Panormítis, on Sými *(see p183)*. This is an important name day throughout Greece.

Eisódia tis Theotókou, Presentation of the Virgin in the Temple *(21 Nov)*. A religious feast day, and one of the most important for the Orthodox church. Name day for María, Máry.

Strings of tomatoes hanging out to dry in the autumn sunshine

Diving for the cross at Epiphany, 6 Jan

Winter

Lashed by wild winds and high seas, the islands can be bleak in winter. *Kafeneía* are steamed up and full of men playing cards or backgammon. Women often embroider or crochet, and cook warming stews and soups. Fishermen celebrate Agios Nikólaos, their patron saint, and then preparations get under way for Christmas. The 12-day holiday begins on Christmas Eve, when the wicked goblins, *kallikántzaroi*, are about causing mischief, until the Epiphany in the new year, when they are banished. Pigs are slaughtered for Christmas pork, and cakes representing the swaddling clothes of the infant Christ are made. The Greek Father Christmas comes on New Year's Day and special cakes, called *vasilópita*, are baked with coins inside to bring good luck to the finder.

December

Agios Nikólaos *(6 Dec)*. This is a celebration for the patron saint of sailors. *Panigýria* (religious ceremonies) are held at harbourside churches, and decorated boats and icons are paraded on beaches.

Agios Spyrídon *(12 Dec)*, Corfu *(see pp78–83)*. A celebration for the patron saint of the island, with a parade of his relics.

Christoúgenna, Christmas *(25 Dec)*. A national holiday. Though less significant than Easter in Greece, Christmas is still an important feast day.

Sýnaxis tis Theotókou, meeting of the Virgin's entourage *(26 Dec)*. A religious celebration nationwide, and a national holiday. The next day *(27 Dec)* is a popular name day for Stéfanos and Stefanía, commemorating the Saint Agios Stéfanos.

January

Agios Vasíleios, also known as *Protochroniá (1 Jan)*.

Almond biscuits eaten at Christmas and Easter

A national holiday to celebrate this saint. The day combines with festivities for the arrival of the new year. Gifts are exchanged and the new year greeting is *Kalí Chroniá*.

Theofánia, or Epiphany *(6 Jan)*. A national holiday and an important feast day throughout Greece. There are special ceremonies to bless the waters at coastal locations throughout many of the islands. A priest at the harbourside throws a crucifix into the water. Young men then dive into the sea for the honour of retrieving the cross.

February

Ypapantí, Candlemas *(2 Feb)*. An important Orthodox feast day throughout Greece. This festival celebrates the presentation of the infant Christ at the temple.

Priests in ceremonial robes at Ypapantí, 2 February

Name Days

In the past, most Greeks did not celebrate their birthdays past the age of about 12. Instead they celebrated their name days, or *giortí*, the day of the saint after whom they were named at their baptism. Choice of names is very important in Greece, and children are usually named after their grandparents – though it has also become fashionable to give children names from Greece's history and mythology. On St George's day or St Helen's day (21 May) the whole nation seems to celebrate, with visitors dropping in, bearing small gifts of beautifully wrapped sweets, preserves or flowers, and being given cakes and liqueurs in return. On a friend's name day you may be told *Giortázo símera* (I'm celebrating today) – the traditional reply is *chrónia pollá* (many years). Today, most people also celebrate their birthdays, regardless of their age.

The Climate of the Greek Islands

Throughout the islands, the tendency is for long, dry summers and mild but rainy winters. The Dodecanese, Cyclades and the Cretan coast are buffeted by a dry north wind called the *meltémi*, which can blow up at any time between June and September, moderating the high temperatures.

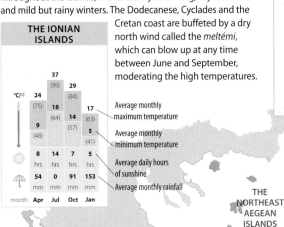

THE NORTHEAST AEGEAN ISLANDS

°C/°F	Apr	Jul	Oct	Jan
max	23 (73)	34 (93)	26 (79)	17 (63)
min	5 (41)	17 (63)	10 (50)	0 (32)
☼	8 hrs	12 hrs	7 hrs	3 hrs
☂	28 mm	11 mm	50 mm	96 mm
month	Apr	Jul	Oct	Jan

THE IONIAN ISLANDS

°C/°F	Apr	Jul	Oct	Jan
max	24 (75)	37 (99)	29 (84)	17 (63)
min	9 (48)	18 (64)	14 (57)	5 (41)
☼	8 hrs	14 hrs	7 hrs	5 hrs
☂	54 mm	0 mm	91 mm	153 mm
month	Apr	Jul	Oct	Jan

Average monthly maximum temperature
Average monthly minimum temperature
Average daily hours of sunshine
Average monthly rainfall

THE SPORADES AND EVVOIA

°C/°F	Apr	Jul	Oct	Jan
max	34 (93)	45 (113)	33 (91)	25 (77)
min	1 (34)	14 (57)	4 (39)	-3 (27)
☼	7 hrs	11 hrs	6 hrs	3 hrs
☂	32 mm	2 mm	36 mm	41 mm
month	Apr	Jul	Oct	Jan

THE NORTHEAST AEGEAN ISLANDS

THE IONIAN ISLANDS

THE SPORADES AND EVVOIA

THE ARGO-SARONIC ISLANDS

°C/°F	Apr	Jul	Oct	Jan
max	32 (90)	42 (108)	37 (99)	21 (70)
min	0 (32)	16 (61)	7 (45)	-4 (25)
☼	8 hrs	12 hrs	6 hrs	4 hrs
☂	23 mm	6 mm	51 mm	62 mm
month	Apr	Jul	Oct	Jan

Athens

THE CYCLADES

THE ARGO-SARONIC ISLANDS

THE DODECANESE

0 km 100
0 miles 100

CRETE

THE CYCLADES

°C/°F	Apr	Jul	Oct	Jan
max	27 (81)	33 (91)	29 (84)	19 (66)
min	10 (50)	20 (68)	13 (55)	6 (43)
☼	6 hrs	13 hrs	6 hrs	3 hrs
☂	19 mm	2 mm	45 mm	91 mm
month	Apr	Jul	Oct	Jan

CRETE

°C/°F	Apr	Jul	Oct	Jan
max	30 (86)	35 (95)	31 (88)	21 (70)
min	8 (46)	18 (64)	12 (54)	5 (41)
☼	8 hrs	13 hrs	6 hrs	3 hrs
☂	26 mm	1 mm	64 mm	95 mm
month	Apr	Jul	Oct	Jan

THE DODECANESE

°C/°F	Apr	Jul	Oct	Jan
max	31 (88)	40 (104)	33 (91)	22 (72)
min	5 (41)	15 (59)	7 (45)	-4 (25)
☼	8 hrs	12 hrs	8 hrs	4 hrs
☂	25 mm	3 mm	61 mm	149 mm
month	Apr	Jul	Oct	Jan

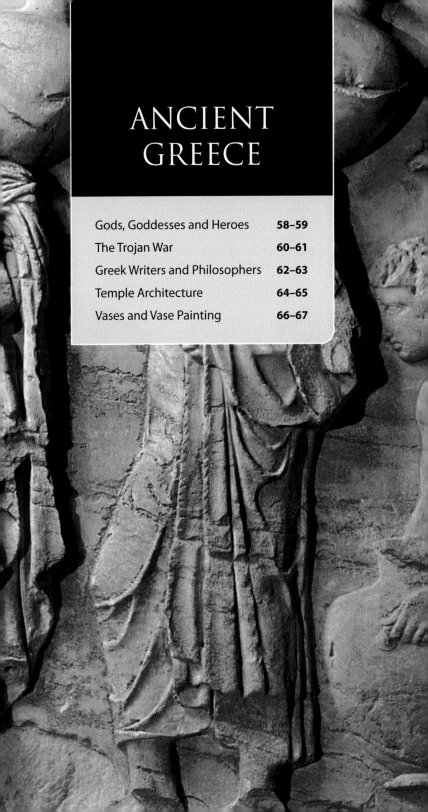

ANCIENT GREECE

Gods, Goddesses and Heroes

The Greek myths that tell the stories of the gods, goddesses and heroes date back to the Bronze Age, when they were told aloud by poets. They were first written down in the early 6th century BC and have lived on in Western literature. Myths were closely bound up with Greek religion and gave meaning to the unpredictable workings of the natural world. They tell the story of the creation and the "golden age" of gods and mortals, as well as the age of semimythical heroes, such as Theseus and Herakles, whose exploits were an inspiration to ordinary men. The gods and goddesses were affected by human desires and failings and were part of a divine family presided over by Zeus. He had many offspring, both legitimate and illegitimate, each with a mythical role.

Hades and Persephone were king and queen of the Underworld (land of the dead). Persephone was abducted from her mother Demeter, goddess of the harvest, by Hades. She was then only permitted to return to her mother for three months each year.

Aris was the goddess of strife.

Zeus was the father of the gods and ruled over them and all mortals from Mount Olympos.

Clymene, a nymph and daughter of Helios, was mother of Prometheus, creator of mankind.

Poseidon, one of Zeus's brothers, was given control of the seas. The trident is his symbol of power, and he married the sea goddess Amphitrite, to whom he was not entirely faithful. This statue is from the National Archaeological Museum in Athens *(see p290).*

Hera, sister and wife of Zeus, was famous for her jealousy.

Athena was born from Zeus's head in full armour.

Paris was asked to award the golden apple to the most beautiful goddess.

Paris's dog helped him herd cattle on Mount Ida, where the prince grew up.

Dionysos, god of revelry and wine, was born from Zeus's thigh. In this 6th-century BC cup, painted by Exekias, he reclines in a ship whose mast has become a vine.

A Divine Dispute

This vase painting shows the gods on Mount Ida, near Troy. Hera, Athena and Aphrodite, quarrelling over who was the most beautiful, were brought by Hermes to hear the judgment of a young herdsman, the Trojan prince, Paris. In choosing Aphrodite, he was rewarded with the love of Helen, the most beautiful woman in the world. Paris abducted her from her husband Menelaos, King of Sparta, and thus the Trojan War began (see pp60–61).

◀ Relief depicting hydria carriers from the north frieze of the Parthenon, c.447–432 BC

Artemis, the chaste goddess of the hunt, was the daughter of Zeus and sister of Apollo. She can be identified by her bow and arrows, hounds and group of nymphs with whom she lived in the forests. Artemis was also the goddess of childbirth.

Happiness, here personified by two goddesses, waits with gold laurel leaves to garland the winner. Wreaths were the prizes in Greek athletic and musical contests.

Helios, the sun god, drove his four-horse chariot (the sun) daily across the sky.

Hermes was the gods' messenger.

Aphrodite, the goddess of love, was born from the sea. Here she has her son Eros (Cupid) with her.

Apollo, son of Zeus and brother of Artemis, was god of healing, plague and also music. Here he is depicted holding a lyre. He was also famous for his dazzling beauty.

The Labours of Herakles

Herakles (Hercules to the Romans) was the greatest of the Greek heroes, and the son of Zeus and Alkmene, a mortal woman. With superhuman strength he achieved success, and immortality, against seemingly impossible odds in the "12 Labours" set by Eurystheus, King of Mycenae. For his first task he killed the Nemean lion, and wore its hide ever after.

Killing the Lernaean hydra was the second labour of Herakles. The many heads of this venomous monster, raised by Hera, grew back as soon as they were chopped off. As in all his tasks, Herakles was helped by Athena.

The huge boar that ravaged Mount Erymanthus was captured next. Herakles brought it back alive to King Eurystheus, who was so terrified that he hid in a storage jar.

Destroying the Stymfalían birds was the sixth labour. Herakles rid Lake Stymfalía of these man-eating birds, which had brass beaks, by stoning them with a sling, having first frightened them off with a pair of bronze castanets.

The Trojan War

The story of the Trojan War, first narrated in the *Iliad*, Homer's 8th-century BC epic poem, tells how the Greeks sought to avenge the capture of Helen, wife of Menelaos, King of Sparta, by the Trojan prince, Paris. The Roman writer Virgil takes up the story in the *Aeneid*, where he tells of the sack of Troy and the founding of Rome. Recent archaeological evidence of the remains of a city identified with ancient Troy in modern Turkey suggests that the myth may have a basis in fact. Many of the ancient sites in the Peloponnese, such as Mycenae and Pylos, are thought to be the cities of some of the heroes of the Trojan War.

Achilles binding up the battle wounds of his friend Patroklos

Gathering of the Heroes

When Paris *(see p58)* carries Helen back to Troy, her husband King Menelaos summons an army of Greek kings and heroes to avenge this crime. His brother, King Agamemnon of Mycenae, leads the force; its ranks include young Achilles, destined to die at Troy.

At Aulis their departure is delayed by a contrary wind. Only the sacrifice to Artemis of Iphigeneia, the youngest of Agamemnon's daughters, allows the fleet to depart.

Fighting at Troy

The *Iliad* opens with the Greek army outside Troy, maintaining a siege that has already been in progress for nine years. Tired of fighting, yet still hoping for a decisive victory, the Greek camp is torn apart by the fury of Achilles over Agamemnon's removal of his slave girl Briseis. The hero takes to his tent and refuses adamantly to fight. Deprived of their greatest warrior, the Greeks are driven back by the Trojans. In desperation, Patroklos persuades his friend Achilles to let him borrow his armour. Achilles agrees and Patroklos leads the Myrmidons, Achilles' troops, into battle. The tide is turned, but Patroklos is killed in the fighting by Hector, son of King Priam of Troy, who mistakes him for Achilles. Filled with remorse at the news of his friend's death, Achilles returns to battle, finds Hector, and kills him in revenge.

King Priam begging Achilles for the body of his son

Patroklos Avenged

Refusing Hector's dying wish to allow his body to be ransomed, Achilles instead hitches it up to his chariot by the ankles and drags it round the walls of Troy, then takes it back to the Greek camp. In contrast, Patroklos is given the most elaborate funeral possible with a huge pyre, sacrifices of animals and Trojan prisoners and funeral games. Still unsatisfied, for 12 days Achilles drags the corpse of Hector around Patroklos's funeral mound until the gods are forced to intervene over his callous behaviour.

Priam Visits Achilles

On the instructions of Zeus, Priam sets off for the Greek camp holding a ransom for the body of his dead son. With the help of the god Hermes he reaches Achilles' tent undetected. Entering, he pleads with Achilles to think of his own father and to show mercy. Achilles relents and allows Hector to be taken back to Troy for a funeral and burial. Although the Greek heroes were greater than mortals, they were portrayed as fallible beings with human emotions who had to face universal moral dilemmas.

Greeks and Trojans, in bronze armour, locked in combat

Achilles Kills the Amazon Queen

Penthesileia was the Queen of the Amazons, a tribe of warlike women reputed to cut off their right breasts to make it easier to wield their weapons. They come to the support of the Trojans. In the battle, Achilles finds himself face to face with Penthesileia and deals her a fatal blow. One version of the story has it that, as their eyes meet at the moment of her death, they fall in love. The Greek idea of love and death would be explored 2,000 years later by the psychologists Jung and Freud.

An early image of the Horse of Troy, from a 7th-century-BC clay vase

Achilles killing the Amazon Queen Penthesileia in battle

The Wooden Horse of Troy

As was foretold, Achilles (see p87) is killed at Troy by an arrow in his heel from Paris's bow. With this weakening of their military strength, the Greeks resort to guile.

Before sailing away they build a great wooden horse, in which they conceal some of their best fighters. The rumour is put out that this is a gift to the goddess Athena and that if the horse enters Troy the city can never be taken. After some doubts, but swayed by supernatural omens, the Trojans drag the horse inside the walls. That night, the Greeks sail back, the soldiers creep out of the horse and Troy is put to the torch. Priam, with many others, is murdered. Among the Trojan survivors is Aeneas, who escapes to Italy and founds the race of Romans: a second Troy. The next part of the story (the *Odyssey*) tells of the heroes' adventures on their way home (see p91).

Death of Agamemnon

Klytemnestra, the wife of Agamemnon, had ruled Mycenae in the 10 years that he had been away fighting in Troy. She was accompanied by Aigisthos, her lover. Intent on vengeance for the death of her daughter Iphigeneia, Klytemnestra receives her husband with a triumphal welcome and then brutally murders him, with the help of Aigisthos. Agamemnon's fate was a result of a curse laid on his father, Atreus, which was finally expiated by the murder of both Klytemnestra and Aigisthos by her son Orestes and daughter Elektra. In these myths, the will of the gods both shapes and overrides that of heroes and mortals.

Greek Myths in Western Art

From the Renaissance onwards, the Greek myths have been a powerful inspiration for artists and sculptors. Kings and queens have had themselves portrayed as gods and goddesses with their symbolic attributes of love or war. Myths have also been an inspiration for artists to paint the nude or Classically draped figure. This was true of the 19th-century artist Lord Leighton, whose depiction of the human body reflects the Classical ideals of beauty. His tragic figure of Elektra is shown here.

Elektra mourning the death of her father Agamemnon at his tomb

Greek Writers and Philosophers

The literature of Greece began with long epic poems, accounts of war and adventure, which established the relationship of the ancient Greeks to their gods. The tragedy and comedy, history and philosophical dialogues of the 5th and 4th centuries BC became the basis of Western literary culture. Much of our knowledge of the Greek world is derived from Greek literature. Pausanias's *Guide to Greece*, written in the Roman period and used by Roman tourists, is a key to the physical remains.

In contrast, the fragments of poems discovered by the poet Sappho, who lived on the island of Lésvos, are exceptional for showing a woman competing in a literary area in the male-dominated society of ancient Greece, and for describing with great intensity her passions for other women.

Hesiod with the nine Muses who inspired his poetry

Epic Poetry

As far back as the 2nd millennium BC, before even the building of the Mycenaean palaces, poets were reciting the stories of the Greek heroes and gods. Passed on from generation to generation, these poems, called *rhapsodes*, were never written down but were changed and embellished by successive poets. The oral tradition culminated in the *Iliad* and *Odyssey (see p91)*, composed around 700 BC. Both works are traditionally ascribed to the same poet, Homer, of whose life nothing reliable is known. Hesiod, whose most famous poems include the *Theogony*, a history of the gods, and

the *Works and Days*, on how to live an honest life, also lived around 700 BC. Unlike Homer, Hesiod is thought to have written down his poems, although there is no firm evidence available to support this theory.

Passionate Poetry

For private occasions, and particularly to entertain guests at the cultivated drinking parties known as *symposia*, shorter poetic forms were developed. These poems were often full of passion, whether love or hatred, and could be personal or, often, highly political. Much of this poetry, by writers such as Archilochus, Alcaeus, Alcman, Hipponax and Sappho, survives only in quotations by later writers or on scraps of papyrus that have been preserved by chance from private libraries in Hellenistic and Roman Egypt. Through these frag-ments we can gain glimpses of the life of a very competitive elite. Since *symposia* were an almost exclusively male domain, there is a strong element of misogyny in much of this poetry.

History

Until the 5th century BC little Greek literature was composed in prose – even early philosophy was in verse. In the latter part of the 5th century, a new tradition of lengthy prose histories, looking at recent or current events, was established with Herodotus's account of the great war between Greece and Persia (490– 479 BC). Herodotus put the clash between Greeks and Persians into a context, and included an ethnographic account of the vast Persian Empire. He attempted to record objectively what people said about the past. Thucydides took a narrower view in his account of the long years of the Peloponnesian war between Athens and Sparta (431– 404 BC). He concentrated

Herodotus, the historian of the Persian Wars

on the political history, and his aim was to work out the "truth" that lay behind the events of the war. The methods of Thucydides were adopted by later writers of Greek history, though few could match his acute insight into human nature.

An unusual vase painting of a *symposion* for women only

The orator Demosthenes in a Staffordshire figurine of 1790

Oratory

Public argument was basic to Greek political life even in the Archaic period. In the later part of the 5th century BC, the techniques of persuasive speech began to be studied in their own right. From that time on some orators began to publish their speeches. In particular, this included those wishing to advertise their skills in composing speeches for the law courts, such as Lysias and Demosthenes. The texts that survive give insights into both Athenian politics and the seamier side of Athenian private life. The verbal attacks on Philip of Macedon by Demosthenes, the 4th-century BC Athenian politician, became models for Roman politicians seeking to defeat their opponents. With the 18th-century European revival of interest in Classical times, Demosthenes again became a political role model.

Drama

Almost all the surviving tragedies come from the hands of the three great 5th-century-BC Athenians: Aeschylus, Sophocles and Euripides.

The latter two playwrights developed an interest in individual psychology (as in Euripides' *Medea*). While 5th-century comedy is full of direct references to contemporary life and dirty jokes, the "new" comedy developed in the 4th century BC is essentially situation comedy employing character types.

Vase painting of two costumed actors from around 370 BC

Greek Philosophers

The Athenian Socrates was recognized in the late 5th century BC as a moral arbiter. He wrote nothing himself, but we know of his views through the "Socratic dialogues", written by his pupil, Plato, examining the concepts of justice, virtue and courage. Plato set up his academy in the suburbs of Athens. His pupil, Aristotle, founded the Lyceum, to teach subjects from biology to ethics, and helped to turn Athens into one of the first university cities. In 1508–11 Raphael painted this vision of Athens in the Vatican.

Plato saw "the seat of ideas" in heaven.

Aristotle, author of the Ethics, had a genius for scientific observation.

Euclid laid the rules of geometry in around 300 BC.

Epicurus advocated the pursuit of pleasure.

Socrates taught by debating his ideas.

Diogenes, the Cynic, lived like a beggar.

Temple Architecture

Temples were the most important public buildings in ancient Greece, largely because religion was a central part of everyday life. Often placed in prominent positions, temples were also statements about political and divine power. The earliest temples, in the 8th century BC, were built of wood and sun-dried bricks. Many of their features were copied in marble buildings from the 6th century BC onwards.

Pheidias, sculptor of the Parthenon, at work

Temple Construction

This drawing is of an idealized Doric temple, showing how it was built and used.

The cella, or inner sanctum, housed the cult statue.

The pediment, triangular in shape, often held sculpture.

The cult statue was of the god or goddess to whom the temple was dedicated.

Fluting on the colum was carved in situ, gui by that on the top bottom dru

The column drums were initially carved with bosses for lifting them into place.

A ramp led up to the temple entrance.

The stepped platform was built on a stone foundation.

700 BC	600 BC	500 BC	400 BC	300 BC

522 Temple of Hera, Sámos (Ionic; see p160)

477–390 Athenian Temple of Apollo, Delos (see pp222–3)

447–405 Temples of the Acropolis, Athens: Athena Nike (Ionic), Parthenon (Doric), Erechtheion (Ionic) (see pp292–4)

Detail of the Parthenon pediment

4th century BC Temple of Lindian Athena, Líndos Acropolis, Rhodes (Doric; see pp200–201)

490 Temple of Aphaia, Aígina (Doric; see pp102–3)

Late 4th century BC Sanctuary of the Great Gods, Samothráki (Doric; see pp136–7)

The gable ends of the roof were surmounted by statues, known as *akroteria*, in this case of a Nike or "Winged Victory". Almost no upper portions of Greek temples survive.

The roof was supported on wooden beams and covered in rows of terracotta tiles, each ending in an upright antefix.

Stone blocks were smoothly fitted together and held by metal clamps and dowels: no mortar was used in the temple's construction.

The ground plan was derived from the megaron of the Mycenaean house: a rectangular hall with a front porch supported by columns.

Caryatids, or figures of women, were used instead of columns in the Erechtheion at Athens' Acropolis. In Athens' Agora (*see p291*), tritons (half-fish, half-human creatures) were used.

The Development of Temple Architecture

Greek temple architecture is divided into three styles, which evolved chronologically, and are most easily distinguished by the column capitals.

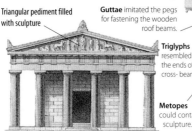

Doric temples were surrounded by sturdy columns with plain capitals and no bases. As the earliest style of stone buildings, they recall wooden prototypes.

Triangular pediment filled with sculpture

Guttae imitated the pegs for fastening the wooden roof beams.

Triglyphs resembled the ends of cross-beams.

Metopes could contain sculpture.

Doric capital

Ionic temples differed from Doric in their tendency to have more columns, of a different form. The capital has a pair of volutes, like ram's horns, front and back.

Akroteria, at the roof corners, could look Persian in style.

The Ionic architrave was subdivided into projecting bands.

The frieze was a continuous band of decoration.

The Ionic frieze took the place of Doric *triglyphs* and *metopes*.

Ionic capital

Corinthian temples in Greece were built under the Romans and only in Athens. They feature columns with slender shafts and elaborate capitals decorated with acanthus leaves.

The pediment was decorated with a variety of mouldings.

Akroterion in the shape of a griffin

The cella entrance was at the east end.

The entablature was everything above the capitals.

Acanthus leaf capital

Vases and Vase Painting

The history of Greek vase painting continued without a break from 1000 BC to Hellenistic times. The main centre of production was Athens, which was so successful that by the early 6th century BC it was sending its high-quality black- and red-figure wares to every part of the Greek world. The Athenian potters' quarter of Kerameikós, in the west of the city, can still be visited today. Beautiful works of art in their own right, the painted vases are the closest we can get to the vanished wall paintings with which ancient Greeks decorated their houses. Although vases could break during everyday use (for which they were intended), a huge number still survive intact or in reassembled pieces.

This 6th-century-BC black-figure vase shows pots being used in an everyday situation. The vases depicted are *hydriai*. It was the women's task to fill them with water from springs or public fountains.

The white-ground *lekythos* was developed in the 5th century BC as an oil flask for grave offerings. They were usually decorated with funeral scenes, and this one, by the Achilles Painter, shows a woman placing flowers at a grave.

The naked woman holding a *kylix* is probably a flute girl or prostitute.

The *Symposion*

These episodes of mostly male feasting and drinking were also occasions for playing the game of kottabos. On the exterior of this 5th-century-BC kylix are depictions of men holding cups, ready to flick out the dregs at a target.

The Development of Painting Styles

Vase painting reached its peak in 6th- and 5th-century BC Athens. In the potter's workshop, a fired vase would be passed to a painter to be decorated. Archaeologists have been able to identify the varying styles of many individual painters of both black-figure and red-figure ware.

The body of the dead man is carried on a bier by mourners.

The geometric design is a prototype of the later "Greek key" pattern.

Chariots and warriors form the funeral procession.

Geometric style characterizes the earliest Greek vases, from around 1000 to 700 BC, in which the decoration is in bands of figures and geometric patterns. This 8th-century-BC vase, placed on a grave as a marker, is over 1 m (3 ft) high and depicts the bier and funeral rites of a dead man.

Eye cups were given an almost magical power by the painted eyes. The pointed base suggests that they were passed around during feasting.

This *kylix* is being held by one handle by another woman feaster, ready to flick out the dregs at a *kottabos* target.

The *rhyton*, such as this one in the shape of a ram's head, was a drinking vessel for watered-down wine. The scene of the *symposion* around the rim indicates when it would have been used.

This drinker holds aloft a branch of a vine, symbolic of Dionysos's presence at the party.

Striped cushions made reclining more comfortable.

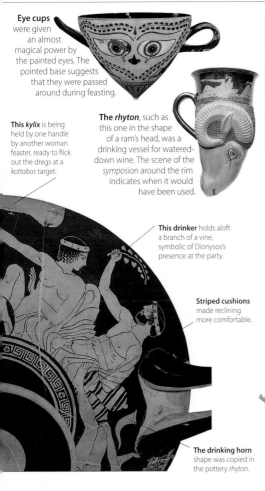

The drinking horn shape was copied in the pottery *rhyton*.

Vase Shapes

Almost all Greek vases were made to be used; their shapes are closely related to their intended uses. Athenian potters had about 20 different forms to choose from. Below are some of the most commonly made shapes and their uses.

The amphora was a two-handled vessel used to store wine, olive oil and foods pre-served in liquid such as olives. It also held dried foods.

This *krater* with curled handles or "volutes" is a wide-mouthed vase in which the Greeks mixed water with their wine before drinking it.

The *hydria* was used to carry water from the fountain. Of the three handles, one was vertical for holding and pouring, two horizontal for lifting.

The *lekythos* could vary in height from 3 cm (1 in) to nearly 1 m (3 ft). It was used to hold oil both in the home and as a funerary gift to the dead.

The *oinochoe*, the standard wine jug, had a round or trefoil mouth for pouring, and just one handle.

The *kylix*, a two-handled drinking cup, was one shape that could take interior decoration.

Black-figure style was first used in Athens around 630 BC. The figures were painted in black liquid clay on to the iron-rich clay of the vase which turned orange when fired. This vase is signed by the potter and painter Exekias.

Red-figure style was introduced in c.530 BC. The figures were left in the colour of the clay, silhouetted against a black glaze. Here a woman pours from an *oinochoe* (wine jug).

THE GREEK ISLANDS AREA BY AREA

The Greek Islands at a Glance

The Greek islands range in size from tiny uninhabited rocks to the substantial islands of Crete and Evvoia. Over the centuries, the sea has brought settlers and invaders and provided the inhabitants with their way of life; it now attracts millions of visitors. Each island has developed its own character through a mix of landscape, climate and cultural heritage. As well as the scattered historical sites, there is enough remote, rugged terrain to satisfy the most discerning walker and, of course, the variety of beaches is extraordinary.

Skópelos
The capital of this rugged island (see pp116–17), Skópelos town, spills down from the hilltop kástro to the sea.

Corfu
The most visited of the Ionians, Corfu (see pp76–87) is a green, fertile island. Corfu town, its capital, contains a maze of narrow streets overlooked by two Venetian fortresses.

Key

- The Ionian Islands *pp72–95*
- The Argo-Saronic Islands *pp96–107*
- The Sporades and Evvoia *pp108–27*
- The Northeast Aegean Islands *pp128–161*
- The Dodecanese *pp162–207*
- The Cyclades *pp208–47*
- Crete *pp248–85*

Corfu
Corfu

THE IONIAN
ISLANDS
(See pp72–95)

Lefkáda

Kefalloniá

Zákynthos

MAINLAND
GREECE

Skiáthos

Chalki

Ath

Spétses

THE AR
SARON
ISLAN
(See pp96–

Kýthira

Aígina
Home to the spectacular and well-preserved ancient Temple of Aphaia, Aígina (see pp100–03) has a rich history due to its proximity to Athens.

Crete
The largest Greek island, Crete (see pp248–85) encompasses historic cities, ancient Minoan palaces, such as Knosós, and dramatic landscapes, including the Samaria Gorge (right).

◄ Palace of the Grand Masters and Mandráki Harbour, Rhodes

Delos
This tiny island *(see pp222–3)* is scattered with the ruins of an important ancient city. From its beginnings as a centre for the worship of Apollo in 1000 BC until its sacking in the 1st century AD, Delos was a thriving cultural and religious centre.

Chíos
The Byzantine monastery of Néa Moní in the centre of the island *(see pp150–57)* contains beautiful mosaics, which survived a severe earthquake in 1881. The mastic villages in the south of the island prospered from the wealth generated by the medieval trade in mastic gum.

Thásos

Samothráki

Límnos

THE NORTHEAST AEGEAN ISLANDS
(See pp128–161)

Lésvos *Mytilíni*

Skýros

E SPORADES ND EVVOIA
ee pp108–127)

Chíos

Andros

Sámos

Ikaría

HE CYCLADES
(See pp208–245)

Léros

Náxos

Ios

Kos

Mílos

Sými

● *Rhodes*

THE DODECANESE
(See pp162–207)

Rhodes

Santoríni

Pátmos
The "holy island" of Pátmos *(see pp166–9)* is where St John the Divine wrote the book of *Revelation*. Pilgrims still visit the Monastery of St John, a fortified complex of churches and courtyards.

Kárpathos

Rethymno

● *Irakleio*

CRETE
(See pp248–285)

Rhodes
Rhodes town is dominated by its walled medieval citadel founded by the crusading Knights of St John. The island has many fine beaches and, inland, some unspoiled villages and remote monasteries *(see pp184–201)*.

0 kilometres   100

0 miles   50

THE IONIAN ISLANDS

CORFU · PAXOS · LEFKADA · ITHACA · KEFALLONIA · ZAKYNTHOS

The Ionian Islands are the greenest and most fertile of all the island groups, characterized by olive groves and cypresses. Lying off the west coast of mainland Greece, these islands have been greatly influenced by Western Europe, in part because the Turks never managed to gain control here, except on the island of Lefkáda.

Famous as the homeland of Homer's Odysseus, these islands were colonized by the Corinthians in the 8th century BC and flourished as a wealthy trading post. In the 5th century BC Corfu defeated Corinth and joined the Athenians, instigating the Peloponnesian War. The Ionians first became a holiday spot during the Roman era.

The islands were not politically grouped together until Byzantine times. They were later occupied by the Venetians whose rule began in 1363 and lasted until 1797. After a brief period of French rule the British took over in 1814. The islands were finally ceded to the Greek state in 1864.

Evidence of the various periods of occupation can be seen throughout the islands, especially in Corfu town, which contains a mixture of Italian, French and British architecture.

Each island has its own distinct character, from tiny Paxós covered in olive trees that play a major role in its economy, to rocky Ithaca and the rugged beauty of

Kefalloniá, with its beautiful beaches, most notably Mýrtou Bay. There is also the quietly serene and historic Zákynthos and mountainous Corfu. The group historically includes Kýthira, which lies just off the southernmost coast of the Peloponnese, but in this guide it is included under the Argo-Saronic Islands due to easier transport connections.

The islands lie on a fault line, which runs south down Greece's west coast, and have been subjected to much earthquake damage. Kefalloniá and Zákynthos in particular suffered massive destruction in the summer of 1953.

Summers are hot and dry, but for the rest of the year the islands have a mild climate; the above-average rainfall supports the lush greenery. There is a huge variety of beaches throughout the Ionians, from resorts providing lively nightlife to quieter stretches, virtually untouched by tourism. The Ionians are popular with sailing enthusiasts, and the larger islands have well-equipped marinas.

Watching from the shade as a ship comes into Sámi town, Kefalloniá

◀ Tripitos Arch on the island of Paxós

Exploring the Ionian Islands

The widely scattered Ionian islands are not particularly well connected with each other, though most are easily reached from the mainland. Corfu is the best base for the northern islands and Kefalloniá for the southern islands. There are few archaeological remains, and museums tend to concentrate on folklore, culture and historical European links. Today's Europeans come mostly for beach holidays. The main islands are large enough to cater for those who like bars and discos, as well as those who prefer a quieter stay, in a family resort or simply in a small fishing village. Traditional Greek life does exist here, inland on the larger islands and on islands such as Meganísi off Lefkáda, or Mathráki, Othonoí and Ereikoússa off northern Corfu.

Islands at a Glance

Corfu *pp76–87*
Ithaca *pp90–91*
Kefalloniá *pp92–3*
Lefkáda *p89*
Paxós *p88*
Zákynthos *pp94–5*

Looking down on Plateía Dimarcheíou in Corfu town with the Town Hall on the left

0 kilometres   25

0 miles   25

A typical house by the roadside in Stavrós village on Ithaca

Key

- ▬▬▬ Motorway
- ▬▬▬ Main road
- ▭▭▭ Minor road
- ▬▬▬ Scenic route
- ‑‑‑ High-season, direct ferry route
- △ Summit

The mountain landscape of Lefkáda

Locator Map

Getting Around

Aside from Paxós, all the main Ionians can be reached by air. Préveza airport serves Lefkáda, which is also connected to the mainland by a road bridge. Larger ferries often travel via the mainland but smaller boats offer direct connections between the islands. Islands often have several ports, so check specific destinations. Buses in the capitals provide services around the islands. Car and bike hire is widespread, but road standards vary, as do local road maps.

An islander working on his boat in Gáïos harbour on Paxós

Préveza

Vónitsa

Lefkáda Town

Agios Nikítas

efkáda

Katoúna

Nydrí

Vathý

Spartochóri

Vasilikí   *Meganísi*   *Kálamos*

Arkoudi   *Kastos*

Fiskárdo   *Atokos*   Astakós

Fríkes

Stavrós   *Ithaca*

Asos

Divaráta   Vathy

a   Aetós   *Echinádes*

Agía Efthymía

Sámi

Kefallonía   Patra

stóli   Kástro

Ainos
1628m   Póros

Pessáda

Markópoulo   Skála

Korithi   Kyllíni

Agios Nikólaos

Volimes

Vrachíonas   Alykés
756m

Zákynthos Town

Argási

Zákynthos   Vasilikós

Laganás

Keri

Holiday apartments at Fiskárdo on Kefalloniá

For keys to symbols *see back flap*

Corfu
Κέρκυρα

Corfu is a green island offering the diverse attractions of secluded coves, stretches of wild coast, bands of coast given over totally to resorts and traditional hill-villages. In 229 BC it became a colony of the Roman Empire, remaining so until AD 337. Byzantine rule then began, intermittently broken by the Goths, the Normans and Angevin rule. Situated between Italy and the Greek mainland, its strategic importance continued under Venetian rule (1386–1797). French rule (1807–14) saw the Greek language restored and the founding of the Ionian Academy, set up for the development of the arts. A period of British rule (1814–64) was followed by unification with Greece.

Perouládes • ⑤ Sidári · Róda • Acha
Avliótes • Karousádes
Kavvadádes • Episk
Episkopí
Valaneió • Nym
Afiónas •
Ano Korakiana •
Skriperó •
Lákones •
⑥ Palaiokastrítsa
Liapádes •
COR
Giannádes •
Ermones •
Vátos ⑦
Glyfá

Angelókastro is a ruined 13th-century fortress, which stands across the bay from Palaiokastrítsa (*see p85*).

Myrtiótissa is one of Corfu's finest beaches (*see p86*).

⑤ Sidári
Unusual rock formations, produced by the effect of sea on sandstone, give the resort of Sidári its appeal. Legend has it that any couple swimming through the Canal d'Amour will stay together forever.

⑦ Vátos
This traditional Greek hill-village is set above the fertile Ropa plain.

⑧ Korisíon Lagoon
This lake is a haven for wildlife and is separated from the Ionian Sea only by some beautiful beaches.

⑥ Palaiokastrítsa
Three main coves cluster around a thickly wooded headland at Palaiokastrítsa. It is now one of the most popular spots on the island and is an ideal base for families, with watersports available and a friendly atmosphere.

| 0 kilometres | 5 |
| 0 miles | 3 |

For hotels and restaurants in this region see p306 and pp322–3

❹ Kassiópi
The unspoiled bay at Kassiópi is overlooked by an attractive quayside lined with tavernas, shops and bars.

VISITORS' CHECKLIST

Practical Information
🗺 120,000. ℹ Corfu town (26610 37520, eotcorfu@otenet.gr). 🎭 For cultural events see
🌐 corfu.gr

Transport
✈ 3 km (2 miles) S of Corfu town. 🚌 Xenofóntos Stratigoú, Corfu town. 🚌

❸ Mount Pantokrátor
This is the highest point on Corfu and offers excellent views over the island and, on a clear day, as far as Italy.

❶ ★ Corfu Town
Corfu town is a delightful blend of European influences. The Liston, focus of café life, was built during the brief French rule. It overlooks the Esplanade that dates to Venetian rule in the town.

❷ Kalámi
Made famous by the author Lawrence Durrell, this remains an attractive coastal village.

Ptichia

Corfu Town ❶

Igoumenitsa, Paxós, Pátra

Potamós

Kanóni

Vlachérna

Pontikonísi

❿ Achílleion Palace
The Empress Elizabeth of Austria built this palace (1890–91).

❿ Achílleion Palace

❾ Benítses

Strongylí

Agios Matthaíos

Moraïtika

Mesongí

Chlomós

❽ Korision Lagoon

Argyrádes   Alykés   *Igoumenitsa*

Perivóli   Lefkímmi

Gardíki Castle was built in the 13th century on the site of Paleolithic remains *(see p86)*.

Dragótina   Kávos

❾ Benítses
An archetypal package holiday resort, Benítses appeals to a young crowd. There is plenty of nightlife and the beach offers every conceivable watersport.

For keys to symbols *see back flap*

❶ Street-by-Street: Corfu Old Town

Πόλη της Κέρκυρας

The 21st century has not spoiled Corfu town, and it continues to be a delightful blend of European influences. The Venetians ruled here for over four centuries, and elegant Italianate buildings, with balconies and shutters, can be seen above French-style colonnades. British rule left a wealth of monuments, public buildings, and the cricket pitch, which is part of the Esplanade, or Spianáda *(see pp80–81)*. This park is a focus for both locals and tourists, with park games and good walks. On its eastern side is the Old Fortress *(see p82)* standing guard over the town, a reminder that Corfu was never conquered by the Turks. The town was designated a UNESCO World Heritage Site in 2007.

New Fortress
(see p82)

The Old Fortress, seen from Corfu old town

THERMISTOKLEOUS

FILELLINON

AGIOU

SPYRIDONOS

KAPODISTRI

N THEOTOKI

N THEOTOKI

ELEFTHE

The Mitrópoli was built in 1577, and became Corfu's Orthodox cathedral in 1841. It is dedicated to St Theodora, whose remains are housed here along with some impressive gold icons.

The Paper Money Museum has a collection of Greek notes and tells Corfu's history through its changes of currency. There is also a display on modern bank-note production *(see p81)*.

Town Hall
(see p82)

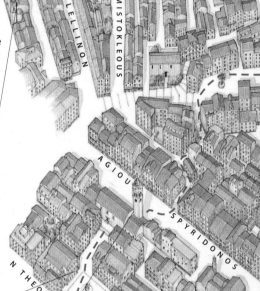

Archaeological Museum
(see pp82–3)

★ **Agios Spyrídon**
The red-domed belfry of this church is the tallest on Corfu. It was built in 1589 and dedicated to the island's patron saint, whose sarcophagus is just to the right of the altar *(see p80)*.

| 0 metres | 25 |
| 0 yards | 250 |

For hotels and restaurants in this region see p306 and pp322–3

Key

— Suggested route

The Corfu Reading Society is housed in this building. The society was founded in 1836 and was modelled on the Reading Society of Geneva. It is the oldest cultural institution in modern Greece.

Byzantine Museum *(see p81)*

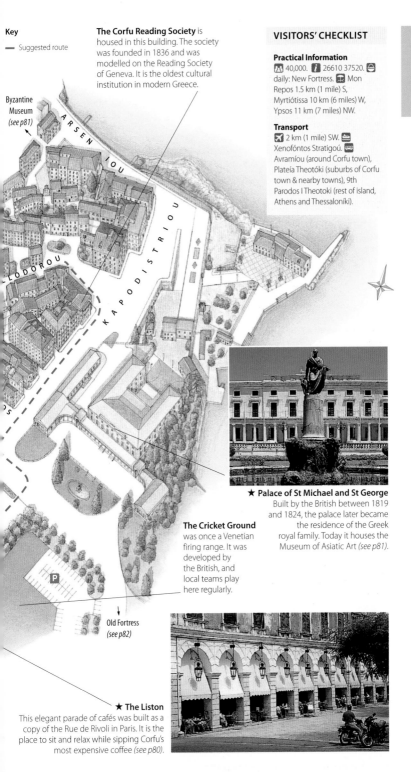

★ **Palace of St Michael and St George**
Built by the British between 1819 and 1824, the palace later became the residence of the Greek royal family. Today it houses the Museum of Asiatic Art *(see p81)*.

The Cricket Ground was once a Venetian firing range. It was developed by the British, and local teams play here regularly.

Old Fortress *(see p82)*

★ **The Liston**
This elegant parade of cafés was built as a copy of the Rue de Rivoli in Paris. It is the place to sit and relax while sipping Corfu's most expensive coffee *(see p80)*.

Exploring Corfu Town

In midsummer the narrow streets of Corfu's old town may be packed with visitors, but there are always quiet places to be found down alleyways and shady cobbled squares. The Corfiot housewives string washing across the streets from their balconies and, below, silversmiths and wood-carvers' shops are hidden away in the maze of alleys. On Nikifórou Theotóki, the southern boundary of the old town, there are several elegant arcaded sections. Built by the French, they are now home to souvenir shops, chapels and churches. Parts of the surrounding new town are quite modern, but many of the buildings date back to French and British rule.

Agios Spyrídon
Agíou Spyrídonos.
Open 6:30am–8pm daily.
The distinctive red-domed tower of Agios Spyrídon guides the visitor to this church, the holiest place on the island. Inside, in a silver casket, is the mummified body of the revered saint, after whom many Corfiot men are named.

Spyrídon himself was not from Corfu but from Cyprus, where he was raised as a shepherd. Later he entered the church and rose to the rank of bishop. He is believed to have performed many miracles before his death in AD 350, and others since – not least in 1716, when he is said to have helped drive the Turks from the island after a six-week siege. His body was smuggled from Constantinople just before the Turkish occupation of 1453. It was only by chance that it came to Corfu, where the present church was built in 1589 to house his coffin. He is regarded as the patron saint of Corfu.

The church is also worth seeing for the immense amount of silverware brought by the constant stream of pilgrims. On four occasions each year (Palm Sunday, Easter Saturday, 11 August and the first Sunday in November) the saint's remains are carried aloft through the streets.

Esplanade
This mixture of park and town square is one of the reasons Corfu town remains such an attractive place. Known as the Esplanade, or Spianáda, it offers relief from the packed streets in summer, either on a park bench or in one of the elegant cafés lining the square on **The Liston**, overlooking the cricket pitch.

The Liston was designed by a Frenchman, Mathieu de Lesseps, who built it in 1807. The name Liston comes from the Venetian practice of having a "List" of noble families in the *Libro d'Oro* or Golden Book – only those on this list were allowed to promenade here.

There are a number of monuments in and around the Esplanade. Near the fountain is the **Enosis Monument**: the word *énosis* means "unification", and this celebrates the 1864 union of the Ionian Islands with the rest of Greece, when British rule came to an end. The marble monument has carvings of the symbols of each of the Ionian Islands.

A statue of **Ioánnis Kapodí-strias**, modern Greece's first president in 1827 and a native of Corfu, stands at the end of the street that flanks the Esplanade and bears his name. He was assassinated in Náfplio in the

Agios Spyrídon can be spotted easily above surrounding buildings with its distinctive red-domed tower

A game of cricket on the pitch by the Esplanade

Peloponnese in 1831 by two Cretans whose uncle he had imprisoned.

Facing this is the **Maitland Rotunda** (1816), a memorial to Sir Thomas Maitland, who became Britain's first Lord High Commissioner to Corfu after the island became a British Protectorate in 1814, though neither he nor his policies were much liked.

🏛 Palace of St Michael and St George

Plateía Spianáda. **Tel** 26610 30443. **Open** Tue–Sun (gardens daily). **Closed** main public hols. 🅿 (gardens free)

The Palace of St Michael and St George was built by the British between 1819 and 1824 from Maltese limestone. It served as the residence of Sir Thomas Maitland, the High Commissioner, and as such is the oldest official building in Greece. When the British left Corfu in 1864 the palace was used for a short time by the Greek royal family, but it was later abandoned and left to fall into disrepair.

The palace, which also once housed the island's treasury, was carefully renovated in the 1950s by Sir Charles Peake, British Ambassador to Greece, and now houses the traffic police, a library and some government offices. Conferences and exhibitions are also held in the palace from time to time.

The Palace of St Michael and St George also houses the **Museum of Asiatic Art**. The core of the museum's collection is the 11,000 items that were collected by a Corfiot diplomat,

Grigórios Mános (1850–1929), during his travels overseas. He offered his vast collection to the state on condition that he could retire and become curator of the museum. Unfortunately he died before he could realize his ambition, though saw its opening in 1927. The exhibits include statues, screens, armour, silk and ceramics from China, Japan, India and other Asiatic countries.

In front of the building is a statue of **Sir Frederick Adam**, the British High Commissioner to Corfu from 1824 to 1831. He built the Mon Repos Villa *(see p83)*, to the south of town and was also responsible for popularizing the west coast resort of Palaiokastrítsa *(see p85)*, one of his favourite spots on the island.

Statue of Sir Frederick Adam

🏛 Byzantine Museum

Prosfórou 30 & Arseníou. **Tel** 26610 38313. **Open** 8:30am–3pm Tue–Sun. **Closed** main public hols. 🅿

The Byzantine Museum opened in 1984 and is housed in the renovated church of Panagía Antivouniótissa, which provided some of the ecclesiastical exhibits. The church is one of the city's oldest buildings. The museum contains about 90 icons dating back to the 15th century. It also has work by artists from the Cretan School. Many of these artists worked and lived on Corfu, as it was a convenient stopping-off point on the journey between Crete and Venice from the 13th to the 17th centuries during the period of Venetian rule.

🏛 Paper Money Museum

Ioniкí Trápeza, Plateía Iróon Kypriakoú Agóna. **Tel** 26610 41552. **Open** Tue, Thu. **Closed** main public hols.

This complete collection of Greek bank notes traces the way in which the island's currency has altered as Corfu's society and rulers changed. The first bank note on the island was issued in British pounds, while later notes show the German and Italian currency of the war years. Another intriguing display shows the process of producing a note from the artistic design to engraving and printing. The museum is housed on the first floor of an imposing 19th-century bank building.

Maitland Rotunda situated in the Esplanade

The Old Fortress towering above the sea on the eastern side of Corfu town

Old Fortress

Tel 26610 48310. **Open** 8am–7:30pm daily (Nov–Mar: to 3pm). **Closed** main public hols. except Sun. limited.

The ruined Old Fortress, or Palaió Froúrio, stands on a promontory believed to have been fortified since at least the 7th or 8th century AD; archaeological digs are still under way. The Old Fortress itself was constructed by the Venetians between 1550 and 1559. The very top of the fortress gives magnificent views of the town and along the island's east coast. Lower down is the church of St George, a British garrison church built in 1840. The fortress is also a venue for concerts and musical events, which are held in the summer months. The fortress is linked to the town by an iron bridge.

New Fortress

Plateía Solomoú. **Tel** 26610 27370. **Open** 8am–7:30pm daily (Nov–Mar: to 3pm).

The Venetians began building the New Fortress, or Néo Froúrio, in 1576 to further strengthen the town's defences. It was not completed until 1589, 30 years after the Old Fortress, hence their respective names. The fortress is used by the Greek navy as a training base, while the surrounding moat is the setting for the town's market.

Mitrópoli

Mitropóleos. **Tel** 26610 39409. **Open** daily.

The Greek Orthodox church of the Panagía Spiliótissa, or Virgin Mary of the Cave, was built in 1577. It became Corfu's cathedral in 1841, when the nave was extended. It is

dedicated to St Theodora, a Byzantine empress whose remains were brought to Corfu at the same time as those of St Spyrídon. Her body is in a silver coffin near the altar.

Plateía Dimarcheíou

Town Hall. **Tel** 26613 62786. **Open** daily. **Closed** main public hols. Agios Iákovos. **Open** daily.

Within this elegant square stands the **Town Hall**. It is a grand Venetian building, which began life in 1663 as a single-storey *loggia* or meeting place for the nobility. It was then converted into the San Giacomo Theatre in 1720, which made it the first modern theatre in Greece. The British added the second floor in 1903, when it became the Town Hall.

Adjacent to it is the Catholic cathedral **Agios Iákovos**, also known by its Italian name of San Giacomo. Built in 1588 and consecrated in 1633, it was badly damaged by bombing in 1943 with only the bell tower surviving intact. Services are held every day, with three Masses on Sundays.

Archaeological Museum

Vraïla 1. **Tel** 26610 30680. **Open** Call for times.

The Archaeological Museum is situated a pleasant stroll south from the centre of town, along the seafront. The museum's collection is not large but a visit is worthwhile to see the centrepiece, the stunning Gorgon frieze.

The frieze, dating from the 6th century BC, originally formed part of the west pediment of the Temple of Artemis near Mon

The 17th-century Catholic cathedral Agios Iákovos in Plateía Dimarcheíou

Corfu Town Centre

① New Fortress
② Mitrópoli
③ Plateía Dimarcheíou
④ Paper Money Museum
⑤ Agios Spyrídon
⑥ Esplanade
⑦ Palace of St Michael
   and St George
⑧ Byzantine Museum
⑨ Old Fortress
⑩ Archeological Museum

```
0 metres        250
0 yards         250
```

The Gorgon frieze in Corfu town's Archaeological Museum

Repos Villa. The layout ensures that the frieze, a massive 17 m (56 ft) long, is not seen until the final room. The museum also displays other finds from the Temple of Artemis and the excavations at Mon Repos Villa.

Environs

Garítsa Bay sweeps south of Corfu town, with the suburb of Anemómilos visible on the promontory. Here, in the street named after it, is the 11th-century church of **Agion Iásonos kai Sosipátrou** (saints Jason and Sossipater). These disciples of St Paul brought Christianity to Corfu in the 1st century AD. Inside are faded wall paintings, including an 11th-century fresco.

South of Anemómilos is **Mon Repos Villa**. It was built in 1824 by Sir Frederick Adam, the second High Commissioner of

the Ionian state, as a present for his wife, and was later passed to the Greek royal family. The remains of the **Temple of Artemis** lie nearby. Opposite the villa are the 5th-century ruins of **Agía Kerkýra**, the church of the old city.

An hour's walk or a short bus ride south of Corfu town is **Kanóni**, with the islands of

Vlachérna and Pontikonísi just off the coast. Vlachérna, with its tiny white convent, is a famous landmark and can be reached by a causeway. In summer boats go to Pontikonísi, or Mouse Island, said to be Odysseus's ship turned to stone by Poseidon. This caused Odysseus to be shipwrecked on Phaeacia, the island often identified with Corfu in Homer's *Odyssey*.

Mon Repos Villa
Tel 26610 41369. **Open** 8am–4pm daily (Nov–Mar: to 3pm).

The church of Agion Iásonos kai Sosipátrou

For keys to symbols *see back flap*

Around Northern Corfu

Northern Corfu, in particular the northeast coast, is emphatically holiday Corfu, with a string of resorts along the main coastal road. These include popular spots such as Kassiópi and Sidári, though there are also quieter villages like Kalámi. In the northwest is one of Corfu's prettiest areas, Palaiokastrítsa, a jigsaw of bays and beaches. Inland stands Mount Pantokrátor, a reminder that there is also a rugged interior to explore.

Looking southwards over the beach at Kalámi Bay

❷ Kalámi

Καλάμι

26 km (16 miles) NE of Corfu town. 🚌 18. 🚤 to Kassiópi.

Kalámi village has retained its charm despite its popularity with visitors. A handful of tavernas line its sand and shingle beach, while behind them cypress trees and olive groves climb up to the lower slopes of Mount Pantokrátor. The hills of Albania are a little over 2 km (1 mile) across Kalámi Bay.

Kalámi's obvious appeal attracted the author Lawrence Durrell to the village in 1939.

Only during the day in high season, when visitors from holiday resorts throng his "peaceful fishing village", might Durrell fail to recognize the place. In the evenings and outside the months of July and August, normality returns.

❸ Mount Pantokrátor

Όρος Παντοκράτωρ

29 km (18 miles) N of Corfu town. 🚌 to Petáleia.

Mount Pantokrátor, whose name means "the Almighty", dominates the northeast bulge of Corfu. It rises so steeply that its peak, at 906 m (2,972 ft), is less than 3 km (2 miles) from the beach resorts of Nisáki and Barbati. The easiest approach is from the north, where a rough road goes all the way to the small monastery at the top. The mountain has great appeal to naturalists as well as walkers, but exploring its slopes is not something to be undertaken lightly, as Corfu's weather can change suddenly. However, the reward is a view to Albania and Epirus in the east, of Corfu town to the south, and even west to Italy when weather conditions are clear.

❹ Kassiópi

Κασσιόπη

37 km (23 miles) N of Corfu town. 🚌 600. 🚌 🚤 Avláki 2 km (1 mile) S.

Kassiópi has developed into one of Corfu's busiest holiday centres without losing either its charm or character. It is set around a harbour that lies between two wooded headlands. Although there is plenty of nightlife to attract younger holiday-makers, there are no high-rise hotels to spoil the setting. The heart of the town is at its harbour, with tavernas and souvenir shops overlooking fishing boats moored alongside motorboats from the many watersports schools.

In the 1st century AD the Emperor Nero is said to have visited a Temple of Jupiter, which was situated on the western side of the harbour, where the church of **Kassiopítissa** now stands. The ruins of a 13th-century castle are a short walk further west.

Fishing boats moored in Kassiópi harbour, east of the castle ruins

For hotels and restaurants in this region see p306 and pp322–3

Caretaker monk at Moní Theotókou, Palaiokastrítsa

The British High Commissioner, Sir Frederick Adam *(see p81)*, loved to picnic here but did not like the awkward journey from Corfu town, so he had a road built between the two.

On the main headland stands **Moní Theotókou**, which dates from the 17th century, although the first monastery stood here in 1228. The church's ceiling features a fine carving of the *Tree of Life*.

Views from the monastery include **Angelókastro**, the ruined 13th-century fortress of Michaíl Angelos Komninós II, the Byzantine despot of Epirus. Situated above the cliffs west of Palaiokastrítsa, the fortress was never taken, and in 1571 it sheltered locals from another failed Turkish attempt to conquer Corfu. The remains include a hilltop chapel and some hermit cells and caves.

Outlying Islands

Corfu has three offshore islands. **Mathráki** offers the simplest Greek island life, with two villages and only a few rooms to rent. **Ereikoússa** is the most popular island, largely because of its glorious sandy beaches. **Othonoí** is the largest island and has the best facilities but lacks the finer beaches.

❺ Sidári

Σιδάρι

31 km (20 miles) NW of Corfu town.
🏠 300. 🚌 🚆 Róda 6 km (4 miles) E.

One of Corfu's first settlements, the village of Sidári has pre-Neolithic remains dating back to about 7000 BC. Today it is a bustling holiday centre with the twin attractions of sandy beaches and unusual rock formations. Erosion of the sandstone has created a number of caves and tunnels, the most famous being a channel between two rocks known as the Canal d'Amour *(see p76)*.

❻ Palaiokastrítsa

Παλαιοκαστρίτσα

26 km (16 miles) NW of Corfu town.
🏠 600. 🚌

Palaiokastrítsa is one of Corfu's most popular spots. Three main coves cluster around a wooded headland, dividing into numerous other beaches which are popular with families because swimming is safe.

Watersports are available as well as boat trips out to see the nearby grottoes. Until the early 19th century the place was noted for its beauty but access was difficult.

Writers and Artists in Corfu

The poet Dionýsios Solomós lived on Corfu from 1828 until his death in 1857. He is best known for his poem "Hymn to Freedom", part of which was adopted as the national anthem after Independence. Other writers have also found inspiration on Corfu, including the British poet and artist Edward Lear, who visited the island in the 19th century, and the Durrell brothers, who both wrote about Corfu. Gerald described his idyllic 1930s childhood in *My Family and Other Animals*, while Lawrence produced *Prospero's Cell* in 1945. He wrote this while staying in Kalámi, where he was visited by Henry Miller, whose 1941 book *The Colossus of Maroussi* is one of the most accurate and endearing books about Greece.

A view from the Benítses road near Gastoúri, by Edward Lear

Around Southern Corfu

Less mountainous but more varied than the north, southern Corfu encompasses Benítses' wild nightlife and the shy wildlife of the Korisíon Lagoon. Much of Corfu's produce grows in the fertile Rópa Plain north of Vátos. To the south lies Myrtiótissa, once described as the world's most beautiful beach. Bus services are good, but to explore off the beaten track you will need your own car.

View inland over the freshwater Korisíon Lagoon

❼ Vátos

Βάτος

24 km (15 miles) W of Corfu town. 🏠 480. 🚌 🚐 Myrtiótissa 2 km (1 mile) S, Ermones 2 km (1 mile) W.

In the hillside village of Vátos, the whitewashed houses with flower-bedecked balconies offer a traditional image of Greece. Vátos has two tavernas and a handful of shops and has mostly remained untainted by the impact of tourism. From the village, a steep climb leads up the mountainside to the top of Agios Geórgios (392 m/1,286 ft). Below lies the fertile Rópa Plain and a beach at Ermones.

Environs
The glorious beach at **Myrtiótissa**, 2 km (1 mile) south of Vátos, is named after the 14th-century monastery behind it dedicated to Panagía Myrtiótissa (Our Lady of the Myrtles). The beach is a long golden sweep of sand backed with cypress and olive trees. Lawrence Durrell was fond of the area and, in his book *Prospero's Cell*, referred to Myrtiótissa as "perhaps the loveliest beach in the world".

South of Vátos lies **Pélekas**, another picturesque and unspoiled hillside village. Its traditional houses tumble down wooded slopes to the small and secluded beach below. Above this is the **Kaiser's Throne**, the hilltop from which Kaiser Wilhelm II of Germany loved to watch the sunset while staying at the Achílleion Palace.

❽ Korisíon Lagoon

Λίμνη Κορισσίων

42 km (26 miles) S of Corfu town. 🚐 Gardíki 1 km (0.5 mile) N.

The Korisíon Lagoon is a 5-km (3-mile) stretch of water, separated from the sea by some of the most beautiful dunes and beaches on Corfu. The lake remains a haven for wildlife, despite the Greek love of hunting. At the water's edge are a variety of waders such as sandpipers and avocets, egrets and ibis. Flowers include sea daffodils and Jersey orchids.

Almost 2 km (1 mile) north lies **Gardíki Castle**, built in the 13th century by Mihaíl Angelos Komninós II *(see p85)*, with the ruined towers and outer castle walls still standing. The site is also known for a find of Paleolithic remains, now removed.

❾ Benítses

Μπενίτσες

14 km (9 miles) S of Corfu town. 🏠 1,400. 🚌 🚐 Benítses.

Benítses has become the archetypal package holiday resort. Its appeal is to young people, and not to those seeking peace and quiet or a real flavour of Greece.

The beaches offer every conceivable watersport, and at the height of the season are extremely busy. The nightlife is also very lively: the bars and discos close about the same time as the local fishermen return from their night at sea.

There are few sights of interest in Benítses other than the remains of a Roman bathhouse near the harbour square.

A whitewashed house in the attractive village of Vátos

⑩ Achílleion Palace

Αχίλλειον

19 km (12 miles) SW of Corfu town. 🚌
Tel 26610 56245. Palace & gardens:
Open 8am–2:30pm Sat–Tue (to 7pm
Mon & Tue). 🎫

A popular day trip from any of Corfu's resorts, the Achílleion Palace was built in 1890–91 by the Italian architect Raphael Carita for the Empress Elizabeth of Austria (1837–98), formerly Elizabeth of Bavaria and best known as Princess Sissy. She used it as a personal retreat from her problems at the Habsburg court: her health was poor and her husband, Emperor Franz Josef, notoriously unfaithful. After the assassination of the Empress Elizabeth by an Italian anarchist in 1898, the palace lay empty for nearly a decade, until it was bought by Kaiser Wilhelm II in 1907. It is famous as the set used for the casino in the James Bond film *For Your Eyes Only*.

The outer entrance to the Achílleion's gardens

A 19th-century painting of
Elizabeth of Bavaria
by Franz Xavier

The Gardens

The lush green gardens below the palace are terraced on a slope which drops 150 m (490 ft) to the coast road. The views along the rugged coast both north and south are spectacular. In the grounds the walls are draped with colourful bougainvillea and a profusion of palm trees. The gardens are also dotted with numerous statues, especially of Achilles, who was the empress's hero, after whom the palace is named. One moving bronze of the *Dying Achilles* is by the German sculptor Ernst Herter. The statue is thought to have appealed to the unhappy empress following the tragic suicide of her second son, the Archduke Rudolph, at Mayerling. Another impressive statue of the hero Achilles is the massive 15 m- (49 ft-) high, cast-iron figure, which was commissioned by Kaiser Wilhelm II.

The Palace

There have been numerous attempts to describe the Achílleion's architectural style, ranging from Neo-Classical to Teutonic, although Lawrence Durrell was more forthright, and declared it "a monstrous building". The empress was not particularly pleased with the finished building, but her fondness for Corfu made her decide to stay.

The palace does, however, contain a number of interesting artifacts. Inside, some original furniture is on display and on the walls there are some fine paintings of Achilles, echoing the bronze and stone statues seen in the gardens. Another exhibit is the strange saddle-seat that was used by Kaiser Wilhelm II whenever he was writing at his desk.

Visitors requiring a pick-me-up after touring the palace can try the Vasilákis Tastery, opposite the entrance, and sample this local distiller's many products, which include a number of Corfiot wines, oúzo and the speciality kumquat liqueur.

The Legend of Achilles

Shortly after his birth, Achilles was immersed in the River Styx by his mother Thetis. This left him invulnerable apart from the heel where she had held him. Achilles' destiny lay at Troy *(see pp60–61)*; Helen, the wife of King Menelaos of Sparta, was held by Paris at Troy, where Menelaos and his allies laid siege. As the Greeks' mightiest warrior, it was Achilles who killed the Trojan hero Hector. However, he did not live to see Troy fall, since he was struck in the heel by a fatal arrow from Paris's bow.

Achilles victoriously dragging the body of Hector around the walls of Troy

Local fishing boats moored at the eastern end of the harbour at Gáïos

Paxós
Παξοί

🏔 2,700. 🚢 Gáïos, Lákka. 🚌 Gáïos.
🛈 Gáïos (26623 60301). 🚤 Mogonísi 3 km (2 miles) SE of Gáïos.

Paxós is green and wooded, with a few farming and fishing villages. The thick groves of olive trees are still a major part of the island's economy. In mythology, Poseidon created Paxós for his mistress, and its small size has saved it from the turbulent history of its larger neighbours. Paxós became part of the Greek state along with the other Ionians in 1864.

Gáïos
Gáïos is a lively, if small-scale, holiday town with two harbours: the main port where ferries dock and, a short walk away, the small harbour, lined with 19th-century houses with Venetian-style shutters and balconies. At the waterfront

stands Pyropolitís, a statue of Constantínos Kanáris, hero in the Greek Revolution (see pp46–7). The grandest house was once residence of the British High Commissioner of Corfu. Behind it are narrow old streets, bars and tavernas.

Around the Island
One main road goes from the south to the north of the island. There are few cars and the best way to get about is by bicycle or moped. Many pleasant tracks lead through woods to high cliffs or secluded coves. At the end of a deep, almost circular inlet on Paxós' northern coast lies the town of **Lákka**. This pretty coastal town is backed by olive groves and pine-covered hills. Lákka is popular with

day-trippers from Corfu, but at night it returns to being a quiet fishing village, with a few rooms to rent and only a scattering of restaurants and cafés.

To the east is the small village of **Pórto Longós**, which is the most attractive of the island's settlements. It has a pebble beach, a handful of houses, a few shops, and tavernas whose tables stand at the water's edge. Pórto Longós is a peaceful place where the arrival of the boat bringing fruit and vegetables every few days is a major event. Paths from the village lead through olive groves to several quiet coves, good for swimming.

Outlying Islands
Around 100 people live on **Antípaxos**, south of Paxós, and mostly in Agrapidiá, although there are a few hamlets inland. The island is unusual in that olive trees are easily outnumbered by vines, which produce Antípaxos's potent and good-quality wine. There is little tourism and no accommodation available, although the sandy beaches do fill up in summer with visitors from Paxós. Offshore from Gáïos lie the two islets of Panagiá and Agios Nikólaos.

Statue of Pyropolitís on the waterfront in Gáïos

The coastal town of Lákka, on the northern coast of Paxós

Houses on a hillside near Kalamítsi

Lefkáda
Λευκάδα

🏘 25,000. 🚢 Nydrí, Vasilikí.
🚌 Dimitroú Golémi, Lefkáda town.
ℹ Lefkáda town (26450 29379).
🚤 Lefkáda town: daily.

Lefkáda offers variety, from mountain villages to beach resorts. It has had a turbulent history, typical of the Ionian Islands, since the Corinthians took control of the island from the Akarnanians in 640 BC, right up until the British left the island in 1864.

Lefkáda Town
The town has suffered repeated earthquakes, but there are interesting backstreets and views of the beautiful ruins of the 14th-century **Sánta Mávra fortress**. Situated on the mainland opposite, the fortress is connected to Lefkáda by a causeway. The main square, Plateía Agíou Spyrídona, is named after the 17th-century church with its rare metal bell towers. Nearby, the **Phonograph Museum** houses a private collection of records and old phonographs. The small **Folk Museum** has local costumes and old photographs of island life. Above the town, **Moní Faneroménis** was founded in the 17th century, though the present buildings date from the 19th century. Its icon of the Panagía is also 19th century.

🏛 **Phonograph Museum**
Konstantínou Kalkáni 10. **Tel** 26450 21088. **Open** Apr–Oct: 10am–2pm & 7–10pm. **Closed** main pub hols. ♿

🏛 **Folk Museum**
Stefanitis 2. **Tel** 26450 22778. **Open** Apr–Oct: Tue–Sun; Nov–Mar: Sun. **Closed** main pub hols.

A bell at Moní Faneroménis

Around the Island
The best way to see the island is to hire a moped or bike, although bus services operate from Lefkáda town. **Agios Nikítas** is a traditional small resort with a harbour and beach. To the south, **Kalamítsi** is a typical Lefkáda mountain village. In the south, the main hill-village is **Agios Pétros**, still a rural community despite the nearby resort of **Vasilikí**, a windsurfer's paradise. **Nydrí** is the main resort on the east coast, with splendid views of the offshore islands.

Outlying Islands
Meganísi has retained its rural lifestyle. Most boats from Nydrí stop at Vathý, the main port. Uphill, the small village of Katoméri has the island's only hotel. **Skorpios** is a private island owned by Aristotle and now, Athina, Onassis.

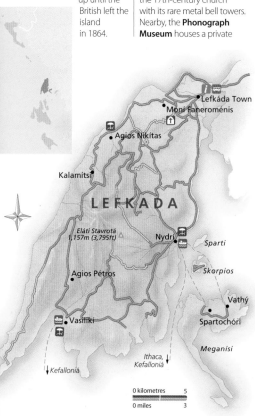

Sailing boats off the white-sand beach at Vasilikí

For keys to symbols *see back flap*

The pebble beach of Pólis Bay on the northwest coast of Ithaca

Ithaca
Ιθάκη

4,000. 🚢 Vathý. 🚌 ℹ️ Vathý (26740 26740). 🚤 Pólis Bay 20 km (12 miles) NW of Vathý.

Small and rugged, Ithaca is famous, according to Homer's epic the *Odyssey*, as the home of Odysseus. Finds on Ithaca date back as far as 4000–3000 BC, and by Mycenaean times it had developed into the capital of a kingdom that included its larger neighbour, Kefalloniá.

Vathý
The capital, also known as Ithaca town, is an attractive port, its brown-roofed houses huddled around an indented bay. The surrounding hills were the site for the first settlement, but the harbour itself was settled in the medieval period, and Vathý became the capital in the 17th century. Destroyed by an earthquake in 1953, it was reconstructed and declared a traditional settlement, which requires all new buildings to match existing styles.

The **Archaeological Museum** contains a collection mainly of vases and votives from the Mycenaean period. In the church of **Taxiárchis** is a 17th-century icon of Christ, believed to have been painted by El Greco (*see p272*).

🏛 **Archaeological Museum**
Behind OTE office. **Tel** 26740 32200.
Open 8:30am–3pm Tue–Sun.
Closed main public hols. ♿ 📷

Around the Island
With just one main town, high hills, a few pebble beaches and little development, Ithaca is a pleasant island to explore. A twice-daily bus (four in season) links Vathý to villages in the north and there are some taxis.

Stavrós, the largest village in northern Ithaca, has only 300 inhabitants but is a thriving hill community and market centre. Nearby Pólis Bay is thought to have been the old port of ancient Ithaca, and site of an important cave sanctuary to the Nymphs. **Odysseus' Palace** may have stood above Stavrós on the hill known as Pilikáta. To find it, ask for directions at the **Archaeological Museum**, whose curator gives guided tours in several languages. Among the local finds is a piece of a terracotta mask from Pólis cave bearing the inscription "Dedicated to Odysseus".

🏛 **Archaeological Museum**
Stavrós. **Tel** 26740 26740.
Open 8:30am–3pm Tue–Sun.
Closed main public hols. ♿

[Map of Ithaca showing: Exogi, Platreithiás, Kefalloniá, Frikes, Lefkáda, Pilikáta, Pólis Bay, Stavrós, Kióni, ITHACA, Léfki, Anogí, Agios Ioánnis, Astakós, Kefalloniá Pátras, Piso Aetós, Vathy, Perachóri, Kefalloniá, Taxiárchis, Filiatró]

0 kilometres 5
0 miles 2

The red-domed roof of a church in Stavrós

The Legend of Odysseus' Return to Ithaca

Odysseus, the king of Ithaca, had been unwilling to leave his wife Penelope and infant son Telemachos and join Agamemnon's expedition against Troy *(see pp60–61)*. But once there his skills as warrior and speaker, and his cunning, ensured he played a vital role. However, his journey home was fraught with such perils as the monstrous one-eyed Cyclops, the witch Circe, and the seductive Calypso. His blinding of the Cyclops angered the god Poseidon, who ensured that, despite the goddess Athena's support, Odysseus lost all his companions, before the kindly Phaeacians brought him home, 10 years after he left Troy. On Ithaca, Odysseus found Penelope besieged by suitors. Disguising himself as a beggar, and aided by his loyal swineherd Eumaios and his son, he killed them all and returned to his marriage bed and to power.

Odysseus' homecoming is depicted in this 15th-century painting attributed to Coracelli. Odysseus had been washed ashore on Phaeacia (Corfu), where King Alkinoös took pity and ferried him back to Ithaca.

Penelope wove a shroud for Odysseus' father Laertes, shown in this 1920 illustration by A F Gorguet. She refused to remarry until the shroud was finished; each night she would unpick the day's weaving.

Eumaios, Odysseus' faithful swineherd, gave his disguised master food and shelter for the night on his arrival in Ithaca. Eumaios then demonstrated his loyalty by praising his absent king while describing the situation on Ithaca to Odysseus. Their meeting is shown on this 5th-century-BC Athenian vase.

Telemachos had challenged Penelope's suitors to string Odysseus' bow and thereby to win his mother's hand in marriage. The suitors all failed the test. Odysseus locked them in the palace hall, strung the bow, and revealed his identity before slaughtering them.

Argus, Odysseus' aged dog, recognized his master without prompting, a feat matched only by Odysseus' old nurse, Eurykleia. Immediately after their meeting, Argus died.

Kefalloniá
Κεφαλλονιά

Archaeological finds date Kefalloniá's first inhabitants to about 50,000 BC. In Mycenaean times the island flourished and remained Greek until the 2nd century BC, when it was captured by the Romans. It was squabbled over by many powers but from 1500 to 1700 it shared the Ionians' history of Venetian occupation. Kefalloniá's attractions range from busy beach resorts to Mount Aínos National Park, which surrounds the Ionians' highest peak.

A church tower in the countryside between Argostóli and Kástro

Argostoli

A big, busy town with lush surrounding countryside, Kefalloniá's capital is situated by a bay with narrow streets rising up the headland on which it stands. Its traditional appearance is deceptive, as Argostóli was destroyed in the 1953 earthquake and rebuilt with donations from emigrants. The destruction and rebuilding is shown in a photographic collection at the **Historical and Folk Museum**. Other exhibits range from rustic farming implements to traditional folk costumes.

The nearby **Archaeological Museum** includes finds from the Sanctuary of Pan, based at the Melissáni Cave-Lake and an impressive 3rd-century-AD bronze head of a man found at Sámi. From the waterfront you can see the **Drápanos Bridge**, built during British rule in 1813.

🏛 **Historical and Folk Museum**
Ilía Zervoú 12. **Tel** 26710 28835.
Open 9am–2pm Mon–Sat.
Closed main public hols. 🔲

🏛 **Archaeological Museum**
Rókkou Vergotí. **Tel** 26710 28300.
Closed for renovation.

Around the Island

It takes time to travel around Kefalloniá, the largest of the Ionian Islands. Despite this, driving is rewarding, with some beautiful spots to discover. The island's liveliest places are **Lássi** and the south coast resorts; elsewhere there are quiet villages and the scenery is stunning. A bus service links Argostóli with most parts of the island.

Capital of Kefalloniá until 1757, the whitewashed village of **Kástro** still flourishes outside the Byzantine fortress of Agios Geórgios. The Venetians renovated the fortress in 1504 but it was damaged by earthquakes in 1636 and 1637, and the 1953 earthquake finally ruined it. The large and overgrown interior is a haven for swallowtail butterflies.

Lefkáda

Fiskárdo

Ithaca (Píso Aetós)

Asos

Agios Spyrídon

Myrtou Bay

Zóla

Sinióri

Kardakáta

Agía Efthimía

Ithaca (Vathý)

Ithaca (Píso Aetós)

Ithaca (Vathý)

Agía Thékla

KEFALLONIA

Agios Dimítrios

Fársa

Melissáni Cave-Lake

Agriliá

Lixoúri

Dilináta

Sámi

Pátra

Drogkaráti Cave

Argostoli

Fragkáta

Lássi

Kástro

Kyllíni

Miniés

Peratáta

Forest Station

Moni Agíou Andréa

Vlacháta

Póros

Pessáda

Mount Aínos 1,630 m (5,350 ft)

Zákynthos

Atsoupádes

Pástra

Kyl

Markópoulo

Néa Skála

0 kilometres — 10
0 miles — 5

Visitors to the blue waters of the subterranean Melissáni Cave-Lake

VISITORS' CHECKLIST

Practical Information

🏠 36,000. 🛈 Waterfront, Argostóli (26710 22248). 🚍 daily, Argostóli. 🎭 Panagía or Snake Festival at Markópoulo: 15 Aug; Wine Festival at Fragáta: 1st Sat after 15 Aug. 🌐 kefallonia.gov.gr

Transport

✈ 9 km (5.5 miles) S of Argostóli. 🚌 Argostóli, Fiskárdo, Agía Efthimía, Sámi, Póros, Pessáda. 🚌 Ioánnou Metaxá, Argostóli.

In 1264 there was a convent on the site of **Moní Agíou Andréa**. The original church was damaged in 1953, but has been restored as a museum to house icons and frescoes made homeless by the earthquake. The new church houses the monastery's holiest relic, supposedly the foot of the apostle Andrew.

There was once a sanctuary to Aenios Zeus at the summit of **Mount Aínos**, which is 1,630 m (5,350 ft) high. Wild horses live in the Mount Aínos National Park, and the slopes of the mountain are covered with the native fir tree, *Abies cephalonica*. A road leads up towards the mountain's summit, but soon becomes a very rough track.

On the east coast, **Sámi** has ferry services to the Peloponnese

Apostle Andrew from the Moní Agíou Andréa

and Ithaca. Nearby are two caves, Drogkaráti Cave, 3.5 km (2 miles) southwest and the Melissáni Cave-Lake, 2 km (1 mile) to the north. **Drogkaráti** drips with stalactites. It is the size of a large concert hall and is sometimes used as such due to its fine acoustics.

The subterranean **Melissáni Cave-Lake** was a sanctuary of Pan in Mycenaean times. Part of its limestone ceiling has collapsed, creating a haunting place with deep, blue water. A channel leads to the enclosed section, where legend says that the nymph Melissáni drowned herself when she was spurned by Pan.

Fiskárdo is Kefaloniá's prettiest village. Its pastel-painted 18th-century Venetian houses cluster by the harbour, which is a popular berth for

yachts. It is also busy in the summer with daily ferry services and day trips from elsewhere on Kefaloniá. Despite the crowds and gift shops Fiskárdo retains its charm.

Asos is an unspoiled village on Kefaloniá's west coast. The surrounding hilly terrain is noted for its stone terracing, which once covered the island. On the peninsula across the isthmus from Asos is a ruined Venetian fortress, built in 1595, which has seen occupation by Venetians, and stays by the French and Russians in the 19th century. Now Asos sees mostly day trippers, as there is little accommodation in the village. South of Asos is **Mýrtou Bay**, a lovely cove with the most beautiful beach on the island.

🛈 Moni Agíou Andréa

Peratáta village. **Tel** 26710 69700. **Open** daily (museum open 8am–2pm Mon–Sat). 🎫 museum only.

Overlooking Asos in the northwest of the island

Zákynthos
Ζάκυνθος

Zákynthos was inhabited by Achaians until Athens took control in the 5th century BC. They were followed by a succession of rulers, including the Spartans, Macedonians, Romans and Byzantines. The Venetians ruled from 1484 until 1797, and Zákynthos finally joined the rest of Greece in 1864. An attractive and green island, there are mountain villages, monasteries, fertile plains and beautiful views to reward exploration.

Statue of the poet Solomós in the main square, Zákynthos town

Zákynthos Town
Completely destroyed in the 1953 earthquake that hit the Ionian Islands, Zákynthos town has now been rebuilt with efforts to recapture its former grace. The traditional arcaded streets run parallel to the waterfront, where fishing boats arrive each morning to sell their catch. Further down the waterfront the ferry boats dock alongside grand Mediterranean cruise ships.

At the southern end of the harbour is the impressive church of **Agios Dionýsios**, the island's patron saint (1547–1622). The church, which houses the body of St Dionýsios in a silver coffin, was built in 1925 and survived the earthquake. The **Byzantine Museum** has a scale model of the pre-earthquake town, an elegant city built by the Venetians. It also houses a breathtaking collection of icons and frescoes rescued from the island's destroyed churches and monasteries.

North of here is the **Solomós Museum**, which contains the tomb of the poet Dionýsios Solomós (1798–1857), author of the Greek national anthem. The collection details lives of prominent Zákynthiot citizens.

A short walk north from the town centre, **Stráni hill** offers good views, while the Venetian kástro, above the town, has even more impressive views of the mainland. The ruined walls contain remnants of several churches and an abundance of plants and wildlife.

🏛 **Byzantine Museum**
Tel 26950 42714. **Open** Jul–Oct: 8:30am–3pm Tue–Sun; Nov–Jun: call for times. **Closed** main public hols.

🏛 **Solomós Museum**
Tel 26950 28982. **Open** 9am–2pm daily. **Closed** main public hols.

Cape Skinári
Blue Caves
Kefalloniá
Korithi
Agios Nikólaos
Volímes
Alykés
Plános
Tsiliví
Kyllíni
Katastári
Zákynthos Town
Moni tis Panagías tis Anafonítrias
ZAKYNTHOS
Argási
Kampí
Lagópodo
Vasilikós
Mouzáki
Geráki
Laganás
Agalás
Kerí

0 kilometres 5
0 miles 5

Loggerhead Turtles

The Mediterranean green loggerhead turtle *(Caretta caretta)* has been migrating from Africa to Laganás Bay, its principal nesting site, for many thousands of years. These giant sea creatures can weigh up to 180 kg (400 lb). They lay their eggs in the sand, said to be the softest in Greece, at night. However, disco and hotel lights disorientate the turtles' navigation and few now nest successfully. Of the eggs that are eventually laid, many are destroyed by vehicles

or by the poles of beach umbrellas. The work of environmentalists has led to some protection for the turtles, with stretches of beach now off-limits, in an attempt to give the turtles a chance to at least stabilize their numbers.

VISITORS' CHECKLIST

Practical Information
🏠 41,000. 🛈 Lomvardou St, Zákynthos town (26950 25428). 🎭 Zákynthos Town Festival: Jul.

Transport
✈ 4 km (2 miles) S of Zákynthos town. 🚌 Zákynthos town; Agios Nikólaos. 🚌 Zákynthos town.

Around the Island

Outside the main resorts there is little tourist development on Zákynthos. It is possible to drive around the island in a day as most of the roads are in good condition. Hiring a car or a powerful motorbike is the best idea, though buses from Zákynthos town are frequent to resorts such as Alykés, Tsiliví and Laganás.

The growth of tourism on Zákynthos has been heavily concentrated in **Laganás** and its 14 km (9 mile) sweep of soft sand. This unrestricted development has decimated the population of loggerhead turtles that nests here – only an estimated 800 remain. Efforts are now being made to protect the turtles and to ensure their future survival. Visitors may take trips out into the bay in glass-bottomed boats to see the turtles, and all sorts of turtle souvenirs fill the large number of trinket shops.

An equally large number of bars and discos ensure the nightlife here continues till dawn.

Head to the north coast for the busy beach resorts of **Tsiliví** and **Alykés**, the latter being especially good for windsurfing.

The 16th-century **Moní tis Panagías tis Anafonítrias** in the northwest has special appeal for locals, as it was here the island's patron saint, Dionýsios, spent the last years of his life as an abbot. During his time here, it is said that Dionýsios heard a murderer's confession; the murderer received the saint's forgiveness, never knowing that

Coat of arms at Moní tis Panagías tis Anafonítrias

his victim was the abbot's brother. When questioned by the authorities, Dionýsios denied seeing the man, which was the only lie he ever told. Dionýsios lived in a cell here which still stands and contains many of the saint's revered possessions. The three-aisled church and the tiny chapel alongside are rare in that they survived the 1953 earthquake. At the northernmost tip of the island are the unusual **Blue Caves**, formed by the relentless action of the sea on the coastline. The principal cave, the Blue Grotto, lies directly underneath the lighthouse on Cape Skinári. It was discovered in 1897 and has become well-known for its stunningly blue and clear water through which the caves' natural stone floors can be seen. The caves can be visited by boat from the resort of Agios Nikólaos, and round-the-island boat trips from the main resorts also stop here.

The Blue Caves of Zákynthos on the northern tip of the island

THE ARGO-SARONIC ISLANDS

SALAMINA · AIGINA · POROS · YDRA · SPETSES · KYTHIRA

Although still supporting fishing and farming communities, the Argo-Saronic Islands have succumbed to a degree of tourism. Many Athenians visit the islands at weekends, when the beaches can become very busy. Kýthira, off the tip of the Peloponnese, shares its history of Venetian and British rule with the Ionians, but is today administered with the Argo-Saronics.

The islands' location close to Athens has given them a rich history. Aígina was very prosperous in the 7th century BC as a maritime state that minted its own coins and built the magnificent temple of Aphaia. Salamína is famed as the site of the Battle of Salamis (480 BC), when the Greek fleet defeated the Persians. Wealth gained from maritime trading also assured the Argo-Saronics' cultural and social development, seen today in the architectural beauty of Ydra and in the grand houses and public buildings of Aígina. Ydra and Spétses were important in the War of Independence *(see pp46–7)*, both islands producing brave fighters, including the notorious Laskarína Bouboulína and Admiral Andréas Miaoúlis.

Salamína and Aígina are so easy to reach from the capital that they are often thought of as island suburbs of Athens. Póros hardly seems like an island at all, divided from the Peloponnese by a narrow channel. However, despite modern colonization peaceful spots can still be found. Póros and Spétses are lush and green, covered with pine forests and olive groves, in contrast to the other more barren and mountainous islands. Scenically, Kýthira's rugged coastline has more in common with the Ionians than the Argo-Saronics. The island's position on ancient shipping routes has led to some major finds, such as the bronze *Youth of Antikýthira*, now on display in the National Archaeological Museum in Athens *(see p290)*.

The chapel of Agios Nikólaos on Aígina

◄ The harbour at Ydra, with the town rising up in the background

Exploring the Argo-Saronic Islands

Close proximity to Athens makes the Argo-
Saronic Islands suitable for short visits as
well as longer stays. The islands have a lush
landscape, with pine forests and crystal-clear
waters in secluded bays. Aígina is an ideal base
and, like the other islands, has picturesque
ports with cobbled streets and Neo-Classical
buildings. Packed with smart bars and shops,
the cosmopolitan atmosphere of the Argo-
Saronics is tempered by harbourside caïques
selling vegetables and horse-drawn carriages
driving along the seafront. Horse power is
particularly evident in Póros, Ydra and Spétses,
where no cars are allowed. Kýthira remains a
well-kept secret. This large island has beautiful
villages and deserted beaches to explore.

The harbour in Aígina town

Islands at a Glance

Key

— Main road
⸱⸱⸱ Minor road
— Scenic route
--- High season, direct ferry route
▲ Summit

The rugged scenery of Palaióchora on Kýthira

0 kilometres   20
0 miles   10

Ermiór
○ Kranídi
Portochéli ○
○ Kósta
Spétses Town
Agia Paraskeví ○   ○ Spétses
Agioi Anárgyroi ○   🏠🏛
Spetsopoula

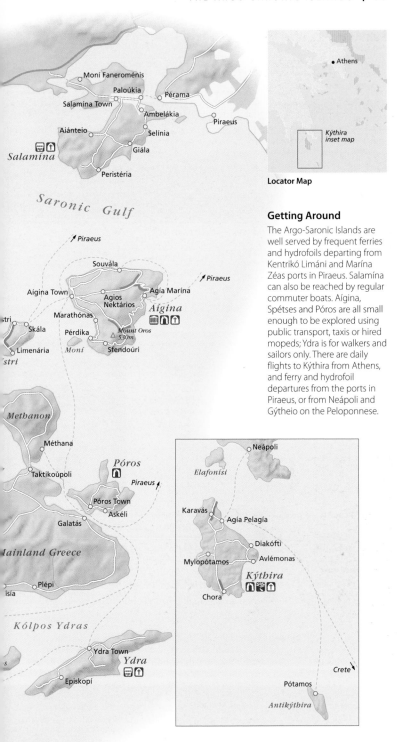

Locator Map

Getting Around

The Argo-Saronic Islands are well served by frequent ferries and hydrofoils departing from Kentrikó Limáni and Marína Zéas ports in Piraeus. Salamína can also be reached by regular commuter boats. Aígina, Spétses and Póros are all small enough to be explored using public transport, taxis or hired mopeds; Ydra is for walkers and sailors only. There are daily flights to Kýthira from Athens, and ferry and hydrofoil departures from the ports in Piraeus, or from Neápoli and Gýtheio on the Peloponnese.

Salamína
Σαλαμίνα

🏠 40,000. 🚢 Paloúkia & Selínia.
🚌 Salamína town. 🚢 Thu at
Salamína town, Sat at Aiánteio.

Salamína is the largest of the
Saronic Gulf islands, and so close
to Athens that most Greeks
consider it part of the mainland.
The island is famed as the site of
the decisive Battle of Salamis in
480 BC, when the Greeks
defeated the Persians. The king of
Persia, Xerxes, watched the
humiliating sight of his
cumbersome ships being
destroyed in Salamis Bay, trapped
by the faster triremes of a smaller
Greek fleet under Themistokles.
The island today is a cheerful
medley of holiday homes,
immaculately whitewashed
churches and cheap tavernas,
although its east coast is lined
with a string of marine
scrapyards and naval bases.

The west coast capital of
Salamína town is a charmless
place, straddling an isthmus of
flat land filled with vineyards.
Both the town and the island are
known as Koúlouri, nicknamed
after a biscuit that resembles the
island's shape.

East of Salamína town **Agios
Nikólaos** has far more
character, with 19th-century
mansions lining the
quayside and small
caïques offloading their
catch of fish. A road from
Paloúkia meanders
across the south of the
island to the villages of
Selínia, Aiánteio and
Peristéria.

In the northwest of the
Salamína, the 17th-century
Moní Faneroménis looks
across a narrow gulf to
Ancient Eleusis on the Attic
coast. The monastery was
used during the War of
Independence *(see pp46–
7)* as a hiding place for
Greek freedom fighters. Its
Byzantine church was restored
by the Venetians, and has fine
18th-century frescoes vividly
depicting the Last Judgment.
Today nuns welcome visitors, and
tend the gardens, home to a
number of peacocks.

Shrine opposite
Moní
Faneroménis

Fishing boats sailing into Aígina harbour

Aígina
Αίγινα

🏠 14,000. 🚢 🚌 Aígina town. ℹ️
Leonárdou Ladá, Aígina town (22970
27777). 🌐 aeginagreece.com

Only 20 km (12 miles) southwest
of the port of Piraeus, Aígina has
been inhabited for over 4,000
years, and has remained an
important settlement throughout that time. According to
Greek mythology, the island's
name was changed from Oinóni
to Aígina, who was the daughter
of the river god Asopós, after
Zeus installed her on the
island as his mistress.

By the 7th century BC
the second-largest Saronic
island was the first
place in Europe to
mint its own silver
coins, which became
accepted currency
throughout the Greek-
speaking world. Plying
the Mediter-ranean and
the Black Sea, the people
of Aígina controlled most
foreign trade in Greece.
However, their legendary
nautical skills and vast
wealth finally incurred
the wrath of neighbouring
Athens, who settled the
long-term rivalry by
conquering the island in
456 BC. Aígina's most famous
site is the well-preserved
Temple of Aphaia *(see pp102–
3)*, built in about 490 BC, prior
to Athenian control. Later,
the island declined during the
centuries of alternating Turkish

and Venetian rule and the
constant plague of piracy.
However, Aígina enjoyed fame
again for a brief period in 1828,
when Ioánnis Kapodístrias
(1776–1857) declared it the first
capital of modern Greece.

The ruinous Venetian Pýrgos Markéllou in
Aígina town

Aígina Town
This picturesque island town
is home to many churches,
including the pretty 19th-
century **Agía Triáda**, next to the
fish market overlooking the
harbour. At the quayside, horse-
drawn carriages take visitors
through narrow streets of
Neo-Classical mansions to
the Venetian tower **Pýrgos
Markéllou** near the cathedral.
Agios Nektários cathedral,
inaugurated in 1994, is said to
be the second-biggest Greek
Orthodox church after Agía
Sofía in Istanbul. Octopuses are
hung out to dry at tavernas in
the street leading to the fish

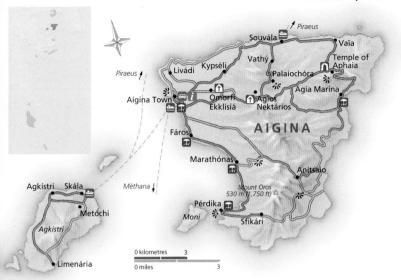

Agios Nektários cathedral in Aígina town

market. To the northwest, past shops selling pistachio nuts and earthenware jugs, are the remains of the 6th-century BC **Temple of Apollo**. The 6th-century Sphinx of Aígina, now in the **Aígina Museum**, was discovered here.

🏛 **Aígina Museum**
Kolóna 8. **Tel** 22970 22248. **Open** 8am–3pm Tue–Sun. **Closed** main public hols. 🈲 🛗

Environs
North of Aígina town, in Livádi, a plaque marks the house where Níkos Kazantzákis wrote *Zorba the Greek (see p280)*.

Around the Island
Aígina, at only 8 km (5 miles) across, is easy to explore by bicycle. Just off the main road east from Aígina town is the

13th-century Byzantine church, **Omorfi Ekklisiá**, which has some fine frescoes. Pilgrims take this road to pay homage at **Agios Nektários**. Archbishop Nektários (1846–1920) was the first man to be canonized in modern times (1961) by the Orthodox Church. Visitors can see his quarters and the chapel where he rests.

On the opposite hillside are the remains of the deserted town of **Palaiochóra**. Populated since Byzantine times, it was destroyed by Barbarossa, the general of Sultan Suleiman I, in 1537. The area around the town was abandoned in 1826.

The scattered ruins of Byzantine chapels around the deserted town of Palaiochóra

South from Aígina town, the road hugs the shore, beneath the shadow of **Mount Oros** at 530 m (1,750 ft). Passing the pistachio orchards and the fishing harbour of Fáros, this scenic route ends at **Pérdika** at the southwestern tip of he island. Overlooking the harbour, this small, picturesque fishing village has some excellent fish tavernas that are packed at weekends with Athenians over for a day trip.

Outlying Islands
Just 15 minutes by caïque from Pérdika is the island of **Moní**, popular for its emerald-green waters, secluded coves and hidden caves.

Agkístri is easily accessible by caïque from Aígina town or by ferry from Piraeus. Originally settled by Albanians, today this island is colonized by Germans who have bought most of the houses in the village of Metóchi, just above Skála port. Although many hotels, apartments and bars have been built in Skála and Mílos, its other main port, the rest of this hilly, pine-clad island remains largely unspoiled. Limenária, in the south of the island, is a more traditional, peaceful community of farmers and fishermen.

For keys to symbols *see back flap*

Aígina: Temple of Aphaia
Ναός της Αφαίας

Surrounded by pine trees, on a hilltop above the busy resort of Agía Marína, the Temple of Aphaia is one of the best-preserved Doric temples in Greece *(see pp64–5)*. The present temple dates from around 490 BC, but the site is known to have been a place of worship from the 13th century BC. In 1901 the German archaeologist Adolf Furtwängler found an inscription to the goddess Aphaia, disproving theories that the temple was dedicated to Athena. Although smaller, the building is similar to the temple of Zeus at Olympia, built 30 years later.

Inner Walls
The inner wall was built with a thickened base and a minimal capital to correspond with the capitals of the colonnade.

KEY

① **Ramp from altar to temple**

② **Architrave**

③ **Metope**

④ **Triglyph**

⑤ **The east pediment sculptures**, with Athena at the centre, were replacements for an earlier set. The west pediment sculptures are Archaic in style.

⑥ **The roof** was made of terracotta tiles with Parian marble tiles at the edges.

⑦ **Opisthodomos, or rear porch**

⑧ **Cult statue of the goddess Aphaia**

⑨ **The pool of olive oil** was a collection of the many libations (offerings) made to the goddess.

Corner Columns
These columns were made thicker for emphasis and to counteract the appearance of thinness in a column that was seen against the sky.

Corner Architraves
Still in good condition, the stonework above the capitals consists of a plain architrave surmounted by a narrow band of plain metopes alternating with ornate triglyphs.

Inner Columns
The cella is enclosed by two storeys of Doric columns, one on top of the other. The taper of the upper columns is continuous with that of the lower.

VISITORS' CHECKLIST

Practical Information
12 km (7 miles) E of Aígina town.
Tel 22970 32398. **Open** 8:30am–7:30pm daily. **Closed** main
public holidays.

View of the Cella
The cella was the inner room of the temple, and the home of the cult statue. Some temples had more than one, the back cella being reserved for the priestess alone.

Reconstruction of the Temple of Aphaia
Viewed from the northwest, this reconstruction shows the temple as it would have been in c.490 BC. Built of local limestone covered in stucco and painted, it was highly colourful.

Temple Pediments
The famous sculptures from the pediments of the Temple of Aphaia were discovered by a group of British and German architects and artists, including John Foster, C R Cockerell and Baron Haller von Hallerstein, in April 1811. They were later sold to the Crown Prince of Bavaria at auction and are now housed in the Glyptothek in Munich. They portray the struggles of various mythological heroes. The sculptures from the west pediment date from around 490 BC and are in the late Archaic style. Those from the east, with their more fluid movements and serious expressions, date from approximately 480 BC and foreshadow the Classical style.

Reconstruction of the *Warriors* sculpture from the west pediment

Póros
Πόρος

🏛 4,500. 🚢 🚌 Póros town. 🛈 Póros town (22980 22462). 🛥 Fri (am) at Paidikí Chará.

Póros takes its name from the 400 m- (1,300 ft-) passage *(póros)* separating it from the mainland at Galatás. Póros is in fact two islands, joined by a causeway: pine-swathed Kalávria to the north, and the smaller volcanic islet of Sfairía in the south over which **Póros town** is built. In spite of much tourist development, the town is an appealing place, extending along the narrow straits, busy with shipping. Its 19th-century houses climb in tiers to its apex at a clock tower.

The **National Naval Academy**, northwest of the causeway and Póros town, was set up in 1849. An old battleship is usually at anchor there for training naval cadets.

The attractive 18th-century **Moní Zoödóchou Pigís** can be found on Kalávria, built around the island's only spring. There are the ruins of a 6th-century hilltop **Temple of Poseidon** near the centre of Kalávria, next to which the orator Demosthenes poisoned himself in 323 BC rather than surrender to the Macedonians. In antiquity the site was linked to ancient Troezen in the Peloponnese. The temple has unlimited access.

The waterfront on Ydra

Ydra
Ύδρα

🏛 2,700. 🚢 Ydra town. 🛈 Ydra town (22980 52210). 🚤 Mandráki 1.5 km (1 mile) NE of Ydra town; Vlychós 2 km (1 mile) SW of Ydra town.

A long, narrow mass of barren rock, Ydra had little history before the 16th century, when it was settled by Orthodox Albanians, who then turned to the sea for a living. Ydra town was built in a brief period of prosperity in the late 18th and early 19th centuries, boosted by blockade-running during the Napoleonic wars. After Independence, Ydra lapsed into obscurity again, until foreigners

Bell tower of Ydra's Panagía church

rediscovered it after World War II. By the 1960s, the trickle had become a flood of outsiders who set about restoring the old houses, transforming Ydra into one of the most exclusive resorts in Greece. Yet the island has retained its charm, thanks to an architectural preservation order which has kept the town's appearance as it was in the 1820s, along with a ban on motor vehicles. Donkey caravans perform all haulage on steep-stair streets.

Ydra Town
More than a dozen three- or four-storey mansions *(archontiká)* survive around the port, though none are regularly open to the public. Made from local stone,

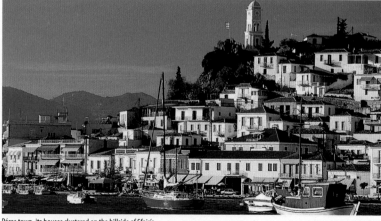
Póros town, its houses clustered on the hillside of Sfairía

they were built by itinerant craftsmen between 1780 and 1820. On the east side of the harbour the **Tsamadoú mansion** is now the National Merchant Marine Academy. On the west, the **Tompázi mansion** is a School of Fine Arts. Just behind the centre of the marble-paved quay is the monastic church of the **Panagía**, built between 1760 and 1770 using masonry from Póros' Temple of Poseidon. The marble belfry is thought to have been erected by a master stonemason from Tínos.

The old harbour of Báltiza on Spétses

Around the Island

Visitors must walk virtually everywhere on Ydra, or hire water taxis to go along the coast. **Kamíni**, 15 minutes' walk southwest along the shore track, has been Ydra's main fishing port since the 16th century. The farm hamlet of **Episkopí**, in the far southwest of the island, used to be a summer refuge and a hunting resort for the upper classes. An hour's steep hike above the town is the convent of **Agía Efpraxía**, which still houses nuns who are keen to sell you handicrafts. The adjacent 19th-century **Profítis Ilías** functions as a monastery. In the island's eastern half, visible from Profítis Ilías, are three uninhabited monasteries, dating from the 18th and 19th centuries. They mark the arduous 3-hour-long route to **Moní Panagía**, situated out near Cape Zoúrvas to the northeast of the island.

Spétses

Σπέτσες

🏠 3,900. 🚢 🚌 Spétses town. **ℹ** Spétses town (22980 27777). 🛒 Wed at Kokinária.

Spétses is a corruption of Pityoússa, or "Piney", the ancient name for this round, green island. Occupied by the Venetians in 1220, by the Turks in 1460, and then by Albanians during the 16th century, the island developed as a naval power, and supplied a fleet for the Greek revolutionary effort.

Possibly the most famous Spetsiot was Laskarína Bouboulína, the admiral who menaced the Turks from her flagship *Agamemnon* and reputedly seduced men at gunpoint. She was shot in 1825 by the father of a girl her son had eloped with. During the 1920s and 1930s, Spétses was a fashionable resort for British expatriates and anglophile Greeks. The ban on vehicles is not total: mopeds and horsecabs can be hired in town, and there are buses to the beaches.

Statue of Bouboulína in Spétses town

Spétses Town

Spétses town runs along the coast for 2 km (1 mile). Its centre lies at Ntápia quay, fringed by cafés. The *archontiká* of Chatzi-Giánnis Méxis, dating from 1795, is now the **Chatzi-Giánnis Méxis Museum**. Bouboulína's coffin is on display as well as figureheads from her ship. Her former resplendent mansion home is now the privately run **Bouboulína Museum**. Southeast from here lies the old harbour at **Báltiza** inlet, where wooden boats are still built using the traditional methods. Above the harbour is the 17th-century church of **Agios Nikólaos**, with fine pebble mosaics and a belfry made by craftsmen from Tínos.

🏛 **Chatzi-Giánnis Méxis Museum**
300 m (985 ft) from the port.
Tel 22980 72994. **Open** 8:30am–3pm Tue–Sun. **Closed** main public hols. 🏛

🏛 **Bouboulína Museum**
Behind Plateía Ntápia. **Tel** 22980 72416. **Open** 25 Mar–31 Oct: 9:45am–8:15pm daily. 🏛 📷
🌐 **bouboulinamuseum-spetses.gr**

Around the Island

A track, only partly concreted, runs all the way round the island, and the best way to get around is by bicycle or moped. East of the town stands the Anargýreios and Korgialéneios College, which is now closed. British novelist John Fowles taught there briefly in the early 1950s. He later used Spétses as the setting for *The Magus*. The pebble beaches on Spétses are the best in the Argo-Saronic group, including **Ligonéri**, **Vréllas** and **Agía Paraskeví**. **Agioi Anárgyroi** is the only sandy one.

Pebble mosaic from the church of Agios Nikólaos, Spétses town

Kýthira
Κύθηρα

Called Tserigo by the Venetians, Kýthira is one of the legendary birthplaces of Aphrodite. Historically, the island shared Venetian and British rule with the Ionian islands; today it is governed from Piraeus with the other Argo-Saronics. Clumps of eucalyptus seem emblematic of the island's modern alias of "Kangaroo Island"; return visits from 60,000 Australian Kythirans are central to Kythiran life. The island is also popular with Athenians seeking unspoiled beaches and holiday homes, many of which are the typical mix of Aegean and Venetian architecture.

Platiá Ammos
Karavás
Neápoli
Agía Pelagía
Gýtheio, Kalamáta
Potamós
Palaióchora
Douriánika
Moní Agíou Theodórou
Friligkiánika
Diakófti
Crete
Agía Sofía Cave
KYTHIRA
Mitáta
Agios Geórgios
Káto Chóra
Mylopotamos
Fónissa
Avlémonas
Palaiópoli
Kastrí
Limniónas
Frátsia
Kaladí
Limnária
Kalokairinés
Komponáda
Livádi
Moní Agios Ioánnis sto Gkremó
Kálamos
Fyrí Ammos
Melidóni
Chóra
Kapsáli
Chalkós

KYTHIRA

0 kilometres   5
0 miles   3

The houses of Chóra clustered on the hillside at dusk

For keys to symbols see back flap

Kapsáli harbour seen from Chóra

Chóra

Chóra has been Kýthira's capital only since the destruction of *Palaióchora* in 1537. Its magnificent **kástro** was built in two phases during the 13th and 15th centuries. A multidomed cistern lies intact near the bottom of the castle; at the summit, old cannons surround the church of **Panagía Myrtidiótissa**. The steepness of the drop to the sea below and Avgó islet, thought to be the birthplace of Aphrodite, is unrivalled throughout the Greek islands. A magnet for wealthy Athenians, the appealing lower town with its solid, flat-roofed mansions dates from the 17th to 19th centuries. The **Archaeological Museum** just outside Chóra has finds from Mycenaean and Minoan sites, plus gravestones dating from the British occupation of 1809–64.

🏛 **Archaeological Museum**
Tel 27360 31739. **Open** Currently closed for renovation; call for opening times. **Closed** main public hols.

Environs

Yachts, hydrofoils and large ferries drop anchor at the harbour of Kapsáli, just east of Chóra. The beach is mediocre, but most foreigners stay here. In the cliff above the pine wood is the 16th-century Moní Agios Ioánnis sto Gkremó, built on to the cliff edge. The nearest good beaches are pebbly Fyrí Ammos, 8 km (5 miles) northeast via Kálamos, with sea caves at its south end; and sandy Chalkós, 7 km (4 miles) south of Kálamos.

Whitewashed house in Mylopótamos

Around the Island

Like many Greek islands, the best way to get around Kýthira is by car, particularly as it is quite mountainous. A bus runs to the main towns once a day during the summer from Agía Pelagía to Kapsáli.

Avlémonas, with its vaulted warehouses and double harbour, forms an attractive fishing port at the east end of a stretch of rocky coast. Just offshore the *Mentor*, carrying many of the Elgin Marbles, sank in 1802. Excellent beaches extend to either side of Kastrí point. The 6th-century hilltop church of **Agios Geórgios**, which has a mosaic floor, sits high above Avlémonas.

Roadside shrine on Kýthira

On the other side of the island is **Mylopótamos**. From here a track leads west to the small Fónissa waterfall, downstream from which is a millhouse, and a tiny stone bridge.

In its blufftop situation with steep drops to the north and west, and a clutch of locked chapels, the Venetian *kástro* at **Káto Chóra** superficially resembles Palaióchora. It was not a military stronghold but a refuge prepared in 1565 for the peasantry in unsettled times. The Lion of St Mark presides over the entrance; nearby an English-built school of 1825 has been restored.

Agía Sofía Cave, 2.5 km (1.5 miles) from Káto Chóra and 150 m (490 ft) above the sea, has formed in black limestone strata. At the entrance, a frescoed shrine, painted by a 13th-century hermit, depicts Holy Wisdom and three attendant virtues. Palaiochóra, the Byzantine "capital" of Kýthira after 1248, was sited so as to be nearly invisible from the sea, but the pirate Barbarossa detected and destroyed it in 1537. The ruins of the town perch on top of a sheer 200 m (655 ft) bluff. Among six churches in Palaióchora, the most striking and best preserved is the 14th-century **Agía Varvára**.

To the south, **Moní Agíou Theodórou** is the seat of Kýthira's bishop. The church, originally 12th century, has been much altered, and the Baroque relief plaque over the door is a rarity in Greece.

To the north, the main port, **Agía Pelagía**, has a handful of hotels. **Karavás**, 5 km (3 miles) northwest is, in contrast, an attractive oasis village, with clusters of houses overhanging the steep banks of a stream valley.

🗺 Agía Sofía Cave

Mylopótamos. **Tel** 27360 31213. **Open** Apr–Oct (call for opening times). 🗐 🞖 Jul & Aug.

Outlying Islands

Directly north of Kýthira, the barren islet of **Elafonísi** is visited mostly by Greeks for its fantastic desert-island beaches. The better of the two is Símos on the east side of a peninsula 5 km (3 miles) southeast of the port town. The remote island of **Antikýthira**, southeast of Kýthira, has a tiny population and no beaches.

VISITORS' CHECKLIST

Practical Information
🏘 4,000. 🛈 27360 31213. 🔲 kythera.gr 🔲 Sun at Potamós.

Transport
✈ 22 km (14 miles) NE of Chóra. ⛴ Agía Pelagía & Kapsáli. 🚌 runs between Agía Pelagía & Kapsáli and between Diakófti & Kapsáli.

Looking east across a gorge from Palaióchora

THE SPORADES AND EVVOIA

SKIATHOS · SKOPELOS · ALONNISOS · SKYROS · EVVOIA

The lush landscape of Evvoia, Greece's second-largest island after Crete, and the Sporades archipelagos comes as a surprise after barren and arid islands such as those found in the Cyclades. Since ancient times, settlers and pirates alike have been lured by the pine-clad mountains now dotted with villages, abundant springs and rivers, endless beaches and dramatic coastlines rich with hidden coves that are found throughout these islands.

Being close to the mainland, the Sporades and Evvoia have been easily conquered throughout history. They were colonized in the prehistoric era by nearby Iolkos (Vólos), and also by the Minoans, who introduced vine and olive cultivation. More than any other island, Evvoia reveals its diverse history in the large number of buildings remaining from the long periods of Venetian and Turkish occupation. Susceptible to pirate raids, the inhabitants of the Sporades lived in the safety of fortified towns until as late as the 19th century. Even in Evvoia, when life proved too difficult in coastal villages such as Límni, the residents simply migrated to Skiáthos for a few generations. The islanders have a rich heritage of maritime trading around the Aegean and are still noted today as sailors. The islands' patchworked interiors of fertile fields and orchards, watered by ample springs and rivers, also encouraged agricultural self-sufficiency and wealth. Particularly on remote and rugged Skýros, such insularity has nurtured some unique folk art and colourful traditions. Its inaccessible coastline enables it to remain relatively unaffected by the numerous tourist hotel complexes that have sprung up on Skiáthos and Skópelos.

The size of Evvoia also means it is one of the few places in the Greek islands where life carries on during the summer, undeterred by the annual invasion of holiday-makers.

Castel Rosso near Kárystos on Evvoia

◄ The blue waters of Cape Amarantos, on the island of Skópelos

Exploring the Sporades and Evvoia

The rich and famous first flocked in their yachts to the deserted beaches of Skiáthos, Skópelos and Alónnisos in the 1960s and 1970s. Although no longer so exclusive, the beautiful coastlines of these islands still lure Greek and foreign holiday-makers alike. There are facilities for windsurfing and boats for hire on most beaches. Skópelos and Skiáthos have a sophisticated array of nightclubs and bars. Quieter Skýros and Evvoia, offering a varied culture and landscape, are perfect for rambling holidays, punctuated by visits to local folk art museums and lingering days on the fine beaches.

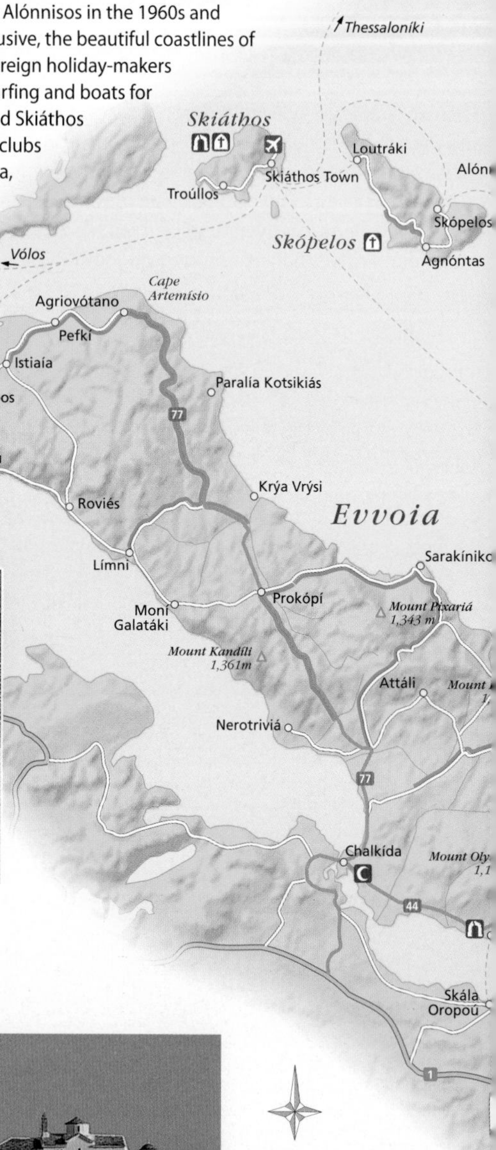

↗ *Thessaloníki*

Skiáthos

Loutráki

Skiáthos Town

Troúllos

Alónn

Skópelos 🏠

Agnóntas

Vólos

Cape Artemísio

Agriovótano

Pefkí

Istiaía

Glýfa

Paralía Kotsikiás

Agiókampos

77

Loutrá Aidipsoú

Agios Geórgios

Roviés

Krýa Vrýsi

Evvoia

Límni

Sarakíniko

Moní Galatáki

Prokópi

Mount Pixariá 1,343 m

Mount Kandíli 1,361 m

Attáli

Mount

Nerotriviá

77

A house in Stení on Evvoia, with Mount Dírfys in the background

Chalkída

Mount Oly 1,1

44

Skála Oropoú

1

The harbour of Skópelos town

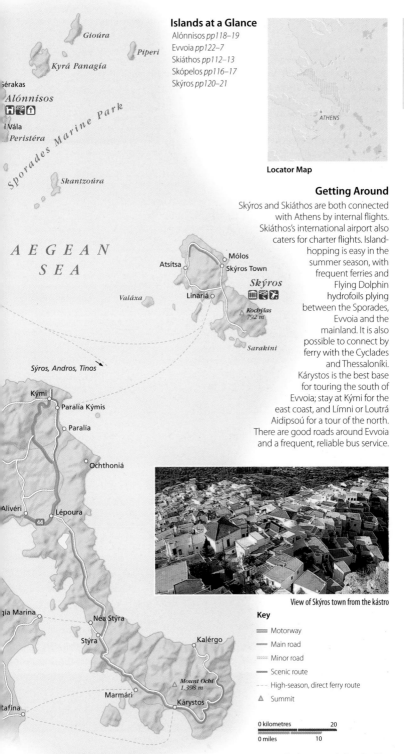

Islands at a Glance

Locator Map

Gioúra

Kyrá Panagía

Piperi

Gérakas

Alónnisos

Vála

Peristéra

Sporades Marine Park

Skantzoúra

Getting Around

Skýros and Skiáthos are both connected with Athens by internal flights. Skiáthos's international airport also caters for charter flights. Island-hopping is easy in the summer season, with frequent ferries and Flying Dolphin hydrofoils plying between the Sporades, Evvoia and the mainland. It is also possible to connect by ferry with the Cyclades and Thessaloníki. Kárystos is the best base for touring the south of Evvoia; stay at Kými for the east coast, and Límni or Loutrá Aidipsoú for a tour of the north. There are good roads around Evvoia and a frequent, reliable bus service.

AEGEAN
SEA

Atsítsa

Valáxa

Linariá

Mólos

Skýros Town

Skýros

Kochýlas
792 m

Sarakíni

Sýros, Andros, Tínos

Kými

Paralía Kýmis

Paralía

Ochthoniá

Alivéri

Lépoura

44

View of Skýros town from the kástro

Néa Stýra

Stýra

Kalérgo

gía Marina

Mount Ochi
1,398 m

Marmári

Kárystos

afína

Key

═══ Motorway

─── Main road

═══ Minor road

─── Scenic route

--- High-season, direct ferry route

△ Summit

0 kilometres		20
0 miles		10

For keys to symbols *see back flap*

Skiáthos
Σκιάθος

Skiáthos has always been an unashamedly hedonistic island from its early tourist development in the 1960s, when it attracted the rich and famous with its legendary beaches, to its current role as bucket-and-spade paradise for family package tours. Although the introduction of direct package flights has diminished Skiáthos' exclusive status, the luxury yachts are still in evidence off Koukounariés beach. In spite of the tourism, the island retains its scenic beauty and a scattering of atmospheric churches and monasteries.

The sweeping bay of Koukounariés

Map showing locations: Kástro, Lalária, Moní Agíou Charalámpou, Panagía Kardási, Moní Evangelismoú, Agios Apóstolos, Taxiárchis, Thessaloníki, Skópelos, Profítis Ilías, Loutráki, Kechriá, Skiáthos Town, Argos, Agios Ioánnis, Asélinos, Fteliá, Andros, Moní Panagías Kounístras, SKIATHOS, Mandráki, Vólos, Syros, Tínos, Tsougkriá, Koukounariés, Máratha, Kalamáki, Tsougriaki, Troúllos

0 kilometres 2
0 miles 1

Skiáthos Town

Still picturesque, the island town is a charming place with its red-tiled roofs and maze of cobbled backstreets. It is built on two small hills, dominated by the large 19th-century churches of **Trión Ierarchón** and **Panagía Limniá**, which offer excellent views of the bustling harbour below. The main street winds up between the two hills to the old quarter of Limniá, a quiet neighbourhood of restored sea captains' houses, covered with trailing bougainvillea and trellised vines. The town is excellent for shopping, full of aromatic bakeries, smart boutiques and antique shops, some of which specialize in genuine folk artifacts, including ceramics, icons, jewellery and embroidery.

The town has twin harbours, separated by **Bourtzi** islet, which is reached by a narrow causeway. The pine-covered islet, once a fortress, is now a cultural centre and hosts the annual Aegean festival of dance, theatre and concert performances each summer. Bourtzi is dominated by a handsome Neo-Classical building, with a statue of the famous Greek novelist Aléxandros Papadiamántis standing guard. Life in Skiáthos town centres on the long, sweeping quaysides lined with numerous *kafeneía*, specializing in *loukoumádes* (small honeyed fritters). In the evenings the waterside attracts many people for a stroll in the cool night air. During the day there is the spectacle of arriving and departing flotilla yachts, ferries and hydrofoils. The western end of the quay has a good fish market, and an *ouzerí* frequented by locals. It is also where small boats and caïques depart for day trips to some of the island's famous beaches, such as Koukounariés and Lalária, or to the nearby islands

Skiáthos town viewed from the church of Profítis Ilías

An ornate fresco in the Christós sto Kástro church

VISITORS' CHECKLIST

Practical Information
6,000. 24270 23172.
skiathos.gr Aegean Festival
of Dance, Skiáthos town: Jul.

Transport
2 km (1 mile) NE of Skiáthos
town. Harbourfront,
Skiáthos town.

of Tsougkriá and Argos.
Behind the harbour is the
Papadiamántis Museum,
former home of the locally
born novelist, whose name it
takes. The museum shows the
simplicity of local island life prior
to the invasion of tourism.

Papadiamántis Museum
Tel 24270 23843. **Open** 9:30am–
1:30pm & 5–8pm daily (subject to
change early and late season).

Moní Agíou Charalámpou, set in the hills
above Skiáthos town

Around the Island
The interior of the northern side
of the island, with its verdant
landscape of pine and olive trees,
reveals deserted monasteries
and churches, springs and plenty
of birdlife. This is in contrast to
the over-developed southern
coast. It is still possible to find
deserted beaches and coves
scattered along the northern
coast. Many of these, such as

Kechriá and **Mandráki**, can
only be visited when the
excursions stop for a few
hours on their day trips around
the island.

The main road south from
Skiáthos town passes Fteliá and
branches to the west just before
Troúllos for Asélinos beach and
Moní Panagías Kounístras. The
monk who founded this 17th-
century monastery, originally
called Panagía Eikonístria,
discovered a miraculous icon in a
nearby tree. The icon is kept in
Trión Ierarchón in Skiáthos town.

The path north from here
leads to **Agios Ioánnis**, where
it is customary to stop and ring
the church bell after completing
the steep walk through
pine trees.

Further north still is the tiny
19th-century chapel of **Panagía
Kechriás**, with its blue ceiling
covered in stars, which perches
high above **Kástro**. Abandoned
in 1829, remains of the 300
houses are still visible in
this deserted town and
three churches have been
restored. The 17th-century
Christós church has a
fine iconostasis.

On the road heading
northwest out of
Skiáthos town lies the
barrel-vaulted
20th-century church of
Profítis Ilías, which has a
good taverna nearby with
stunning views over the town.
Continuing north, past rich
farms and the 20th-century
Agios Apóstolos church, the

track descends through sage
and bracken to **Moní Agíou
Charalámpou**, built in 1809.
Aléxandros Moraïtidis, the
writer, spent his last days here
as a monk in the early 1920s.
Just south of here is **Moní
Evangelistrías**. Founded in 1775
by monks from Mount Athos, it
played a crucial role in the War
of Independence (see pp46–7),
hiding many freedom fighters.

To the south of Moní Agíou
Charalámpou, on the way back
to Skiáthos town, is the beautiful
church of **Taxiárchis**. It is
covered in plates in the shape
of a cross, and the best mineral
spring water on the island flows
out of a tap that is by the church.

Alexandros Papadiamántis

The island's most famous
native is one of Greece's
outstanding literary figures.
Aléxandros Papadiamántis
spent his early childhood on
the island, with five brothers
and sisters, before leaving
to study in Athens, where
he began his career in
journalism. He wrote more
than 100 novellas and short
stories, all set against the
backdrop of island life.
Among his best-
known works are
*The Gypsy, The
Murderess*, a
compulsive
psychological
drama, and *The
Man Who Went to
Another Country*. In
1908 he returned to
Skiáthos where he died a
few years later in 1911 at the
age of 60.

Skópelos
Σκόπελος

Surprisingly, given its close proximity to Skiáthos, Skópelos has not totally succumbed to tourism. It is known to have been colonized by the Minoans as far back as 1600 BC and was used as a place of exile by the Byzantines. The Venetians held power for about 300 years after 1204. Famed for its wine in ancient times, Skópelos is still renowned for its fruit today. It offers many good beaches, and has a beautiful pine-covered interior.

The way up to Panagía tou Pýrgou above Skópelos town

[Map showing locations: Glóssa, Loutráki, Skópelos Town, Skiáthos, Mount Délfi 680 m (2,230 ft), Elios, Glystéri, Loutráki, Skiáthos, Vólos, Alónnisos, SKÓPELOS, Moní Timíou Prodrómou, Skópelos Town, Moní Metamórfosis tou Sotíros, Dasiá, Miliá, Moní Evangelistrías, Adrína, Pánormos, Mount Paloúki 385 m (1,260 ft), Moní Taxiarchón, Limnonári, Velóna, Agnóntas, Stáfylos]

0 kilometres 5
0 miles 3

Skópelos Town
This charming town proudly reveals its rich pedigree with 123 churches, many fine mansion houses and myriad shops selling local delicacies such as honey, prunes and various delicious sweets. The cobbled streets wind up from the waterfront, and are covered with intricate designs made from sea pebbles and shells. There are numerous classic examples of the old Sporadhan town house, with its wooden balcony and fish-scale, slate-tiled roof.

In the upper town the cruciform church of **Panagía Papameletíou** is particularly splendid. Built in 1662, it is also known as Koímisis tis Theotókou. It has a well-kept interior, with an interesting display case of ecclesiastical *objets d'art* and a carved iconostasis by the Cretan

craftsman António Agorastós. Perched on a clifftop above the town, the landmark church of **Panagía tou Pýrgou**, with its shining fish-scale roof, overlooks the harbour.

The old quarter of Skópelos town, the Kástro, sits above the modern town and is topped by the remains of the Venetian castle. Built by the Ghisi family in

the 13th century, the castle stands on the site of the 5th-century BC acropolis of ancient Skópelos. The church nearest the castle is **Agios Athanásios**. It was built in the 11th century, but the foundations date from the 9th century. There are some fine 16th-century frescoes inside.

The **Folk Art Museum** sits behind the harbourfront in a 19th-century mansion. Examples of traditional local costumes and embroidery are on display.

🏛 Folk Art Museum
Chatzistamáti. **Tel** 24240 23494.
Open Nov–May: 10am–2pm Mon–Fri & Sun, 11am–2pm Sat; Jun–Oct: 10am–2pm & 7–9pm Mon, Wed & Fri (to 2pm Tue, Thu & Sat), 11am–2pm Sun. 🚫

Environs
In the hills above Skópelos town there are numerous impressive monasteries. Reached by the road going east out of the town,

The attractive bay of Skópelos town, viewed from the Kástro

Fish tavernas around the bay at Agnóntas in the late afternoon

they all have immaculate churches with carved iconostases and icons. **Moní Evangelistrías** (also known as Evangelismós) was built in 1712 and is one of the largest on the island. The nuns sell their handicrafts, including weavings, embroidery and food. Further up the road is **Metamórfosis tou Sotíros**, one of the oldest monasteries on Skópelos. It was built in the 16th century and is now inhabited by a solitary monk.

Moní Timíou Prodrómou, north of Moní Metamórfosis tou Sotíros, was restored in 1721. It has been inhabited by nuns, who also sell crafts, since the 1920s, and has a commanding view of Skópelos. From here a rough track leads up to **Mount Paloúki**. The deserted **Moní Taxiarchón** is reached by a track from Mount Paloúki that hugs the *sares*, the local name for the steep cliffs facing Alónissos.

Around the Island

The island is easy to explore, with its main road traversing the developed southern coast, and continuing as far as Glóssa to the northwest. It has a beautiful interior, full of plum orchards, pine forests and *kalývia* (farmhouses), but beware of the lack of signposts when travelling inland.

A steep road leads down to the popular beaches south of Skópelos town, Stáfylos and Velóna. **Agnóntas**, which serves as a port for ferries in rough weather, is quieter than Skópelos town. It is popular with locals, who come for the fish tavernas beside its pebble beach. Nearby

Limnonári, with its stunning pebble beach and azure-coloured water, is reached by boat or along the narrow clifftop road. Before reaching the modern village of Elios, there are two thriving resorts at Miliá and Pánormos. For a quieter location, the tiny beach of **Adrína** nearby

Whitewashed houses in Glóssa with colourful doors and shutters

is often deserted. Sitting opposite the beach is wooded Dasía island, named after a female pirate who drowned there long ago.

Glóssa is the other major settlement on the island, and sits directly opposite Skiáthos. Reminders of the Venetian occupation of Skópelos are evident in the picturesque remains of Venetian towers and houses. The small port of **Loutráki** below Glóssa has cafés and restaurants, and most ferries stop here as well as at Skópelos town.

On the north coast, caïques shuttle every half-hour between the pebbled **Glystéri** beach, which lies protected in a rocky cove, and Skópelos town. From Glystéri a good but winding road leads inland to the wooded region just east of the island's highest peak, **Mount Délfi**. A short walk through the enchanting pine forest leads to four mysterious niches, signposted as *sentoúkia*, literally "crates", that are carved in the rocks. Believed to be Neolithic sarcophagal tombs, their position offers fine views over the island.

Kalývia

Skópelos's interior is covered with an unusual array of beautiful *kalývia* (farmhouses). Some of these traditional stone buildings are still occupied all year round, while others are only used as weekend retreats during important seasonal harvests or for celebratory feasts on local saints' days. They all have distinctive outdoor prune ovens – a legacy from the days when Skópelos was renowned for its prunes. The farmhouses provide a rare insight into the rural life that has virtually disappeared on many of the neighbouring islands.

A traditional *kalývia* among olive and cypress trees

Bathers enjoying beautiful weather and clear, calm waters at Kokkinókastro beach

Alónissos
Αλόννησος

🏘 2,800. 🚤 🚌 Patitíri. 🚆
Kokkinókastro 6 km (4 miles) N of
Patitíri. 🌐 alonissos.gr

Sharing a history of attacks by
the pirate Barbarossa with the
other Sporades and having
endured earthquake damage
in 1965, Alónissos has suffered
much over the years.
However, the island is relatively
unspoiled by tourism, and
most of the development is
centred in the main towns of
Patitíri and Palaiá Alónissos.

Patitiri

The port of Patitíri is a centre
of bustling activity. Boats are
available for day trips to the
neighbouring islands, and there
is excellent swimming off the
rocks, northeast of the port.
The picturesque backstreets
display typical Greek pride
in the home, evident in the
immaculate whitewashed
courtyards and pots of flowers.
Rousoúm Gialós and Vótsi,
3–4 km (1–2 miles) north of

Fishing vessels and cargo boats moored in
Patitíri harbour

Patitíri, are quieter alternatives
with their natural cliff-faced
harbours and tavernas.

Around the Island

This quiet island has a surfeit of
beaches and coves and the
interior is crisscrossed by dirt
tracks accessible only to
intrepid shepherds and
motorbikes. The old capital of
Palaiá Alónissos, west of
Patitíri, perches precariously on
a clifftop. There are ruins of a
15th-century Venetian castle
and a beautiful small chapel,
Tou Christoú, that has a fish-
scale roof. The town was
seriously damaged by the
earthquake in 1965, and the
inhabitants were forced to leave
their homes. They were
rehoused initially in
makeshift concrete
homes at Patitíri.
Today, the houses of
Palaiá Alónissos have
been bought and
restored by German
and British families,
and the town retains
all the architectural
beauty of a traditional
Sporadhan village.
    The road across
the island, northeast
from Patitíri, reveals
a surprisingly fertile
land of pine, olive
and arbutus trees.
At **Kokkinókastro**, a
popular pebble beach
edged by red cliffs
and pines, there are

scant remains of the site of
ancient Ikos – the old name
of the island.
    Further north lies the seaside
village of **Stení Vála**. From here,
a road snakes towards **Gérakas**,
at the wild northern tip of the
island. In summer, the beach
here is home to the research
centre for the **HSSPMS** (Hellenic
Society for the Study and
Protection of the Monk Seal).
The organisation's main
premises are located in the
harbour area of Patitíri.

🔲 **Hellenic Society for the
Study and Protection of the
Monk Seal (HSSPMS)**
Patitíri. **Tel** 24240 66350. **Open** Apr–
Oct: daily; Nov–Mar: on request. ♿
🌐 mom.gr

Taverna at Stení Vála

Two endangered Mediterranean monk seals

Sporades Marine Park

Θαλάσσιο Πάρκο

🚢 from Skiáthos, Skópelos, Alónissos.

Founded in 1992, the National Marine Park of Alónissos and the Northern Sporades, to give it its full name, is an area of great environmental importance. It is the only such park in the Aegean, and includes not just Alónissos but also its uninhabited outlying islands of Peristéra, Skantzoúra and Gioúra. Day trips by boat are possible but access is limited.

The park was created to protect an important breeding colony of the endangered Mediterranean monk seal and a fragile marine ecosystem of other rare wildlife, flora and fauna. Thanks to the pioneering efforts of marine biologists from the University of Athens, who first formed the Hellenic Society for the Study and Protection of the Monk Seal in 1988, Greece's largest population of the elusive Mediterranean monk seal is now scientifically monitored. Fewer than 500 of these seals exist worldwide, making it one of the world's most endangered species. There is an estimated population of 300 seals around the Aegean, with about 50 in the marine park. A campaign to promote awareness of the endangered status of the seals and restrictions on fishing in the area seems to be paying off.

Sightings of seals are not always guaranteed and there is no longer access for the public to view the wild goats on Gioúra, Audouin's gull or Eleonora's falcons on the islet of Skantzoúra: only scientists are now permitted.

The marine park is also an important route and staging post for many migrant birds during the spring and autumn. Land birds, ranging in size from tiny warblers through to elegant pallid harriers, pass through the region in large numbers to and from their breeding grounds in northeast Europe.

Marine Wildlife in the Sporades

Visitors can observe a wide range of other wildlife in the Sporades while watching out for monk seals. Grey herons and kingfishers are both birds of the coast here, a surprise for many birdwatchers from northern Europe who usually associate them with freshwater habitats. Spring and autumn in particular are good times for seeing several species of gulls and terns and, when venturing close to sea cliffs, keep an eye out for the Eleonora's falcons which nest on the inaccessible ledges; in the air, they are breathtakingly acrobatic birds.

Further out to sea, look for jellyfish in the water and the occasional group of common dolphins which may accompany the boat for a while. Cory's shearwaters fly with rigid wings close to the waves and head towards the shore in high winds and as dusk approaches. If you are at sea after dark, you are likely to see a glowing bioluminescence on the surface of the waves, caused by microscopic marine animals.

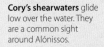
Cory's shearwaters glide low over the water. They are a common sight around Alónissos.

Jellyfish flourish in the seas off the Sporadic islands. This is a *Pelagia noctiluca*.

Mediterranean gulls are easily recognized by the pure white wings and black hood that characterize their summer plumage.

Common dolphins can sometimes be seen in small groups diving in and out of the waves around the boat's wake or swimming alongside.

Skýros
Σκύρος

Renowned in myth as the hiding place of Achilles *(see p87)* and the home-in-exile of the hero Theseus, Skýros has always played an important role in Greek history. A rich Athenian colony from 476 BC, it later became a place of exile for the wealthy from Byzantine Constantinople. Currently one of the homes of the Greek Navy and Air Force, its unique heritage, landscape and architecture bear more resemblance to the Dodecanese than the Sporades.

An example of traditional Skýrian embroidery in the Faltáits Museum

Skýros Town

The main town is architecturally unusual in the Aegean; it has a fascinating mixture of cube-shaped houses, Byzantine churches and spacious squares. Although its main street has been spoiled by loud tavernas and bars, many backstreets give glimpses into Skýrian homes. Traditional ceramics, woodcarving, copper and embroidery are always proudly on display.

Immortal Poetry in Plateía Rupert Brooke

Topping the kástro of the old town with its impressive mansion houses are the remains of the **Castle of Lykomedes**, site of both an ancient acropolis and later a Venetian fortress. It is reached through a tunnel underneath the whitewashed **Moní Agíou Georgíou**, which contains a fine painting of St George killing the dragon. The views from the kástro of the bay below are quite breathtaking. Nearby are the remains of two Byzantine churches, and three tiny chapels, with colourful pastel pink and blue interiors. The town has two good museums. The **Archaeological Museum** displays some bracelets and pottery that were discovered during excavations of minor Neolithic and Mycenaean sites around the island. The museum also presents a traditional Skýrian town house that has been accurately recreated with local furnishings.

Housed in an old mansion owned by the Faltáits family, the excellent **Faltáits Museum** was opened in 1964 by one of their descendants, Manos Faltáits. It has a diverse collection of folk art, including rare books and manuscripts, photographs and paintings, which reveal much about Skýrian history and culture. It not only shows how craftsmen absorbed influences from the Byzantine, Venetian and Ottoman occupations, but also how the development of a wealthy aristocracy actively helped transform the island's woodcarving, embroidery, ceramics and copperware into highly sophisticated art forms.

One place to learn some of these crafts is the **Skýros Centre**, a unique holiday centre which also has courses in such wide-ranging subjects as yoga, reflexology, creative writing and windsurfing. The main branch is in Skýros town, with another branch at Atsítsa, on the west coast of the island.

Plateía Rupert Brooke, above the town, is famous for its controversial statue of a naked man by M Tómpros. Erected in 1930 in memory of the British poet Rupert Brooke who died on the island, the statue is known as *Immortal Poetry*.

🏛 **Archaeological Museum**
Plateía Brooke. **Tel** 22220 91327. **Open** Tue–Sun. **Closed** main public hols. 🗭

🏛 **Faltáits Museum**
Palaiópyrgos. **Tel** 22220 91232. **Open** 10am–noon & 6–9pm daily (winter: 5:30–8pm). 🗭 🗭 🗷 faltaits.gr

🗷 **Skýros Centre**
Tel 01983 865566 (contact London office for bookings). 🗷 skyros.com **Open** Apr–Oct.

Environs

Beneath Skýros town are the resorts of **Mólos** and **Magaziá**. Around these two resorts there are plenty of decent hotels, tavernas and rooms to rent. Further along the coast from Magaziá, there is another sandy stretch of beach at **Pouriá**, which offers excellent spear-fishing and snorkelling. At **Cape Pouriá** itself, the chapel of Agios Nikólaos is built into a cave. Just off the coast are the islets of Vrikolakonísia, where the incurably ill were sent during the 17th century.

The Castle of Lykomedes towering above Skýros town

east coast. Access to Vounó, the mountainous southern part of the island, is through a narrow fertile valley south of **Ormos Achíli** between the island's two halves. The road continues south to **Kalamítsa** bay, and beyond to **Treís Mpoúkes**, a natural deep-water harbour used by pirates in the past and the Greek navy today. Reached by dirt-track road, this is also the site of poet Rupert Brooke's simple marble grave, set in an olive grove. Brooke (1887–1915) died on a hospital ship that was about to set sail to fight at Gallipoli.

The azure waters and tree-lined sand of Péfkos beach

Around the Island

The island divides into two distinct halves bisected by the road from Skýros town to the port of Linariá. Meroí, the northern part of the island, is where most people live and farm on the fertile plains of Kámpos and Trachý.

Skýros is famous for its indigenous ponies, thought by some to be the same breed as the horses that appear on the Parthenon frieze (see p294). It is certainly known that the animals have been bred exclusively on Skýros since ancient times and can still be seen in the wild on the island today, particularly in the south, near the grave of Rupert Brooke.

The road running north from Skýros town leads first to the airport and then west around the island through pine forests to **Kalogriá** and **Kyrá Panagiá**, two leeward

beaches sheltered from the *meltémi* (north wind). From here, the road leads to the small village and pine-fringed beach of **Atsítsa**, where there are rooms to rent and a good taverna. As noted above, Atsítsa is also home to the other branch of the Skýros Centre, the island retreat offering alternative holidays. A little way south are the two beaches of **Agios Fokás** and **Péfkos**. The road loops back from Péfkos to the port of **Linariá**. Caïques depart from here to the inaccessible sea caves at Pentekáli and Diatrýpti on the

The Skýros Goat Dance

This famous goat dance is one of Greece's few rites that have their roots in pagan festivals. It forms the centrepiece of the pre-Lenten festivities in Skýros, celebrated with dancing and feasting. Groups of masquerading men parade noisily around the narrow streets of Skýros town. Each group is led by three central characters: the *géros* (old man), wearing a traditional shepherd's outfit and a goatskin mask and weighed down with noisy bells; the *koréla*, a young man in Skýrian women's clothing; and the *frángos*, or foreigner, a comic figure wearing dishevelled clothes.

The *géros* in full costume

Evvoia
Εύβοια

After Crete, Evvoia is Greece's largest island. It is generally unspoiled by tourism, and its diverse landscape and history make it a microcosm of the whole country. From Macedonian rule in 338 BC, to Turkish government until 1833, the island has suffered many occupations. Traces of Evvoia's mixed history are widely evident, from the range of religious cultures in Chalkída to the descendants of 15th-century Albanian immigrants who still speak their own dialect of Arvanitika.

❽ Cape Artemísio
This is the site of the Battle of Artemisium, which took place in 480 BC.

Istiaía is the main town in the northern part of the island. It is a pretty market town with sleepy squares *(see p127)*.

Map labels: Cape Artemísio ❽, Agriovótano, Glýfa, Istiaía, Psaropoúli, Paralía Kotsikiás, Agiókampos, Giáltra, Loutrá Giáltron, Agios Geórgios, Loutrá Aidipsoú ❾, Arkitsa, Roviés, Agios Vasíleios, Krýa Vrýsi, Mantoúdi, Sara..., Límni ❿, Prokópi ❼, Moní Galatáki, Mount Kandíli 1,361m (4,464ft), Mour... 1,343 r..., Nerotriviá, Néa Arták..., Chalkída ❶

❾ ★ Loutrá Aidipsoú
Old-fashioned, this charming resort has attracted visitors for centuries with its warm spa waters. Local fishermen still continue their trade in the wide bay.

❿ Límni
This picturesque fishing town is full of narrow streets lined with white houses, and colourful flowers that pour out on to the pavement.

❼ Prokópi
The large Kandíli estate, belonging to the English Noel-Baker family, sits just outside the quiet village of Prokópi.

0 kilometres   15
0 miles   10

❶ Chalkída
A modern town, Chalkída is the capital of the island, and has a mixed populace of Muslims, Jews and Orthodox Greeks. By the waterfront is a flourishing market.

❻ Stení
Nestling in the green hills of Mount Dírfys, Stení's cool climate makes it a pleasant escape from the summer heat and a popular place for a day trip.

❺ Kými
A wealthy port in the 1880s, Kými is quieter today, with a fine Folk Museum displaying traditional crafts such as this embroidered picture frame.

Mount Dírfys, the highest point on Evvoia, is a trekker's paradise (see p126).

❹ Ochthoniá
The wild and exposed beaches surrounding Ochthoniá are quiet and often deserted, offering a relaxing break from the busy village.

Skýros

Kými
Mount Dírfys ❺ Paralía Kýmis
1,745 m (5,720 ft)
Platána
❻ Paralía Mourterí
Stení
ount Olympos Ochthoniá ❹
72 m (3,844 ft) Avlonári
E V V O I A
cient Erétria Lépoura
Erétria Alivéri
Skála Oropoú

❸ ★ Kárystos
The traditional seaside and port town of Kárystos is overlooked by the dramatic slopes of Mount Ochi.

Lake Dýstos is a large swampy area on the road to Néa Stýra (see p125).

Néa Stýra
Agía Marina
Stýra
Kalérgo
Mount Ochi
1,398 m (4,585 ft)
Marmári
Rafina
❸ Kárystos

❷ Ancient Erétria
Finds from Ancient Erétria, such as this statue of the goddess Athena, are displayed in the modern town's Archaeological Museum.

Néa Stýra is one of the minor ports on the island for ferries to the mainland (see p125).

Rafina

Mount Ochi provides a scenic day's trek with excellent views (see p125).

For keys to symbols see back flap

❶ Chalkída
Χαλκίδα

Ancient Chalkis was one of the major independent city-states until it was taken by Athens in 506 BC, and it remained an Athenian ally until 411 BC. Briefly Macedonian, the town was under Roman rule by 200 BC. There followed the same history of Byzantine, Frankish and Venetian rule that exists in the Sporades. A bridge has spanned the fast-flowing Evripos channel since the 6th century BC. According to legend, Aristotle was so frustrated at his inability to understand the ever-changing currents that he threw himself into the water.

VISITORS' CHECKLIST

Practical Information
🗺 65,000. Diákou & Frízi. 🛈 22210 22314. 🏛 Mon–Sat. 🎭 Agía Paraskeví celebrations: 26 Jul–1 Aug. 🖵 **chalkidacity.gr**

Transport
🚉 🚌 Athinón. 🚌 corner of Athanasíou

Chalkída's waterfront market

Exploring Chalkída
Although much of modern Chalkída is dominated by commercial activity, there are two areas of the town that are worth a visit: the waterfront which overlooks the Evripos channel, and the old Kástro quarter, on the slopes overlooking the seafront.

The Waterfront
Lined with old-fashioned hotels, cafés and restaurants, Chalkída's waterfront also has a bustling enclosed market where farmers from the neighbouring villages sell their produce. This often leads to chaotic traffic jams in the surrounding narrow streets, an area still known by its Turkish name of Pazári, where there are interesting shops devoted to beekeeping (No. 6 Neofýtou) and other rural activities.

Kástro
In the old Kástro quarter, southeast of the Evripos bridge, the deserted streets reveal a fascinating architectural history. Many houses still bear the traces of their Venetian and Turkish ancestry, with timbered façades or marble heraldic carving. Now inhabited by Thracian Muslims

who settled here in the 1980s, and the surviving members of the oldest Jewish community in Greece, the Kástro also has an imposing variety of religious buildings. Three examples of these include the 19th-century **synagogue** on Kótsou, the beautiful 15th-century mosque, **Emir Zade**, in the square marking the entrance to the Kástro, and the church of **Agía Paraskeví**. The mosque is usually closed, but outside is an interesting marble fountain with an Arabic inscription.

Agía Paraskeví, situated near the Folk Museum, reveals the diverse history of Evvoia more than any other building in Chalkída. This huge 13th-century basilica is built on the site of a much earlier Byzantine church. Its exterior resembles a Gothic cathedral but the interior is a patchwork of different styles, a result of years of modification by invading peoples, including the Franks and the Turks. It has a marble iconostasis, a carved wooden pulpit, brown stone walls and a lofty wooden ceiling. Opposite the church on a house lintel is a carving of St Mark's winged lion, the symbol of Venice.

Housed in the vaults of the old Venetian fortress at the top of the Kástro quarter, the **Folk Museum** presents a jumble of local costumes, engravings and a bizarre set of uniforms from a brass band. The **Archaeological Museum** is a more organized collection of finds from ancient Evvoian sites such as Kárystos. Exhibits include some 5th-century-BC gravestones and vases.

Roman horse head in the Archaeological Museum

🏛 **Folk Museum**
Skalkóta 4. **Tel** 22210 21817. **Open** 10am–1pm Tue–Sun. 🎫 📷

🏛 **Archaeological Museum**
Venizélou 13. **Tel** 22210 25131. **Open** 8am–3pm Tue–Sun. **Closed** main public hols. 📷

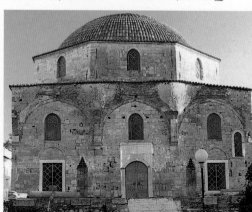

The 15th-century mosque in the Kástro, home to some Byzantine relics

Around Evvoia

The forests of pine and chestnut trees, rivers and deserted beaches in the fertile north contrast dramatically with the dry and scrubby south. Separated by the central mountains, the south becomes rough and dusty with sheep grazing in flinty fields, snaking roads along cliff tops and the scree slopes of Mount Ochi.

Picturesque Kárystos harbour, with Mount Ochi in the background

❷ Ancient Erétria

Αρχαία Ερέτρια

22 km (14 miles) SE of Chalkída. 🚌

Excavations begun in the 1890s in the town of Néa Psará have revealed the sophistication of the ancient city-state of Erétria, which was destroyed by the Persians in 490 BC and the Romans in AD 198. At the height of its power it had colonies in both Italy and Asia Minor. The ancient harbour is silted up, but evidence of its maritime wealth can be seen in the ruined agora, temples, gymnasium, theatre and sanctuary, which still remain around the modern town.

Artifacts from the ancient city are housed in the **Archaeological Museum**. The tomb finds include some bronze cauldrons and funerary urns. There are votive offerings from the Temple of Apollo, gold jewellery and a terracotta gorgon's head, which was found in a 4th-century BC Macedonian villa.

Archaeologists have also restored the **House with Mosaics** (ask for the key at the museum). Its floor mosaics are of lions attacking horses, sphinxes and panthers.

🏛 **Archaeological Museum**
On the road from Chalkída to Alivéri.
Tel 22290 62206. **Open** 8:30am–3pm daily. 🅿 ♿

Environs

Past **Alivéri**, with its medieval castle and ugly power station, the road divides at the village of **Lépoura**. Venetian towers can be seen on the hillside here, and also around the Dýstos plain northwards to Kými and south to Kárystos. A road twists through tiny villages such as **Stýra**, with their surrounding wheat fields. Below lie the seaside resorts of **Néa Stýra** and **Marmári**, both of which provide ferry services to the mainland port of Rafína.

Gorgon's head, Archaeological Museum, Erétria

❸ Kárystos

Κάρυστος

130 km (80 miles) SE of Chalkída. 🏔 5,000. 🚢 🚌

Kárystos, overlooked by the imposing Castel Rosso and the village of Mýloi where plane trees surround the *kafeneía*, is a picturesque town. The modern part of the town dates from the 19th century, and was built during the reign of King Otto. Kárystos has five Neo-Classical municipal buildings, excellent waterfront fish tavernas close to its Venetian Bourtzi fortress and a **Folk Museum**. Set up as a typical Karystian house, the museum contains examples of rural life – copper pots and pans, oil amphorae and ornate 19th-century furniture and embroidery. Kárystos is also famed for its green and white marble and green slate roof and floor tiles.

🏛 **Folk Museum**
50 m (165 ft) from the town square. **Tel** 22240 22240.
Open 8am–10pm Tue & Thu.
Closed main public hols.

Environs

Southeast of Kárystos, remote villages, such as Platonistós and Amigdaliá, hug the slopes of **Mount Ochi**. Caïques from these villages take passengers on boat trips to visit nearby coves where there are prehistoric archaeological sites.

Dragon Houses

Off the main road at Stýra, a signpost points the way to the enigmatic dragon houses, known locally as *drakóspita*. Red arrows mark the trail that leads to these low structures. Constructed with huge slabs of stone, they take their name from the only creatures thought capable of carrying the heavy slabs. There are many theories about the *drakóspita*, but the most plausible links them to two other similar

sites, on the summits of Mount Ochi and Mount Ymittós in Attica. All three are near marble quarries, and it is believed that Carian slaves from Asia Minor (where there are similiar structures) built them as temples in around the 6th century BC.

Scenic road running through olive groves between Ochthoniá and Avlonári

❹ Ochthoniá

Οχθωνιά

90 km (56 miles) E of Chalkída.
🚍 1,140. 🚌

Both Ochthoniá and its
neighbouring village of
Avlonári, with their
Neo-Classical houses
clustered around ruined
Venetian towers, are
reminiscent of
protected Umbrian
hill-towns.

A Frankish castle
overlooks the village of
Ochthoniá, and west of
Avlonári is the
distinctive 14th-century
basilica of Agios Dimítrios,
the largest Byzantine church
in Evvoia. Beyond the fertile
fields that surround these
villages, wild beaches, such
as Agios Merkoúris and
Mourterí, stretch out towards
Cape Ochthoniá.

❺ Kými

Κύμη

90 km (56 miles) NE of Chalkída. 🚍
4,000. 🚢 🚌 🏨 Sat. 🚉 Platána 7 km
(4.5 miles) S.

Four km (2 miles) above Paralía
Kýmis lies the thriving town
of Kými. With a commanding
view of the sea, this remote
settlement had surprisingly rich
resources, derived from silk
production and maritime

trading, in the 19th century. In
the 1880s, 45 ships from Kými
plied the Aegean sea routes.
The narrow streets of elegant
Neo-Classical houses testify
to its past wealth. It is known
today mainly for the
medicinal spring water
from nearby Choneftikó,
and a statue in the main
square of Dr Geórgios
Papanikoláou, Kými's
most famous son and
inventor of the cervical
smear "Pap test". An
extensive and well-
organized **Folk
Museum** contains
exhibits from Kymian life, such
as a fine collection of unique
cocoon embroideries. On the
road north of Kými, the 17th-
century **Moní Metamórfosis
tou Sotíros**, now inhabited by
nuns, perches on the cliff edge.

Dr Papanikoláou
(1883–1962)

🏛 Folk Museum
Tel 22220 22011. **Open** 10:30am–3pm
daily. **Closed** main public holidays.

❻ Steni

Στενή

31 km (20 miles) NE of Chalkída.
🚍 1,250. 🚌

This mountain resort is much
loved by Greeks who come for
the cool climate and fine
scenery. Steni is also popular
with hikers setting their sights
on **Mount Dírfys**, the island's
highest peak at 1,745 m
(5,720 ft), with spectacular
views from the summit. A brisk
walk followed by a lazy lunch
of classic mountain cuisine –
grilled meats and oven-baked
beans – make for a pleasant day.
The main square is also good for
shops selling local specialities,
such as wild herbs and
mountain tea.

The road from Steni to the
northern coast snakes up the
mountain. It passes through
spectacular scenery of narrow
gorges filled with waterfalls and
pine trees, and cornfields that
stretch down to the sea.

Moní Metamórfosis tou Sotíros in the mountains near Kými

❼ Prokópi

Προκόπι

52 km (32 miles) NW of Chalkída. 1,200. 🚌 🚆 Sun. 🚆 Krýa Vrýsi 15 km (9 miles) N.

Sleepy at most hours, Prokópi only wakes when the tourist buses arrive with pilgrims coming to worship the remains of St John the Russian (Agios Ioánnis o Rósos), housed in the modern church of Agíou Ioánnou tou Rósou. Souvenir shops and hotels around the village square cater fully for the visiting pilgrims. In reality a Ukranian, John was captured in the 18th century by the Turks and taken to Prokópi (present-day Ürgüp) in central Turkey. After his death, his miracle-working remains were brought over to Evvoia by the Greeks during the exodus from Asia Minor in 1923.

Prokópi is also famous for the English Noel-Baker family, who own the nearby Kandíli estate. Although the family have done much for the region, local feeling is mixed about the once-feudal status of this estate. Many locals, however, now accept the important role Kandíli plays in its latest incarnation as a specialist holiday centre, by bringing money into the local economy.

Environs

The road between Prokópi and Mantoúdi runs by the River Kiréa, and a path leads to one of the oldest trees in Greece, said to be over 2,000 years old. This huge Oriental plane tree (*Platanus orientalis*) has a circumference of over 11 m (37 ft).

Façade of the mansion on the Noel Baker Kandíli estate, Prokópi

The pebble beach at Cape Artemísio

❽ Cape Artemísio

Ακρωτήριο Αρτεμίσιο

105 km (65 miles) NW of Chalkída. 🚌 to Agriovótano. ℹ️ Agriovótano (22210 76131) 🚆 Psaropoúli 15 km (9 miles) SE.

Below the Picturesque village of Agriovótano sits Cape Artemísio, site of the Battle of Artemisium. Here the Persians, led by King Xerxes, defeated the Greeks in 480 BC. In 1928, local fishermen hauled the famous bronze statue of Poseidon out of the sea at the cape. It is now on show in the National Archaeological Museum in Athens *(see p290)*.

Old Mercedes truck delivering produce

Environs

About 20 km (12 miles) east lies **Istiaía**, a pleasant market town with sleepy squares, white chapels and ochre-coloured houses.

❾ Loutrá Aidipsoú

Λουτρά Αιδηψού

100 km (62 miles) NW of Chalkída. 🚌 3,000. 🚌 ℹ️ 22260 22500. 🚆 Mon–Sat 🚆 Giáltra 15 km (9 miles) SW.

Loutrá Aidipsoú is Greece's largest spa town, popular since antiquity for its cure-all sulphurous waters. These waters bubble up all over the town and many hotels are built directly over hot springs to provide a supply to their treatment rooms. In the rock pools of the public baths by the sea, the steam rises in winter scalding the red rocks.

Dominating the town is the Thérmai Sýlla. Voted one of the world's top spas, this luxurious hotel offers wellness and medical treatments that use the natural, mineral-rich spring waters. These luxuries are reminders of the days when the rich and famous came to take the cure. Other Neo-Classical hotels along the seafront also recall the town's days of glory in the late 1800s. The town has a relaxed atmosphere and is popular with Greek families.

Environs

In the summer a ferry service goes across the bay to **Loutrá Giáltron,** where warm spring water mixes with the shallows of a quiet beach edged by tavernas.

❿ Límni

Λίμνη

87 km (54 miles) NW of Chalkída. 🚌 2,100. 🚌 ℹ️ 22270 32111.

Once a wealthy 19th-century seafaring power, the pleasant town of Límni has elegant houses, cobbled streets and a charming seafront. Just south of the town is the magnificent Byzantine **Moní Galatáki**, the oldest monastery on Evvoia, etched into the cliffs of Mount Kandíli. Inhabited by nuns since the 1940s, its church is covered with beautiful frescoes. The Last Judgment is shown in particularly gory detail, with some souls frantically climbing the ladder to heaven, while others are dragged mercilessly into the leviathan's jaws.

THE NORTHEAST AEGEAN ISLANDS

THASOS · SAMOTHRAKI · LIMNOS · LESVOS · CHIOS · IKARIA · SAMOS

More than any other archipelago in Greece, the seven major islands of the Northeast Aegean defy easy categorization. Though they are neighbours, sharing a common history of rule by the Genoese and lively fishing industries, the islands are culturally distinct, encompassing a range of landscapes and lifestyles.

Although Sámos and Chíos were prominent in ancient times, few traces of that former glory remain. Chíos offers the region's most compelling medieval monuments, including the Byzantine monastery of Néa Moní, considered one of the finest examples of Macedonian Renaissance architecture in the world, and the mastic villages, while vineyard-covered Sámos has a fascinating museum of artifacts from the long-venerated Heraion shrine. In Límnos's capital, Mýrina, you encounter evidence of the Genoese and Ottoman occupations, in the form of its castle and domestic architecture.

Lésvos, the third-largest Greek island, shares the fortifications and volcanic origin of Límnos, though the former's monuments are grander and its topography more dramatic. To the south, the islands of Sámos, Chíos and Ikaría have mountainous profiles and are richly forested with pine, olive and cypress trees. Most of the pines of Thásos were devastated by forest fires in the 1980s, but its rugged coastline of deep bays lined with white beaches makes it a popular holiday island. Samothráki remains unspoiled; its numerous hot springs and waterfalls, as well as the brooding summit of Mount Fengári, are a counterpoint to the long-hallowed Sanctuary of the Great Gods.

Beaches come in all sizes and consistencies, from the finest sand to melon-sized volcanic shingle. Apart from Thásos, Sámos and Lésvos, package tourism is scarce in the north, where summers are shorter. Wild Ikaría, historically a backwater, will appeal mostly to spa-plungers and beachcombers, while its tiny dependency, Foúrnoi, is an ideal do-nothing retreat owing to its convenient beaches and abundant seafood.

Mólyvos harbour, Lésvos, overlooked by the town's 14th-century Genoese castle

◄ The picturesque Agios Isídoros on the island of Chíos

Exploring the Northeast Aegean Islands

For its beaches and ancient ruins, both composed of white marble, Thásos is hard to fault, while Samothráki has long been a destination for hardy nature lovers. Less energetic visitors will find Límnos ideal, with picturesque villages and beaches close to the main town. Olive-rich Lésvos offers the greatest variety of scenery but requires time and effort to tour. For first-time visitors to the eastern isles, Sámos is the best touring base, though the cooler climate of Chíos is more attractive, and its main town offers good shopping. Connoisseurs of relatively unspoiled islands will want to sample a slower pace of life on Ikaría, Psará or Foúrnoi.

Fishing boat in Mólyvos harbour, Lésvos

Islands at a Glance

Chíos *pp150–56*
Ikaría *p157*
Lésvos *pp140–41*
Límnos *pp138–9*
Sámos *pp158–61*
Samothráki *pp136–7*
Thásos *pp132–5*

Key

▬▬ Motorway
— Main road
═══ Minor road
— Scenic route
--- High-season, direct ferry route
▬▬ International border
△ Summit

Byzantine monastery of Néa Moní, Chíos, seen from the southwest

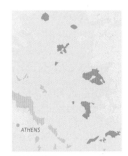

Locator Map

Volcanic landscape near Kontiás, Límnos

0 kilometres 40
0 miles 20

Getting Around

Thásos and Samothráki have no airports, but are served by ferries from Alexandroúpoli and Kavála on the mainland, while Límnos and Lésvos have air and ferry links with Athens and Thessaloníki. Bus services vary from virtually nonexistent on Límnos and Samothráki, or Lésvos's functional schedules, to Thásos's frequent coaches. Chíos, Ikaría and Sámos are served by flights from Athens, and are connected by ferry. Chíos has an adequate bus service but is best explored by car; Sámos has more frequent buses, and is small enough to be toured by motorbike; Ikaría has skeletal public transport and steep roads requiring sturdy vehicles.

Sandy Messaktí beach, Ikaría

For keys to symbols *see back flap*

Thásos
Θάσος

Thásos has been inhabited since the Stone Age, with settlers from Páros colonizing the east coast during the 7th century BC. Spurred by revenues from gold deposits near modern Thásos town, Ancient Thásos became the seat of a seafaring empire, though its autonomy was lost to the Athenians in 462 BC. The town thrived in Roman times, but lapsed into medieval obscurity. Today, the island's last source of mineral wealth is delicate white marble, cut from quarries whose scars are prominent on the hillsides south of Thásos town.

Exterior of the Archaeological Museum, Thásos

🏛 Ancient Thásos
Site & Museum Tel 25930 22180. **Open** 8am–3pm Tue–Sun. **Closed** main public hols. 🖼 ♿

Founded in the 7th century BC, Ancient Thásos is a complex series of buildings, only the remains of which can be seen today. French archaeologists have conducted excavations here since 1911; digs have continued at a number of locations in Thásos town. The **Archaeological Museum**, next to the agora, houses treasures from the site.

Well defined by the ruins of four stoas, the Hellenistic and Roman **agora** covers a vast area behind the ancient military harbour, today the picturesque Limanáki, or fishing port. Though only a few columns

Thásos town harbour, viewed from the agora

❶ Thásos Town
Λιμένας

🏔 3,130. 🚌 🚕 ℹ️ 25930 23111. 🚢 daily. 🚕 Pachýs 9 km (6 miles) W.

Modern Liménas, also known as Thásos town, is an undistinguished resort on the coastal plain which has been settled for nearly three millennia.

Interest lies in the vestiges of the ancient city and the manner in which they blend into the modern town. Foundations of a Byzantine basilica take up part of the central square, while the road to Panagiá cuts across a vast shrine of Herakles before passing a monumental gateway.

Sights at a Glance
❶ Thásos Town
❷ Potamiá
❸ Alykí
❹ Moní Archangélou Michaïl
❺ Theológos
❻ Kástro
❼ Sotíras
❽ Megálo Kazavíti

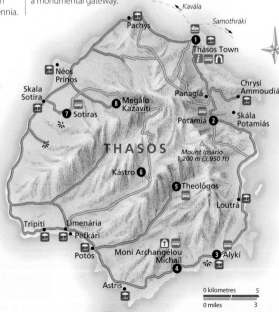

Kavála
Samothráki
Pachýs
❶ Thásos Town
ℹ️ 🏛
Néos
Prínos
Skala
Sotira
Chrysí
Ammoudiá
Panagiá
❽ Megálo
Kazavíti
❼ Sotíras
Potamiá ❷
Skála
Potamiás
THÁSOS
Mount Ipsário 1,200 m (3,950 ft)
Kástro ❻
❺ Theológos
Loutrá
Tripití
Limenária
Pefkári
Potós
Moní Archangélou
Michaïl
❹
❸ Alykí
Astris

| 0 kilometres | | 5 |
| 0 miles | | 3 |

Plan of Ancient Thásos

VISITORS' CHECKLIST

Practical Information
16,000. Panagiá: 15 Aug.

Transport
Thásos town bus station.

The Gate of Parmenon in the south wall of Ancient Thásos

Key to Plan

① Archaeological Museum
② Agora
③ Temple of Dionysos
④ Theatre
⑤ Citadel
⑥ Walls
⑦ Temple to Athena Poliouchos
⑧ Shrine to Pan
⑨ Gate of Parmenon

have been re-erected, it is easy to trace the essentials of ancient civic life, including several temples to gods and deified Roman emperors, foundations of heroes' monuments and the extensive drainage system.

Foundations of a **Temple of Dionysos**, where a 3rd-century BC marble head of the god was found, mark the start of the path up to the acropolis. Partly overgrown by oaks, the Hellenistic **theatre** has spectacular views out to sea. The Romans adapted the stage area for their bloody spectacles; it is now being excavated with the intent of complete restoration.

The ancient **citadel**, once the location of an Apollo temple, was rebuilt during the 13th century by the Venetians and Byzantines. It was then ceded by Emperor Manuel II Palaio-lógos to the Genoese Gatelluzi clan in 1414, who enlarged and occupied it until 1455. Recycled ancient masonry is conspicuous at the south gateway. By the late 5th century BC, substantial **walls** of more than 4 km (2 miles) surrounded the city,

the sections by the sea having been mostly wrecked on the orders of victorious besiegers in 492 and 462 BC.

Foundations of a **Temple to Athena Poliouchos** (Patroness of the City), dated to the early 5th century BC, are just below the acropolis summit; massive retaining walls support the site terrace. A cavity hewn in the rocky outcrop beyond served as a **shrine to Pan** in the 3rd

century BC; he is depicted in faint relief playing his pipes.

Behind the summit point, a steep 6th-century BC stairway descends to the **Gate of Parmenon** in the city wall. The gate retains its lintel and takes its name from an inscription "Parmenon Made Me" (denoting its mason), on a nearby wall slab.

Columns of the agora, with the town church in the background

Around Thásos Island

Thásos is just small enough to explore by motorbike, though the bus service along the coastal ring road is good and daily hydrofoils link Thásos town with the western resorts. The best beaches are in the south and east, though the coastal settlements are mostly modern annexes of inland villages, built after the suppression of piracy in the 19th century.

Boats in the peaceful harbour of Skála Potamiás

❷ Potamiá

Ποταμιά

9 km (6 miles) S of Thásos town. ⛰ 1,000. 🚌 🚢 daily. 🚗 Loutrá 12 km (7 miles) S; Chrysí Ammoudiá 5 km (3 miles) E.

Potamiá is a small village, with one of the most popular paths leading to the 1,200 m (3,950 ft) summit of Mount Ipsário. Following bulldozer tracks upstream brings you to the trailhead for the ascent, which is a 7-hour excursion; although the path is waymarked by the Greek Alpine Club, it is in poor condition.

The sculptor and painter Polýgnotos Vágis (1894–1965) was a native of the town, although he emigrated to America at an early age. Before his death, the artist bequeathed most of his works to the Greek

state and they are now on display at the small **Vágis Museum**, situated in the village centre. His work has a mythic, dreamlike quality; the most compelling sculptures are representations of birds, fish, turtles and ghostly faces which he carved on to boulders or smaller stones.

🏛 **Vágis Museum**
Tel 25930 61400. **Open** 10am–1pm, 6–9pm Tue–Sun. **Closed** main public holidays. 📷

Environs
Many visitors stay and enjoy the traditional Greek food at **Skála Potamiás**, 3 km (2 miles) east of Potamiá, though **Panagía**, 2 km (1 mile) north, is the most visited of the inland villages. It is superbly situated above a sandy bay,

Blue-washed house in Panagía

has a lively square and many of its 19th-century houses have been preserved or restored.

❸ Alykí

Αλυκή

29 km (18 miles) S of Thásos town. 🚌 🚗 Astrís 12 km (7 miles) W.

Perhaps the most scenic spot on the Thasian shore, the headland at Alykí is tethered to the body of the island by a slender spit, with beaches to either side. The westerly cove is fringed by the hamlet of Alykí, which has well-preserved 19th-century vernacular architecture due to its official classification as an archaeological zone. A Doric temple stands over the eastern bay, while behind it, on the headland, are two fine Christian basilicas, dating from the 5th century, with a few of their columns re-erected.

Local marble was highly prized in ancient times; now all that is left of Alykí's quarries are overgrown depressions on the headland. At sea level, "bathtubs" (trenches scooped out of the rock strata) were once used as evaporators for salt-harvesting.

Moní Archangélou Michaïl, perched on its clifftop

❹ Moní Archangélou Michaïl

Μονή Αρχαγγέλου Μιχαήλ

34 km (21 miles) S of Thásos town. **Tel** 25930 31500. 🚌 **Open** daily.

Overhanging the sea 3 km (2 miles) west of Alykí, Moní

Archangélou Michaïl was founded early in the 12th century by a hermit called Luke, on the spot where a spring had appeared at the behest of the Archangel. Now a dependency of Moní Filothéou on Mount Athos in northern Greece, its most treasured relic is a Holy Nail from the Cross. Nuns have occupied the grounds since 1974.

Slate-roofed house with characteristically large chimneypots, Theológos

❺ Theológos

Θεολόγος

50 km (31 miles) S of Thásos town. 🚲 900. 🚌 **Open** daily. 🚕 Potós 10 km (6 miles) SW.

Well inland, secure from attack by pirates, Theológos was the Ottoman-era capital of Thásos. Tiered houses still exhibit their typically large chimneys and slate roofs. Generous gardens and courtyards give the village a green and open aspect. A ruined tower and low walls on the hillside opposite are evidence of Theológos's original 16th-century foundation by Greek refugees from Constantinople.

❻ Kástro

Κάστρο

45 km (28 miles) SW of Thásos town. 🚲 6. 🚕 Tripití 13 km (8 miles) W of Limenária.

At the centre of Thásos, 500 m (1,640 ft) up in the mountains, the village of Kástro was even more secure than Theológos.

Taverna overhung by plane trees in Sotíras village

Founded in 1403 by Byzantine Emperor Manuel II Palaiológos, it became a stronghold of the Genoese, who fortified the local hill which is now the cemetery. Kástro was slowly abandoned after 1850, when a German mining concession created jobs at Limenária, on the coast below.

This inland hamlet has now been reinhabited on a seasonal basis by sheep farmers. The *kafeneío*, on the ground floor of the former school, beside the church, shelters the single telephone; there is no mains electricity.

❼ Sotíras

Σωτήρας

23 km (14 miles) SW of Thásos town. 🚲 12 🚌 🚕 Skála Sotíra 3 km (2 miles) E.

Facing the sunset, Sotíras has the most alluring site of all the inland villages – a fact not lost on the dozens of foreigners who have made their homes here. Under gigantic plane trees

watered by a triple fountain, the tables of a small taverna fill the relaxed balcony-like square. The ruin above the church was a lodge for German miners, whose exploratory shafts still yawn on the ridge opposite.

Traditional stone houses with timber balconies, Megálo Kazavíti

❽ Megálo Kazavíti

Μεγάλο Καζαβίτι

22 km (14 miles) SW of Thásos town. 🚲 1,650. 🚌 🍴 daily. 🚕 Néos Prínos 6 km (4 miles) NE.

Greenery-shrouded Megálo Kazavíti (officially Ano Prínos) surrounds a central square, which is a rarity on Thásos. There is no better place to find examples of traditional domestic Thasian architecture with its characteristic mainland Macedonian influence: original house features include narrow-arched doorways, balconies and overhanging upper storeys, with traces of the indigo, magenta and ochre plaster pigment that was once commonly used across the Balkans.

Samothráki
Σαμοθράκη

🏛 2,700. 🚢 🚌 Kamariótissa. 🚉 Pachiá Ammos 15 km (9 miles) SW of Kamariótissa.

With virtually no level terrain except for the western cape, Samothráki is synonymous with the bulk of Mount Fengári. In the Bronze Age the island was occupied by settlers from Thrace. Their religion of the Great Gods was incorporated into the culture of the Greek colonists in 700 BC, and survived under Roman patronage until the 4th century AD. The rawness of the weather seems to go hand in hand with the brooding landscape, making it easy to see how belief in the Great Gods endured.

Chóra
Lying 5 km (3 miles) east of Kamariótissa, the main port of the island, Chóra is the capital of Samothráki. The town almost fills a pine-flecked hollow, which renders it invisible from the sea.

With its labyrinthine bazaar, and cobbled streets threading past sturdy, tile-roofed houses, Chóra is the most handsome village on the island. A broad central square with two tavernas provides an elegant vantage point, looking out to sea beyond the Genoese castle. Adapted from an earlier Byzantine fort, little other than

The town of Chóra with the remains of its Genoese castle in the background

the castle's gateway remains, though more substantial fortifications can be found downhill at Chóra's predecessor, **Palaiópoli**; here three Gatelluzi

Sanctuary of the Great Gods
Ερείπια του Ιερού των Μεγάλων Θεών

The sanctuary of the Great Gods on Samothráki was, for almost a millennium, the major religious centre of ancient Aeolia, Thrace and Macedonia. There were similar shrines on Límnos and Ténedos, but neither commanded the following or observed the same rites as the one here. Its position in a canyon at the base of savage, plunging crags on the northeast slope of Mount Fengári was perhaps calculated to inspire awe; today, though thickly overgrown, it is scarcely less impressive. The sanctuary was expanded and improved in Hellenistic times by Alexander's descendants, and most of the ruins visible today date from that period.

Nike Fountain
A marble centrepiece, the Winged Victory of Samothráki, once decorated the fountain. It was discovered by the French in 1863 and is now on display in the Louvre, Paris.

The stoa is 90 m (295 ft) long and dates to the early 3rd century BC.

Hall for votive offerings

The Theatre held performances of sacred dramas in July, during the annual festival.

Hieron
The second stage of initiation, *epopteia*, took place here. In a foreshadowing of Christianity, this involved confession and absolution followed by baptism in the blood of a sacrificed bull or ram. Rites took place in an old Thracian dialect until 200 BC.

The Temenos is a rectangular space where feasts were probably held.

(see p142) towers of 1431 protrude above the extensive walls of the ancient town.

Around the Island

Easy to get around by bike or on foot, Samothráki has several villages worth visiting on its southwest flank, lost in olive groves or poplars. The north coast is moister, with plane, chestnut and oak trees lining the banks of several rivers. Springs are abundant, and waterfalls meet the sea at Kremastá Nerá to the south. Stormy conditions compound the lack of adequate harbours.

Thérma has been the island's premier resort since the Roman era, due to its hot springs and lush greenery. You can choose among two rustic outdoor

The Gatelluzi towers at ancient Palaiópoli

pools of about 34° C (93° F), under wooden shelters; an extremely hot tub of 48° C (118° F) in a cottage, only for groups; and the rather sterile modern bathhouse at 39° C (102° C). Cold-plunge fans will find rock pools and low waterfalls 1.5 km (1 mile) east at **Krýa Váthra**. These are not as impressive or cold as the ones

only 45 minutes' walk up the Foniás canyon, 5 km (3 miles) east of Thérma.

The highest summit in the Aegean, at 1,600 m (5,250 ft), is the granite mass of **Mount Fengári**. Although often covered with cloud, it serves year-round as a seafaring landmark and the views from the top are superb. In legend, the god Poseidon watched the Trojan War from this mountain. The peak is usually climbed from Thérma as a 6-hour round trip, though there is a longer and easier route up from Profítis Ilías village on its southwest flank.

Arsinoeion

At over 20 m (66 ft) across, this rotunda is the largest circular building known to have been built by the Greeks. It was dedicated to the Great Gods in the 3rd century BC.

VISITORS' CHECKLIST

Practical Information
6 km (4 miles) NE of Kamariótissa.
Tel 25510 41474. **Open** 8:30am–3pm daily.

Transport
to Palaiópoli.

Sanctuary of Anaktoron

This building was where myesis, the first stage of initiation into the cult, took place. This involved contact with the *kabiri* mediated by prior initiates.

Small theatre

The Propylon

(monumental gate) was dedicated by Ptolemy II of Egypt in 288 BC.

useum

Deities and Mysteries of Samothráki

When Samothráki was colonized by Greeks in 700 BC, the settlers combined later Olympian deities with those they found here. The principal deity of Thrace was Axieros, the Great Mother, an earth goddess whom the Greeks identified with Demeter, Aphrodite and Hekate. Her consort was the fertility god Kadmilos and their twin offspring were the *kabiri* – a Semitic word meaning "Great Ones" which soon came to mean the entire divine family. These two deities were later recognized as the *dioskouri* Castor and Pollux, whose emblems were snakes and a star. The cult was open to allcomers of any age or gender, free or slave, Greek or barbarian. Details of the mysteries are unknown, as adherents honoured a vow of silence.

The twin *kabiri*, Castor and Pollux

Límnos
Λήμνος

The mythological landing place of Hephaistos, the god of metalworking cast out of Olympos by Zeus, Límnos is appropriately volcanic; the lava soil crumbles into broad beaches and grows excellent wine and herbal honey. Controlling the approaches to the Dardanelles, the island was an important outpost to both the Byzantines and the Turks, under whom it prospered as a trading station. The Greek military still controls much of the island, but otherwise it is hard to imagine a more peaceful place.

Map labels: Pláka, Sergitsi, Kabeirió, Propoúli, Ifaisteía, Katálakko, Dáfni, Atsikí, Város, Kondopoúli, Mount Skopiá 430 m (1,410 ft), Sardes, Karpási, Kalliópi, Kavála, Kornós, Livadochóri, Repanídi, Romanó, Roussopoúli, Káspakas, Portianó, Moúdros, Polióchni, Avlónas, Kontiás, Mount Paradeísi 260 m (850 ft), Thessaloníki, Mýrina, Agios Pávlos, Fisíni, Platí, Thános, Mount Fakós 265 m (865 ft), Skandáli, Rafína, Lésvos Agios Efstrátios

0 kilometres 5
0 miles 3

dominated by pottery shards which may only interest a specialist. The most compelling ceramic exhibits are a pair of votive lamps in the form of sirens from the temple at Ifaisteía, while metalwork from Polióchni is represented by bronze tools and a number of decorative articles.

Spread across the headland, and overshadowing Mýrina, the **kástro** boasts the most dramatic position of any North Aegean stronghold. Like others in the region, it was in turn an ancient acropolis and a Byzantine fort, fought over and refurbished by Venetians and Genoese until the Ottomans took the island in 1478. Though dilapidated, the kástro makes a rewarding evening climb for beautiful views over western Límnos.

Mýrina
Successor to ancient Mýrina Límnos's second town in antiquity, modern Mýrina sprawls between two sandy bays at the foot of a rocky promontory. Not especially touristed, it is one of the more pleasant island capitals in the North Aegean, with cobbled streets, an unpretentious bazaar and

imposing, late-Ottoman houses. The most ornate of these cluster behind the northerly beach, Romeíkos Gialós, which is also the centre of the town's nightlife. The south beach, Toúrkikos Gialós, extends beyond the compact fishing port with its half-dozen quayside tavernas. The only explicitly Turkish relic is a fountain on Kída, inscribed with Turkish calligraphy, from which delicious potable water can still be drawn.

Housed in an imposing 19th-century mansion behind Romeíkos Gialós, the **Archaeological Museum** is exemplary in its display of artifacts belonging to the four main ancient cities of Límnos. The most prestigious items have been sent to Athens, however, leaving a collection

Mýrina harbour, overlooked by the kástro in the background

The volcanic landscape of Límnos, viewed from the village of Kontiás

Around the Island

Though buses run from Mýrina to most villages in summer, the best way to travel around Límnos is by car or motorbike; both can be hired at Mýrina. Southeast from Mýrina, the road leads to **Kontiás**, the third-largest settlement on Límnos, sited between two volcanic outcrops supporting the only pine woods on the island. Sturdily constructed, red-tiled houses, including some fine *belle époque* mansions, combine with the landscape to make this the island's most appealing inland village.

The bay of **Moúdros** was Commonwealth headquarters during the ill-fated 1915 Gallipoli campaign. Many casualties were evacuated to hospital here; the unlucky ones were laid to rest a short walk east of Moúdros town on the road to Roussopoúli. With 887 graves, this ranks as the largest Commonwealth cemetery from either world war in the Greek islands; 348 more English-speaking servicemen lie in another graveyard across the bay at **Portianoú**.

Founded just before 3000 BC, occupying a clifftop site near the village of Kamínia, the fortified town of **Polióchni** predates Troy on the coast of Asia Minor just across the water. Like Troy, which may have been a colony, it was levelled in 2100 BC by an earthquake. It was never resettled. The suddenness of the catastrophe gave many people no time to escape – skeletons were unearthed among the ruins. Polióchni was noted for its metalsmiths, who refined and worked raw ore from Black Sea deposits, and shipped the finished objects to the Cyclades and Crete. A hoard of gold jewellery, now displayed in Athens, was found in one of the houses. Italian archaeologists continue the excavations every summer, and have penetrated four distinct layers since 1930.

The patron deity of Límnos was honoured at **Ifaisteía**, situated on the shores of Tigáni Bay. This was the largest city on the island until the Byzantine era. Most of the site has yet to be completely revealed. Currently, all that is visible are outlines of the Roman theatre, parts of a necropolis and

Looking down on the remains of a Roman theatre, Ifaisteía

foundations of Hephaistos's temple. Rich grave offerings and pottery found on the site can be seen in the Mýrina Archaeological Museum.

The ancient site of the **Kabeirio** (*Kavírio* in modern Greek) lies across Tigáni Bay from Ifaisteía and has been more thoroughly excavated. The Kabeirioi, or Great Gods, were worshipped on Límnos in the same manner as on Samothráki (*see pp136–7*), though at this sanctuary little remains of the former shrine and its adjacent stoa other than a number of column stumps and bases.

Below the sanctuary ruins, steps lead down to a sea grotto known as the Cave of Philoctetes. It takes its name from the wounded Homeric warrior who was supposedly abandoned here by his comrades on their way to Troy until his infected leg injuries had healed.

Outlying Islands

Certainly the loneliest outpost of the North Aegean, tiny, oak-covered **Agios Efstrátios** (named after the saint who was exiled and died here) has scarcely a handful of tourists in any summer. The single port town was damaged by an earthquake in 1967, with dozens of islanders killed; some pre-quake buildings survive above the ferry jetty. Deserted beaches can be found an hour's walk to either side of the port.

Lésvos
Λέσβος

Once a favoured setting for Roman holidays, Lésvos, with its thick southern forests and idyllic orchards, was known as the "Garden of the Aegean" to the Ottomans. Following conquest by them, in 1462, much of the Greek population was enslaved or deported to Constantinople, and most physical traces of Genoese or Byzantine rule were obliterated by both the Turks and the earthquakes the island is prone to. Lésvos has been the birthplace of a number of artists, its most famous child being the great 7th-century-BC lyric poet Sappho.

⑥ ★ Mólyvos
The tourist capital of the island, Mólyvos has a harbour overlooked by a Genoese castle with fine views of Turkey.

⑦ Pétra
This popular resort takes its name from the huge perpendicular rock at its heart. Steps in the rock lead to an 18th-century church on the summit.

⑧ Kalloní
Known mainly for the sardines caught off the coast of nearby Skála Kallonís, this is a cross-roads for most of the island's bus routes.

⑨ Antissa
Situated just below a pine grove, this is the largest village in the area. It has several excellent *kafeneía* in its central square, overshadowed by huge plane trees.

⑩ Moní Ypsiloú
Straddling the summit of an extinct volcano on the edge of a fossilized forest, 12th-century Ypsiloú has a museum of ecclesiastical treasures.

Map labels:
Mólyvo
Pétra
Anax
Kámpos
Skalochóri
Moní Perivolís
Moní Ypsiloú
Moní Leimónos
Sígri ⑪
Antissa ⑨ ⑩
Vatoússa
Kallon
Chídira
Skála Kallon
Eresós
Skála Eresoú ⑫
Mesótopos
Kólpos Kallo
Vatera

⑪ Sígri
Near the westernmost point of the island, this small chapel stands at the waterfront on the edge of the remote village of Sígri.

⑫ Skála Eresoú
One of the largest resorts on the island, the beach at Eresós lies only a short walk from the birthplace of the poet Sappho.

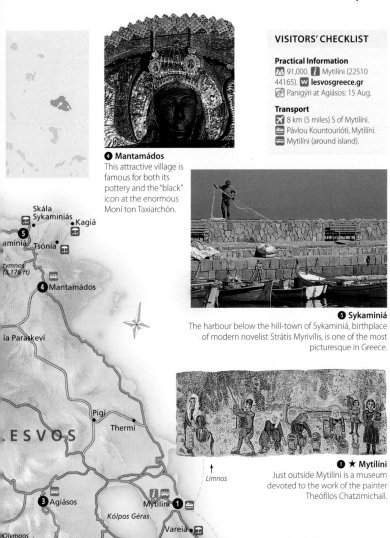

Mantamádos

This attractive village is famous for both its pottery and the "black" icon at the enormous Moní ton Taxiarchón.

VISITORS' CHECKLIST

Practical Information
91,000. Mytilíni (22510 44165). lesvosgreece.gr
Panigýri at Agiásos: 15 Aug.

Transport
8 km (5 miles) S of Mytilíni.
Pávlou Kountourióti, Mytilíni.
Mytilíni (around island).

Sykaminiá

The harbour below the hill-town of Sykaminiá, birthplace of modern novelist Strátis Myrivílis, is one of the most picturesque in Greece.

★ Mytilíni

Just outside Mytilíni is a museum devoted to the work of the painter Theófilos Chatzimichaïl.

Agiásos

Widely regarded as the most beautiful hill-town of the island, Agiásos' main church has an icon supposedly painted by St Luke.

Plomári

This large coastal resort, with its Varvagiánnis distillery, is the oúzo capital of Lésvos.

For keys to symbols *see back flap*

❶ Mytilíni

Μυτιλήνη

Modern Mytilíni has assumed both the name and site of the ancient town. It stands on a slope descending to an isthmus bracketed by a pair of harbours. An examination of Ermoú reveals the heart of a lively bazaar. Its south end is home to a fish market selling species rarely seen elsewhere, while at the north end the roofless shell of the Gení Tzamí marks the edge of the former Turkish quarter. The Turks ruled from 1462 to 1912 and Ottoman houses still line the narrow lanes between Ermoú and the castle rise. The silhouettes of such *belle époque* churches as Agioi Theódoroi and Agios Therápon pierce the tile-roofed skyline.

VISITORS' CHECKLIST

Practical Information
🏠 27,000. 🛈 Aristárchou 6 (22510 44165). 🚌 Agios Ermogénis 12 km (7 miles) S; Charamída 14 km (9 miles) S. 🎭 15 Jul–15 Aug.

Transport
✈ 8 km (5 miles) S. ⛴ 🚌 Pávlou Kountourióti.

The dome of Agios Therápon

🏛 Kástro Mytilónis
Tel 22510 27790. **Open** 8am–3pm daily. **Closed** main public hols. 🎫

Surrounded by pine groves, this Byzantine foundation of Emperor Justinian (527–65) still impresses with its huge curtain walls, but it was even larger during the Genoese era. Many ramparts and towers were destroyed during the Ottoman siege of 1462 – an Ottoman Turkish inscription can be seen at the south gate. Over the inner gate the initials of María Palaiologína and her husband Francesco Gatelluzi – a Genoan who helped John Palaiológos regain the Byzantine throne – complete the resumé of the castle's various occupants. The ruins include those of the Gatelluzi palace, a Turkish *medresse* (theological school) and a dervish cell; a Byzantine cistern stands by the north gate.

🏛 Archaeological Museum
Argýris Eftaliótis. **Closed** for renovation. **New Archaeological Museum:** Corner of 8 Noemvríou & Melínas Merkoúri. **Tel** 22510 28032. **Open** 8am–3pm Tue–Sun. **Closed** main pub hols. 🎫 ♿ (new Museum only).

Lésvos's archaeological collection occupies a *belle époque* mansion and a modern building nearby (New Archaeological Museum). The most famous exhibits are Roman villa mosaics. Neolithic finds from the 1929–33 British excavations at Thermí, just north of town, can also be seen.

🏛 Byzantine Museum
Agios Therápon. **Tel** 22510 28916. **Open** 9am–1pm Mon–Sat. 🎫

This ecclesiastical museum is devoted almost entirely to exhibiting icons. The collection ranges from the 13th to the 18th century and also includes a more recent, folk-style icon by Theófilos Chatzimichaïl.

Environs
The **Theófilos Museum**, 3 km (2 miles) south, offers four rooms of canvases by Theófilos Chatzimichaïl (1873–1934), the Mytilíni-born artist. All were commissioned by his patron Tériade in 1927 and created over the last seven years of the painter's life. Theófilos detailed the fishermen, bakers and harvesters of rural Lésvos and

executed creditable portraits of personalities he met on his travels. For his depictions of historical episodes or landscapes beyond his experience, Theófilos relied on his imagination. The only traces of our age are occasional aeroplanes or steamboats in the background of his landscapes.

Just along the road is the **Tériade Museum**, housing the collection of Stratís Eleftheriádis – a local who emigrated to Paris in the early 20th century, adopting the name Tériade. He became a publisher of avant-garde art and literature. Miró, Chagall, Picasso, Léger and Villon were some of the artists who took part in his projects.

🏛 Theófilos Museum
Vareiá. **Tel** 22510 41644. **Open** 8:30am–3pm daily. 🎫 ♿

🏛 Tériade Museum
Vareiá. **Tel** 22510 23372. **Open** 8:30am–3pm daily. 🎫 ♿

Daphnis and Chloe, by Marc Chagall (1887–1984), in the Tériade Museum

Olive Growing in Greece

The Cretan Minoans are thought to have been the first people to have cultivated the olive tree, around 3800 BC. The magnificent olive groves of modern Greece date back to 700 BC, when olive oil became a valuable export commodity. According to Greek legend, Athena, goddess of peace as well as war, planted the first olive tree in the Athenian Acropolis – the olive has thus become a Greek symbol for peace. The 11 million or so olive trees on Lésvos are reputed to be the most productive oil-bearing trees in the Greek islands; Crete produces more and better-quality oil, but no other island is so dominated by olive monoculture. The fruits can be cured for eating throughout the year, or pressed to provide a nutritious and versatile oil; further crushing yields oil for soap and lanterns, and the pulp is a good fertilizer.

Olive groves on Lésvos largely date from after a killing frost in 1851. The best olives come from the hillside plantations between Plomári and Agiásos, founded in the 18th century by local farmers desiring land relatively inaccessible to Turkish tax collectors.

In myth, the olive is a virgin tree, sacred to Athena, tended only by virgin males. Its abundant harvest has been celebrated in verse, song and art since antiquity. This vase shows three men shaking olives from a tree, while a fourth gathers the harvest into a basket.

Greek olive oil, greenish-yellow after pressing, is believed by the Greeks to be of a higher quality than its Spanish and Italian counterparts, owing to hotter, drier summers which promote low acid levels in olive fruit.

The olive harvest on Lésvos takes place from late November to late December. Each batch is brought to the local *elaiotrıveío* (olive mill), ideally within 24 hours of being picked, pressed separately and tested for quality.

Types of Olive

From the mild fruits of the Ionians to the small, rich olives of Crete, the Greek islands are a paradise for olive lovers.

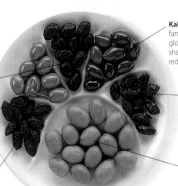

Kalamáta, the most famous Greek olive, is glossy-black, almond-shaped and cured in red-wine vinegar.

Elítses are small, sweetly flavoured olives from the island of Crete.

Thásos olives are salt-cured and have a strong flavour that goes well with cheese.

Tsakistés are picked young and lightly cracked before curing in brine.

Ionian greens are mild, mellow-flavoured olives, lightly brine-cured.

Throúmpes are a true taste of the countryside, very good as a simple *mezés* with olive-oil bread.

Around Eastern Lésvos

Eastern Lésvos is dominated by the two peaks of Lepétymnos in the north and Olympos in the south, both reaching the same height of 968 m (3,176 ft). Most of the island's pine forests and olive groves are found here, as well as the two major resort areas and the most populous villages after the port and capital. There are also several thermal spas, the most enjoyable being at Loutrá Eftaloús, near Mólyvos. With an early start from Mytilíni, which provides bus connections to all main towns and villages, the east of the island can be toured in a single day.

❷ Plomári

Πλωμάρι

42 km (26 miles) SW of Mytilíni. 🏠 3,400. ⓘ 22520 32200. 🚌 ⛴ Mon–Sat. 🚤 Agios Isidoros, 3 km (2 miles) NE; Melínta, 6 km (4 miles) NW.

Plomári's attractive houses spill off the slope above its harbour and stretch to the banks of the usually dry Sedoúntas River, which runs through the central commercial district. The houses date mostly from the 19th century, when Plomári became wealthy through its role as a major shipbuilding centre. Today, Plomári is known as the island's "oúzo capital", with five distilleries in operation, the most famous being Varvagiánnis.

❸ Agiásos

Αγιάσος

28 km (17 miles) W of Mytilíni. 🏠 3,100. ⓘ 22520 22080. 🚌 ⛴ Mon–Sat. 🚤 Vaterá, 31 km (19 miles) S.

Hidden in a forested ravine beneath Mount Olympos, Agiásos is possibly the most beautiful hill-town on Lésvos. It

began life in the 12th century as a dependency of the central monastic church of the **Panagía Vrefokratoússa**, which was constructed to enshrine a miraculous icon reputed to have been painted by St Luke.

After exemption from taxes by the Sultan during the 18th century, Agiásos swelled rapidly with Greeks fleeing hardship elsewhere on the island. The town's tiled houses and narrow, cobbled lanes have changed little since then, except for stalls of locally crafted souvenirs which line the way to the church with its belfry and surrounding bazaar. The presence of shops built into the church's foundations, with rents going towards its upkeep, is an ancient arrangement. It echoes the country-fair element of the traditional religious *panigýria* (festivals), where pilgrims once came to buy and sell as well as perform devotions. Agiásos musicians are hailed as the best on Lésvos – they are out in force during the 15 August festival of the Assumption of the Virgin, considered one of the liveliest in Greece. The

Oúzo

Oúzo is the Greek version of a spirit found throughout the Mediterranean. The residue of grape skins left over from wine-pressing is boiled in a copper still to make a distillate originally called raki. The term oúzo may derive from the Italian *uso Massalia*, used to label early shipments leaving the Ottoman Empire for Marseille. Today it means a base of raki flavoured with star anise or fennel. Oúzo's alcohol content varies from 38–48 per cent, with 44 per cent considered the minimum for a quality product. When water is added, oúzo turns milky white – this results from the binding of anethole, an aromatic compound found in fennel and anise.

pre-Lenten carnival is also celebrated with verve at Agiásos; there is a special club devoted to organizing it.

❹ Mantamádos

Μανταμάδος

36 km (22 miles) NW of Mytilíni. 🏠 1,500. ⓘ 22530 61200. 🚌 ⛴ Mon–Sat. 🚤 Tsónia, 12 km (7 miles) N.

The attractive village of Mantamádos is famous for its pottery industry and the adjacent **Moní Taxiarchón**. The existing monastery dates from the 17th century and houses a black icon of the Archangel Michael, reputedly made from mud and the blood of monks slaughtered in an Ottoman raid. A bull is publicly sacrificed here on the third Sunday after Easter and its meat eaten in a communal stew, the first of several such rites on the island's summer festival calendar. Mantamádos ceramics come in a wide range of sizes and colours, from giant *pythária* (olive oil containers) to smaller *koumária* (ceramic water jugs).

Plomári, viewed from the extended jetty

For hotels and restaurants in this region see p308 and pp324–6

Fishing boats at Mólyvos harbour with the castle in the background

❺ Sykaminiá

Συκαμινιά

46 km (29 miles) NW of Mytilíni. 300. Mon–Sat. Kágia 4 km (2 miles) E; Skála Sykaminiás, 2 km (1 mile) N.

Flanked by a deep valley and overlooking the straits to the Asia Minor coast, Sykaminiá has the most spectacular position of any village on Mount Lepétymnos, which stands at a height of 968 m (3,176 ft). Novelist Efstrátios Stamatópoulos (1892–1969), known as Strátis Myrivílis, was born close to the atmospheric central square. The jetty church, which featured in his novel *The Mermaid Madonna*, can be seen down in Skála Sykaminiás. One of Skála's tavernas is named after the *mouriá* or mulberry tree in which Myrivílis slept on hot summer nights.

❻ Mólyvos (Míthymna)

Μόλυβος (Μήθυμνα)

61 km (38 miles) NW of Mytilíni. 1,500. 22530 71313. Mon–Sat.

Situated in a region celebrated in antiquity for its vineyards, Mólyvos is the most popular and picturesque town on Lésvos. It was the birthplace of Arion, the 7th-century-BC poet, and the site of the grave of Palamedes, the Achaian warrior buried by Achilles. According to legend, Achilles besieged the city until the king's daughter fell in love with him and opened the gates – though he killed her for her treachery. There is little left of the ancient town apart from the tombs excavated near the tourist office, but its ancient name, Míthymna, has been revived and is used as an alternative to Mólyvos (a Hellenization of the Turkish *Molova*). Artifacts from Ancient Míthymna are on display in the Archaeological Museum in Mytilíni town *(see p142)*.

Before 1923 over a third of the population was Muslim, forming a landed gentry who built many sumptuous three-storey town houses and graced Mólyvos with a dozen street fountains, some of which retain original ornate inscriptions. The mansions, or *archontiká*, are clearly influenced by Eastern architecture *(see p26)*; the living spaces are arranged on the top floor around a central stairwell, or *chagiáti* – a design which had symbolic, cosmological meaning in the original Turkish mansions from which it was taken. The picturesque harbour and cobbled lanes of tiered stone houses are all protected by law; any new development must conform architecturally with the rest of the town.

Overlooking the town, and affording splendid views of the Turkish coast, stands a sizeable Byzantine **kástro**. The castle was modified by the Genoese adventurer Francesco Gatelluzi *(see p142)* in 1373, though it fell into Turkish hands during the campaign of Mohammed the Conqueror in 1462. Restored in 1995, the castle still retains its wood-and-iron medieval door and a Turkish inscription over the lintel. During summer, the interior often serves as a venue for concerts and plays.

A boatyard operates at the fishing harbour, a reminder of the days when Mólyvos was one of the island's major commercial ports.

🏰 Kástro

Tel 22530 71803. **Open** May–Oct: 8am–8pm daily; Nov–Apr: 8:30am–3pm daily. **Closed** main public hols.

Colourfully restored Ottoman-style houses in Mólyvos

Windmills on the island of Chíos ▶

Around Western Lésvos

Though mostly treeless and craggy, western Lésvos has a severe natural beauty, broken by inland villages, beach resorts and three of Lésvos's most important monasteries. Many of the island's famous horses are bred in this region, and where the streams draining the valleys meet the sea, reedy oases form behind the sand, providing a haven for bird-watchers during spring. Bus schedules are too infrequent for touring the area, but cars can be hired at Mólyvos.

Tiered houses of the village of Skalochóri

❼ Pétra

Πέτρα

55 km (34 miles) NW of Mytilíni. 🗺 3,700. ℹ 22530 42222. 🚌 🚕 Anaxos 3 km (2 miles) W.

The village of Pétra takes its name (meaning "rock") from the volcanic monolith at its centre. By its base is the 16th-century basilica of **Agios Nikólaos**, still with its original frescoes, while a flight of 103 steps climbs to the 18th-century church of **Panagía Glykofiloúsa** church. The **Archontikó Vareltzídainas**, one of the last of the Ottoman dwellings once widespread on Lésvos (*see p145*), is also 18th century.

🏛 **Archontikó Vareltzídainas**
Sapphous. **Tel** 22530 41510. **Open** Tue–Sun. **Closed** main public hols.

❽ Kalloní

Καλλονή

40 km (25 miles) NW of Mytilíni. 🗺 1,600. ℹ 22530 22288. 🚌 🚕 Mon–Sat. 🚕 Skála Kallonís 2 km (1 mile) S.

An important crossroads and market town, Kalloní lies 2 km (1 mile) inland from its namesake gulf. Sardines are netted at the beach of **Skála Kallonís**.

Environs

In 1527, the abbot Ignatios founded **Moní Leimónos**, the second most important monastery on Lésvos. You can still view his cell, maintained as a shrine. A carved wood ceiling, interior arcades and a holy spring distinguish the central church. Moní Leimónos also has various homes for the infirm, a mini-zoo and two museums: one ecclesiastical and one of folkloric miscellany.

🏛 **Moní Leimónos**
5 km (3 miles) NW of Kalloní. **Tel** 22530 22289. Ecclesiastical Museum: **Open** daily. Folk Museum: **Open** on request.

❾ Antissa

Άντισσα

76 km (47 miles) NW of Mytilíni. 🗺 1,410. ℹ 22530 53600. 🚌 🚕 daily. 🚕 Kámpos 4 km (2.5 miles) S.

The largest village of this part of Lésvos, Antissa merits a halt for its fine central square alone, in which a number of cafés and tavernas stand overshadowed by three huge plane trees. The ruins of the eponymous ancient city, destroyed by the Romans in 168 BC, lie 8 km (5 miles) below by road, near the remains of the Genoese **Ovriókastro**. This castle stands on the shore, east of the tiny fishing port of Gavathás and the long sandy beach of Kámpos.

Environs

Although, unlike Antissa, there is no view of the sea, **Vatoússa**, 10 km (6 miles) east, is the area's most attractive village. Tiered **Skalochóri**, another 3 km (2 miles) north, does overlook the north coast and – like most local villages – has a ruined mosque dating to the days before the 1923 Treaty of Lausanne (*see p47*).

Hidden in a lush river valley, 3 km (2 miles) east of Antissa, stands the 16th-century **Moní Perivolís**, situated in the middle of a riverside orchard. The narthex features three 16th-century frescoes, restored in the 1960s: the apocalyptic *Earth and Sea Yield Up Their Dead*, the *Penitent Thief of Calvary* and the *Virgin* (flanked by Abraham). The interior is lit by daylight only, so it is advisable to visit the monastery well before dusk.

Frescoes adorning the narthex of Moní Perivolís

⑩ Moní Ypsiloú

Μονή Υψηλού

62 km (39 miles) NW of Mytilíni. 🚌
Tel 22530 56259. **Open** daily.

Spread across the 511-m (1,676-ft) summit of Mount Ordymnos, an extinct volcano, Moní Ypsiloú was founded in the 12th century and is now home to just four monks. It has a handsome double gate, and a fine wood-lattice ceiling in its *katholikón* (main church) beside which a rich exhibition of ecclesiastical treasures can be found. In the courtyard outside stand a number of fragments of petrified trees. The patron saint of the monastery is John the Divine (author of the book of *Revelation*), a typical dedication for religious communities located in such wild, forbidding scenery.

Triple bell tower of Moní Ypsiloú

Environs
The main entry to Lésvos's **petrified forest** is just west of Ypsiloú. Some 15 to 20 million years ago, Mount Ordymnos erupted, beginning the process whereby huge stands of sequoias, buried in the volcanic ash, were transformed into stone.

⑪ Sígri

Σίγρι

93 km (58 miles) NW of Mytilíni. 🏔 400. 🚌

An 18th-century Ottoman **castle** and the church of **Agía Triáda** dominate this sleepy port, protected from severe weather by long, narrow Nisópi island. Sígri's continuing status as a naval base has discouraged tourist development, though it has a couple of small beaches; emptier ones are only a short drive away.

The peaceful harbour of Sígri

⑫ Skála Eresoú

Σκάλα Ερεσού

89 km (55 miles) W of Mytilíni town. 🏔 1,500. 🚌

Extended beneath the acropolis of ancient Eresós, the wonderful, long beach at Skála Eresoú supports the island's third-largest resort. By climbing the acropolis hill, you can spot the ancient jetty submerged in the modern fishing anchorage. Little remains at the summit, but the Byzantine era is represented in the ancient centre by the foundations of the basilica of **Agios Andreás**; its 5th-century mosaics await restoration.

Environs
The village of **Eresós**, 11 km (7 miles) inland, grew up as a refuge from medieval pirate raids; a vast, fertile plain extends between the two settlements. Two of Eresós's most famous natives were the philosopher Theophrastos, a pupil of Aristotle (*see p63*), and Sappho, one of the greatest poets of the ancient world.

Sappho, the Poet of Lesvos

One of the finest lyric poets of any era, Sappho (c.615–562 BC) was born, probably at Eresós, into an aristocratic family and a society that gave women substantial freedom. In her own day, Sappho's poems were known across the Mediterranean, though Sappho's poetry was to be suppressed by the church in late antiquity and now survives only in short quotations and on papyrus scraps. Many of her poems were also addressed to women, which has prompted speculation about Sappho's sexual orientation. Much of her work was inspired by female companions: discreet homosexuality was unremarkable in

her time. Even less certain is the manner of her death; legend asserts that she fell in love with a younger man, whom she pursued as far as the isle of Lefkáda. Assured that unrequited love could be cured by leaping from a cliff, she did so and drowned in the sea: an unlikely, and unfortunate, end for a poet reputed to be the first literary lesbian.

Chíos
Χίος

Although Chíos has been prosperous since antiquity, today's island is largely a product of the Middle Ages. Under the Genoese, who controlled the highly profitable trade in gum mastic (see pp152–3), the island became one of the richest in the Mediterranean. It continued to flourish under the Ottomans until March 1822, when the Chians became the victims of one of the worst massacres (see p155) of the Independence uprising. Chíos had only partly recovered when an earthquake in 1881 caused severe damage, particularly in the south.

Shopfront in Chíos town bazaar

❶ Chíos Town
Χίος

🔢 25,000. 🚢 🚌 Polytechníou (around island), Dimokratías (environs). 🛈 Kanári 18 (22710 44389). 🛒 Mon–Sat. 🚉 Karfás 7 km (4 miles) S.

Chíos town, like the island, was settled in the Bronze Age and was colonized by the Ionians from Asia Minor by the 9th century BC. The site was chosen for its convenient position for travelling to the Turkish mainland opposite, rather than good anchorage: a series of rulers have been obliged to construct long breakwaters as a consequence. Though it is a modernized island capital (few buildings predate the earthquake of 1881), there are a number of museums and other scattered relics from the town's eventful past. Besides the kástro, the most interesting sights are the lively bazaar at the top of Roïdou, and an ornate Ottoman fountain dating to 1768 at the junction of Martýron and Dimarchías.

🏰 Kástro
Maggiora. **Tel** 22710 22819.
Open daily. ♿

The most prominent medieval feature of the town is the kástro, a Byzantine foundation improved by the Genoese after they acquired Chíos in 1346. Today the kástro lacks the southeasterly sea rampart, which fell prey to developers after the devastating earthquake in 1881. Its most impressive gate is the southwesterly Porta Maggiora; a deep dry moat runs from here around to the northwest side of the walls. Behind the walls, Ottoman-era houses line narrow lanes of what were once the Muslim and Jewish quarters of the town; after the Ottoman conquest, in 1566, Orthodox and Catholics were required to live outside the walls. Also inside, a disused mosque, ruined Turkish baths and a small Ottoman cemetery can be found. The latter contains

Sights at a Glance

❶ Chíos Town
❷ Mastic Villages
❸ Néa Moní
❹ Avgónyma
❺ Volissós
❻ Moní Moúndon

(Map labels: Agiásmata, Agio Gála, Pelinaío 1,297 m (4,255 ft), Potamiá, Kardámyla, Volissós, Moní Moúndon, Lefkáda, Limniá, Langáda, Mánagros, CHIOS, Vrontádos, Psará, Anávatos, Néa Moní, Elínta, Chíos Town, Avgónyma, Kámpos, Mastic Villages, Karfás, Véssa, Vávyloi, Sámos Piraeus / Sámos, Mestá, Armólia, Ol'y mpoi, Pyrgí, Kalamotí, Chíos Straits, Mávra Vólia, Kómi, Emporeiós)

0 kilometres 5
0 miles 5

For keys to symbols see back flap

Chíos town waterfront with the dome and minaret of the Mecidiye Mosque

VISITORS' CHECKLIST

Practical Information
🏘 45,000. ℹ Chíos town
(22710 44389). 🌐 chios.gr

Transport
✈ 4 km (2 miles) S of Chíos
town. 🚢 Chíos town. 🚌

the grave and headstone of Admiral Kara Ali, who commanded the massacre of 1822. He was killed aboard his flagship when it was destroyed by the Greek captain Kanáris.

Porta Maggiora, the southwesterly entrance to the kástro

🏛 Justiniani Museum
Kástro. **Tel** 22710 22819. **Open** 8am–7pm Tue–Sun (Nov–May: to 2:30). 🎫
This collection is devoted to religious art and includes a 5th-century-AD floor mosaic rescued from a neglected Chian chapel. The saints featured on the icons and frescoes include Isídoros, who is said to have taught the islanders how to make liqueur from mastic *(see pp152–3)*, and Matrona, a martyr of Roman Ankara whose veneration here was introduced by refugees from Asia Minor after 1923.

🏛 Byzantine Museum
Plateía Vounakíou. **Tel** 22710 26866. **Open** 10am–1:30pm Tue–Sun (to 3pm Sun). 🎫
Though called the Byzantine Museum, this is little more than an archaeological warehouse and restoration workshop. It is housed within the only mosque to have survived intact in the East Aegean, the former Mecidiye Cami, which still retains its minaret. A number of Jewish, Turkish and Armenian gravestones stand propped up in the courtyard, attesting to the multiethnic population of the island during the medieval period.

🏛 Argéntis Folklore Museum
Koraïs 2. **Tel** 22710 28256. **Open** 8am–2pm Mon–Fri. 🎫
Endowed in 1932 by a member of a leading Chian family and occupying the floor above the Koraïs library, this collection features rural wooden implements, plus examples of traditional embroidery and costumes. Also on view, alongside a number of portraits of the Argéntis family, are rare engravings of islanders and numerous copies of the *Massacre at Chíos* by Delacroix (1798–1863). This painting, as much as any journalistic dispatch, aroused the sympathy of Western Europe for the Greek revolutionary cause *(see pp46–7)*. The main core of the Koraïs library, situated on the ground floor, consists of a number of books and manuscripts bequeathed by the cultural revolutionary and intellectual Adamántios Koraïs (1748–1833); these include works given by Napoleon.

Environs
The fertile plain known as the **Kámpos** extends 6 km (4 miles) south of Chíos town. The land is crisscrossed by a network of unmarked lanes which stretch between high stone walls that betray nothing of what lies behind. However, through an ornately arched gateway left open, you may catch a glimpse of what were once the summer estates of the medieval Chian aristocracy.

Several of the mansions were devastated by the 1881 earthquake, but some have been restored with their blocks of multicoloured sandstone arranged so that the different shades alternate. Many of them still have their own water-wheels, which were once donkey-powered and drew water up from 30 m- (98 ft-) deep wells into open cisterns shaded by a pergola and stocked with fish. These freshwater pools, which are today filled by electric pumps, still irrigate the vast orange, lemon and tangerine orchards for which the region is widely known.

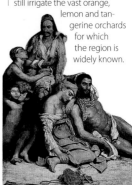

Detail of Delacroix's *Massacres de Chíos* (1824) in the Argéntis Folklore Museum

❷ Mastic Villages

Μαστιχοχώρια

The 20 settlements in southern Chíos known as the *mastichochória*, or "mastic villages", received their name from their most lucrative medieval product. Genoese overlords founded the villages well inland as an anti-pirate measure during the 14th and 15th centuries. Constructed to a design unique in Greece, they share common defensive features made all the more necessary by the island's proximity to the Turks. Though they were the only villages to be spared in the 1822 massacres *(see p155)*, most have had their architecture compromised by both earthquake damage and ill-advised modernization.

Main Mastic Villages

KEY

① **Flat roofs** of adjacent buildings were ideally of the same height to facilitate escape.

② **Streets followed** an intricate grid plan designed to confuse strangers.

③ **Narrow passages** were overarched by flying buttresses, to limit earthquake damage.

④ **Houses** reached three storeys, with vaulted ceilings except on the top floor.

⑤ **Fortification towers** guarded each corner of the village.

⑥ **A square tower** in the centre of the village was the last refuge in troubled times.

⑦ **The outer circuit** of houses doubled as a perimeter wall.

Véssa
This is the one village whose regular street plan can easily be seen from above while descending from Agios Geórgios Sykoúsis or Eláta.

Pyrgí
Pyrgí is renowned for its bright houses, many patterned with *xystá* (grating) decoration. Outer walls are plastered using black sand and coated with white-wash. This is then carefully scraped off in repetitive geometric patterns, revealing the black undercoat. An example of this is the church of Agioi Apóstoloi, which also has medieval frescoes.

Armólia
One of the smallest and simplest of the *mastichochória*, Armólia is renowned for its pottery industry.

Olýmpoi

Olýmpoi is almost square in layout. Its central tower has survived to nearly its original height, and today two cafés occupy its ground floor. Here local men and women can be seen winnowing mastic.

Vávyloi

The 13th-century Byzantine church of Panagía tis Krínis, on the edge of the village, is famed for its frescoes and its alternating courses of stone and brickwork.

Mastic Production

The mastic bush of southern Chíos secretes a resin or gum that, before the advent of petroleum-based products, formed the basis of paints, cosmetics and medicines. Today it is made into chewing gum, liqueur and even toothpaste. About 300 tonnes of gum are harvested each summer through incisions in the bark, which weep resin "tears"; once solidified a day later, the resin is scraped off and spread to air-cure on large trays.

Mastic bush bark and crystals

Crystals separated from the bark

Mestá

Viewed here from the southwest, Mestá is considered the best preserved of the mastic villages. It has the most even roof heights and still retains its perimeter corner towers.

Taxiárchis Church

Mestá's 19th-century church, the largest on Chíos, dominates the central square. The atmospheric interior has a fine carved altar screen.

❸ Néa Moní
Néa Movή

Hidden in a wooded valley 11 km (7 miles) west of Chíos town, the monastery of Néa Moní and its mosaics – some of Greece's finest – both date from the 11th century. It was established by Byzantine Emperor Constantine IX Monomáchos in 1042 on the site where three hermits found an icon of the Virgin. It reached the height of its power after the fall of the Byzantine Empire, and remained influential until the Ottoman reprisals of 1822. Néa Moní has now been a convent for decades, but when the last nun dies it is to be taken over again by monks.

Néa Moní, viewed from the west

Narthex
Seen here with the main church dome in the background, the narthex contains the most complex mosaics. Twenty-eight saints are depicted, including St Anne, the only woman. The *Virgin with Child* adorns the central dome.

KEY

① **Ornate marble inlays** were highly prized in the Byzantine Empire.

② **The belfry** is a modern structure, added after the 1881 earthquake.

③ **St Mark the Evangelist mosaic**

④ **Descent from the Cross mosaic**

⑤ **The dome** was repaired after the 1881 earthquake, though its magnificent Pantokrátor was lost.

⑥ **The main apse** has a mosaic of the Virgin. It is positioned above the walls and represents earthly subjects, while the dome depicts Christ.

⑦ **Altar screen**

⑧ **The floor** is covered with marble segments, which echo the disciplined architecture of the nave.

⑨ **St Joachim mosaic**

★ **Christ Washing the Disciples' Feet**
Here Christ washes the feet of Peter, who indicates he wishes his head and hands also to be bathed.

★ Anástasis
After the Resurrection, Christ rescues Adam and Eve from Hell before entering Heaven.

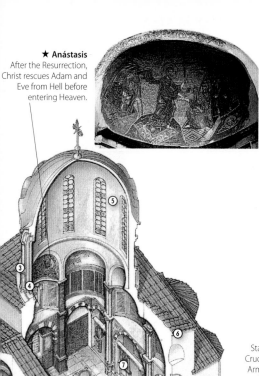

VISITORS' CHECKLIST

Practical Information
11 km (7 miles) W of Chíos town. **Tel** 22710 79391.
Open 9am–1pm Tue–Sun.
🕇 ♿ limited.

Transport
🚌

Byzantine Clock
Standing beneath the Crucifixion mosaic, this Armenian-made clock came from Smyrna after its destruction in 1922.

The Massacre at Chíos

After 250 years of Ottoman rule, the Chians joined the Independence uprising in March 1822, incited by Samian agitators. Enraged, the Sultan sent an expedition that massacred 30,000 Chians, enslaved almost twice that number and brutally sacked most of the monasteries and houses. Many Chians fled to Néa Moní for safety, but they and most of the 600 monks were also killed. Just inside the main gate of the monastery stands a chapel containing the bones of those who died here. The savagery of the Turks is amply illustrated by the axe-wounds visible on many skulls, including those of children.

Cabinet containing the skulls of the Chian martyrs of 1822

Betrayal in the Garden
A detail of this mosaic shows Peter lopping off the ear of Malchus, the High Priest's servant, following the betrayal of Jesus in Gethsemane. Unfortunately, the *Kiss of Judas* has been damaged.

Around Chíos Island

With its verdant, semi-mountainous terrain, edged by rocky cliffs in the south and sandy beaches to the northwest, Chíos is one of the Aegean's most beautiful isles. Roads and public transport radiate in all directions from Chíos town and the best bus service is to be found on the densely populated southeast coast; to explore anywhere else you need to hire a taxi, car or powerful motorbike.

One of the many restored stone houses of Avgónyma

❹ Avgónyma

Αυγώνυμα

20 km (12 miles) W of Chíos town. 🏠 15. 🚌 Elínta 7 km (4 miles) W.

This is the closest settlement to Néa Moní *(see pp154–5)* and the most beautiful of the central Chian villages, built in a distinct style: less labyrinthine and claustrophobic than the mastic villages, and more elegant than the houses of northern Chíos. The town's name means "clutch of eggs", perhaps after its clustered appearance when viewed from the ridge above. Virtually every house has been tastefully restored by Greek-Americans with roots here. The medieval *pýrgos* (tower) on the main square, with its interior arcades, is home to the excellent central taverna.

Environs

Few Chian villages are as striking glimpsed from a distance as **Anávatos**, 4 km (2 miles) north of Avgónyma. Unlike Avgónyma, Anávatos has scarcely changed in recent decades; shells of houses blend into the palisade on which they perch, overlooking occasionally tended pistachio orchards. The village was the scene of a particularly traumatic incident during the atrocities of 1822 *(see p155)*. Some 400 Greeks threw themselves into a ravine from the 300 m (985 ft) bluff above the village, choosing suicide rather than death at the hands of the Turks.

❺ Volissós

Βολισσός

40 km (25 miles) NW of Chíos town. 🏠 500. 🚌 🚤 Mánagros 2 km (1 mile) SW.

Volissós was once the primary market town for the 20 smaller villages of northwestern Chíos, but today the only vestige of its former commercial standing is a single saddlery on the western edge of town. The strategic importance of medieval Volissós is borne out by the crumbled hilltop castle, erected by the Byzantines in the 11th century and repaired by the Genoese in the 14th.

The town's stone houses stretch along the south and east flanks of the fortified hill; many have been bought and restored by Volissós's growing expatriate population.

Environs

Close to the village of **Agio Gála**, 26 km (16 miles) northwest of Volissós, two 15th-century chapels can be found lodged in a deep cavern near the top of a cliff. The smaller, hindmost chapel is the more interesting of the two; it is built entirely within the grotto and features a sophisticated and mysterious fresco of the *Virgin and Child*. The larger chapel, at the entrance to the cave, boasts an intricate carved *témblon* or altar screen. Agio Gála can be reached by bus from Volissós and admission to the churches should be made via the resident warden who holds the keys.

The largely deserted town of Anávatos, with the few inhabited dwellings in the foreground

❻ Moní Moúndon

Μονή Μουνδών

35 km (22 miles) NW of Chíos town.
🚌 to Volissós. **Open** daily (ask for key
at first house in Diefha village).

Founded late in the 16th
century, this picturesque
monastery was once second
in importance to Néa Moní
(see pp154–5). The *katholikón*
(or central church) has a
number of interesting late
Medieval murals, the most
famous being the *Salvation
of Souls on the Ladder to
Heaven*. Although the church
is only open to the public
during the monastery's
festival (29 August), the
romantic setting makes
the stop worthwhile.

Moúndon's *Salvation of Souls on the Ladder
to Heaven* mural

Outlying Islands

Domestic architecture on the
peaceful islet of **Oinoússes**, a
few miles east of Chíos town, is
deceptively humble, for it is the
wealthiest territory in Greece.
Good beaches can be found to
either side of the port, and in
the northwest of the island is
the Evangelismoú convent,
endowed by the Pateras family.

Much of **Psará**, 71 km
(44 miles) to the west, was
ruined in the Greek War of
Independence *(see pp46–7)*;
as a result, the single town,
built in a pastiche of island
architectural styles, is a product
of the last 100 years. The
landscape is still desolate and
infertile, though there are good
beaches to visit east of the
harbour, and Moní Koímisis tis
Theotókou in the far north.

The remains of a Hellenistic tower near Fanári, Ikaría

❼ Ikaría

Ικαρία

🏔 7,500. ✈ 🛥 🚌 Agios Kýrikos.
ℹ 22750 50400. 🚢 Fanári 16 km
(10 miles) NE of Agios Kírykos.

Lying 245 km (150 miles) south
of Chíos, Ikaría is named after
the Ikaros of legend who flew
too near the sun on artificial
wings and plunged to his
death in the sea when his
wax bindings melted.

Agios Kírykos, the capital
and main port, is a pleasant
town flanked by two spas,
one of them dating to Roman
times and still popular with
an older Greek clientele.
A number of hot baths can
be visited at **Thérma**, a short
walk to the northeast, while
at **Thérma Lefkádas**, to the
southwest, the springs still well
up among the boulders in the
shallows of the sea.

About 2 km (1.5 miles) west
of Evdilos, a village port on the
north coast, lies the village of
Kámpos. It boasts a broad,
sandy beach and, beside the
ruins of a 12th-century church,
the remains of a Byzantine
manor house can be seen. The
building recalls a time when
the island was considered
a humane place of exile for
disgraced noblemen; there
was a large settlement of such
officials in Kámpos. A small
museum contains artifacts
from the town of Oinoe,
Kámpos' ancient predecessor.

Standing above Kosoíki
village, 5 km (3 miles) inland,
the Byzantine castle of
Nikariás was built during
the 10th century to guard a
pass on the road to Oinoe.
The only other well-preserved
fortification is a 3rd-century-BC
Hellenistic tower (Drakánou),
once an ancient lighthouse,
near Fanári.

Tiny **Armenistís**, with its
surrounding forests and fine
beaches, such as Livádi and
Messaktí to the east, is Ikaría's
main resort. The foundations
of a temple to the goddess
Artemis Tavropólos (Artemis
incarnated as the patroness
of bulls) lie 4 km (2 miles) west.

Home to the most active
fishing fleet in the East Aegean,
the island of **Foúrnoi**, due east
of Ikaría, is far more populous
and lively than its small size
suggests. The main street
of the port town, lined with
mulberry trees, links the
quay with a square well
inland, where an ancient
sarcophagus sits between the
two cafés. Within walking
distance lie Kampí and Psilí
Ammos beaches.

Coastal town of Agios Kírykos, the capital
of Ikaría

Sámos
Σάμος

Settled early, owing to its natural richness and ease of access from Asia, Sámos was a major maritime power by the 7th century BC and enjoyed a golden age under the rule of Polykrates (538–522 BC). After the collapse of the Byzantine Empire, most of the islanders fled from pirates and Sámos lay deserted until 1562, when Ottoman Admiral Kiliç Ali repopulated it with returned Samians and other Orthodox settlers. The 19th century saw an upsurge in fortunes made in tobacco trading and shipping. Union with Greece occurred in 1912.

Fishermen at Vathý harbour

Sights at a Glance

1. Vathý
2. Efpalíneio Orygma
3. Pythagóreio
4. Moní Megális Panagías
5. Heraion
6. Kokkári
7. Karlóvasi
8. Mount Kerketéfs

Assyrian bronze horse figurine, Vathý Archaeological Museum

❶ Vathý
Βαθύ

🔼 5,700. 🚢 🚌 Ioánnou Lekáti. 🛈 25 Martíou 4 (22730 28582). 🛒 daily. ⛱ Psilí Ammos 8 km (5 miles) SE; Mykáli 6 km (4 miles) S.

Though the old village of Ano Vathý existed in the 1600s, today's town is recent; the harbour quarter grew up only after 1832, when the town became the capital of the island. Just large enough to provide all amenities in its bazaar, lower Vathý caters to tourists, while cobble-laned Ano Vathý carries on oblivious to the commerce in the streets below.

The Sámos **Archaeological Museum** contains artifacts from the excavations at the Heraion sanctuary (*see p160*). Because of the far-flung origins of the pilgrims who visited the shrine, the collection of small votive offerings is one of the richest in Greece – among them are a bronze statuette of an Urartian god, Assyrian figurines and an ivory miniature of Perseus and Medusa. The largest freestanding sculpture to have survived from ancient Greece is the star exhibit: a 5 m- (16 ft-) tall marble *koúros* dating from 580 BC and dedicated to the god Apollo.

🏛 Archaeological Museum
Kapetán Gymnasiárchou Katevéni. **Tel** 22730 27469. **Open** 8:30am–5pm Tue–Sun. **Closed** main public hols. 🎟

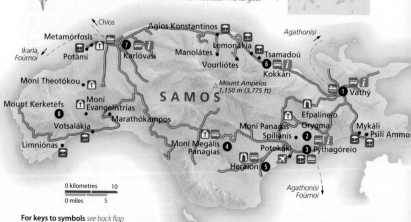

Chíos
Metamórfosis
Ikaría, Fournoí
Potámi
Karlóvasi
Agios Konstantínos
Lemonákia
Manolátes
Vourliótes
Tsamadoú
Kokkári
Agathonísi
Moní Theotókou
Mount Kerketéfs
Moní Evangelistrías
Votsalákia
Marathókampos
Limniónas
Moní Megális Panagías
Mount Ampelos
1,150 m (3,775 ft)
SAMOS
Moní Panagías
Spilianís
Potokáki
Heraion
Efpalíneio Orygma
Pythagóreio
Mykáli
Psilí Ammo
Vathý
Agathonísi
Fournoí

0 kilometres 10
0 miles 5

For keys to symbols *see back flap*

Around Sámos Island

Sámos has a paved road around the island, but buses are frequent only between Pythagóreio and Karlóvasi, via Vathý. Vehicle hire is easy, though many points can be reached only by jeep or foot. In the south and west there are many rough dirt roads where caution is necessary.

❷ Efpalíneio Orygma

Ευπαλίνειο Όρυγμα

15 km (9 miles) SW of Vathý. **Tel** 22730 61400. **Open** Tue–Sun. **Closed** main public hols. 🐾

Efpalíneio Orygma (Eupalinos's tunnel) is a 1,040-m (3,410-ft) aqueduct, ranking as one of the premier engineering feats of the ancient world. Designed by the engineer Eupalinos and built by hundreds of slaves between 529 and 524 BC, the tunnel guaranteed ancient Sámos a water supply in times of siege, and remained in use until this century. Eupalinos' surveying was so accurate that, when the work crews met, having begun from opposite sides of the mountain, their vertical error was nil.

Visitors may walk along the ledge used to remove rubble from the channel far below. Half the total length is open to the public, with grilles to protect you from the worst drops.

❸ Pythagóreio

Πυθαγόρειο

13 km (8 miles) SW of Vathý. 🚟 1,300. 🚌🚕ℹ️ Lykoúrgou Logothéti (22730 61400). 🏖️ Potokáki 3 km (2 miles) W.

Cobble-paved Pythagóreio, named after the philosopher Pythagoras, who was born here in 580 BC, has long been the lodestone of Samian tourism. The extensive foundations and walls of ancient Sámos act as a brake on tower-block construction; the only genuine tower is the 19th-century manor of **Lykoúrgos Logothétis**, the local chieftain who organized a decisive naval victory over the Turks on 6 August 1824, the date of the Feast of the Transfiguration. Next to this stronghold is the church of the **Metamórfosis**, built to celebrate the victory. At the far

Pythagoras statue (1989) by Nikoláos Ikaris, Pythagóreio

western edge of town are the extensive remains of **Roman Baths**, still with a few doorways intact. Further west, the Doryssa Bay luxury complex stands above the silted-in area of the Archaic harbour; all that remains is Glyfáda lake, crossed by a causeway.

🏛️ Roman Baths

W of Pythagóreio. **Tel** 22730 61400. **Open** variable. ♿

Environs

Polykrates protected Pythagóreio by constructing a circuit of walls enclosing Kastrí hill, to a circumference of more than 6 km (4 miles), with 12 gates. The walls were damaged by an Athenian siege of 439 BC, and today are most intact just above Glyfáda, where a fortification tower still stands. Enclosed by the walls, just above the ancient theatre, sits **Moní Panagías Spilianís** with its 100 m (330 ft) cave containing a shrine to the Virgin.

❹ Moní Megális Panagías

Μονή Μεγάλης Παναγίας

27 km (17 miles) W of Vathý. 🚌 **Open** May–Oct: daily.

Founded in 1586 by Nílos and Dionýsios, two hermits from Asia Minor, the monastery of Megális Panagías is the second oldest on Sámos and contains the island's best surviving frescoes from that period. The central church is orientated diagonally within the square compound of cells, now restored, probably built directly above a temple of Artemis which it replaced. Sadly, the area was ravaged by fire in 1990, shortly after the last monk died. Visiting hours depend on the whim of the caretaker.

Fresco of Jesus washing the apostles' feet, Moní Megális Panagías

For hotels and restaurants in this region see p308 and pp324–6

The single remaining column of Polykrates' temple, Heraion

❺ Heraion

Ηραίον

21 km (13 miles) SW of Vathý.
Tel 22730 95277. 🚌 Iraío. **Open** Tue–Sun. **Closed** main public hols. 🅿

A fertility goddess was worshipped here from Neolithic times, though the cult only became identified with Hera after the arrival of Mycenaean colonists *(see pp32–3)*, who brought their worship of the Olympian deities with them. The sanctuary's site on flood-prone ground honoured the legend that Hera was born under a sacred osier (willow tree) on the banks of the

Imvrasos and celebrated her nuptials with Zeus among the osiers here, in the dangerous pre-Olympian days when Kronos still ruled.

A 30 m- (98 ft-) long temple built in the 8th century BC was replaced in the 6th century BC by a stone one of the Ionic order, planned by Rhoikos, a local architect. Owing to earthquakes, or a design fault, this collapsed during the reign of Polykrates, who ordered a grand replacement designed by Rhoikos's son, Theodoros. He began the new temple in 525 BC, 40 m (130 ft) west of his father's, recycling building materials from its predecessor. Building continued off and on for many centuries, but the vast structure was never completed. The interior, full of votive offerings, was described by visitors in its heyday as a veritable art gallery.

Most of the finds on display at the Archaeological Museum in Vathý *(see p158)* date from the 8th to the 6th centuries BC, when the sanctuary was at the height of its prestige. The precinct was walled and contained several temples to other deities, though only Hera herself had a sacrificial altar.

Plinth from Polykrates' temple, Heraion

Pilgrims could approach from the ancient capital along a 4,800 m (15,750 ft) Sacred Way.

Despite diligent 20th-century German excavations, much of the sanctuary is confusing. Byzantine and medieval masons removed ready-cut stone for reuse in their buildings, leaving only one column untouched. Early in the 5th century, Christian masons built a basilica dedicated to a new mother figure: the Virgin Mary. Its foundations lie east of the Great Temple.

❻ Kokkári

Κοκκάρι

10 km (6 miles) W of Vathý. 🏠 1,000.
🚌 ℹ Agíou Nikoláou (22730 92333).
🏖 Tsamadoú & Lemonákia 2 km (1 mile) W.

Built on and behind twin headlands, this charming little port takes its name from the shallot-like onions once cultivated just inland. Today it is the island's third resort after Pythagóreio and Vathý, with its wind-blown location turned to advantage by a multitude of windsurfers.

The town's two beaches are stony and often surf-battered, but the paved

The Cult of Hera

Hera was worshipped as the main cult of a number of Greek cities, including Argos on the mainland, and always at out-of-town sanctuaries. Before the 1st millennium BC, she was venerated in the form of a simple wooden board, which was

Hera, led by peacocks, and depicted on Samian coins

later augmented with a copper statue. One annual rite, the Tonaia, commemorated a foiled kidnapping of the wooden statue by Argive and Etruscan pirates. During the Tonaia, the idol would be paraded to the river mouth, bound on a litter of osiers (sacred to Hera), bathed in the sea and draped with gifts. The other annual festival, the Heraia, when the copper statue was dressed in wedding finery, celebrated Hera's union with Zeus, and was accompanied by concerts and athletic contests. Housed in a special shrine after the 8th century, the statue of Hera was flanked by a number of live peacocks and sprigs from an osier tree. Both are shown on Samian coins of the Roman era, stamped with the image of the richly dressed goddess.

The beach and harbour of Kokkári, flanked by its twin headlands

quay and its waterside tavernas are the busy focus of nightlife.

Environs
Though many of Sámos's hill-villages are becoming deserted, **Vourliótes** is an exception, thriving thanks to its orchards and vineyards. The picturesque central square is one of the most beautiful on the island, with outdoor seating at its four tavernas. Vourliótes is situated at a major junction in the area's network of hiking trails; paths come up from Kokkári, descend to Agios Konstantínos, and climb to Manolátes, which is the trailhead for the ascent of Mount Ampelos, a 5-hour round trip.

❼ Karlóvasi
Καρλόβασι

33 km (20 miles) NW of Vathý. 🚌 5,500. 🛈 22730 32265. 🚌🚌🚌 Potámi 2 km (1 mile) W.

Sprawling, domestic Karlóvasi, gateway to western Sámos and the island's second town, divides into four separate districts. Néo Karlóvasi served as a major leather production centre between the world wars, and abandoned tanneries and ornate mansions built on shoe-wealth can still be seen down by the sea. Meséo Karlóvasi, on a hill across the river, is more attractive, but most visitors stay at the harbour of Limín, with its tavernas and lively boatyard. Above the port, Ano, or Palaió

Karlóvasi, is tucked into a wooded ravine, overlooked by the landmark hilltop church of **Agía Triáda**, the only structure in Ano visible from the sea.

Environs
An hour's walk from Ano Karlóvasi, inland from Potámi beach, is the site of a medieval settlement. Its most substantial traces include the 11th-century church of **Metamórfosis**, the oldest on the island, and a Byzantine castle immediately above.

❽ Mount Kerketéfs
Όρος Κερκετεύς

50 km (31 miles) W of Vathý. 🚌 to Marathókampos. 🚌 Votsalákia, 2 km (1 mile) S of Marathókampos; Limniónas, 5 km (3 miles) SW of Marathókampos.

Dominating the western tip of Sámos, 1,437 m (4,715 ft) Mount Kerketéfs is the second-highest peak in the Aegean after Sáos on Samothráki. On an island otherwise composed of smooth sedimentary rock, the partly volcanic mountain is an anomaly, with jagged rocks and bottomless chasms.

Kerketéfs was first recorded in Byzantine times, when religious hermits occupied some of its caves. Nocturnal glowings at the cave-mouths were interpreted by sailors as the spirits of departed saints, or the aura of some holy icon awaiting discovery. Today, two monasteries remain on Kerketéfs: the 16th-century **Moní Evangelistrías**, perched on the south slope, and **Moní Theotókou**, built in 1887, tucked into a valley on the northeast side.

Despite past forest fires, and the paving of a road to remote villages west of the summit, Mount Kerketéfs still boasts magnificent scenery, with ample opportunities for hiking. At Seïtáni Bay on the north coast, a marine reserve protects the Mediterranean monk seal *(see p119)*.

Mount Kerketéfs, seen from the island of Ikaría

THE DODECANESE

PATMOS · LIPSI · LEROS · KALYMNOS · KOS · ASTYPALAIA · NISYROS
TILOS · SYMI · RHODES · CHALKI · KASTELLORIZO · KARPATHOS

Scattered along the coast of Turkey, the Dodecanese are the most southerly group of Greek islands, their hot climate and fine beaches attracting many visitors. They are the most cosmopolitan archipelago, with an eastern influence present in their architecture. They were the last territories to be incorporated into modern Greece.

Due to their distance from Athens and mainland Greece, these islands have been subject to a number of invasions, with traces of occupation left behind on every island. The Classical temples built by the Dorians can be seen on Rhodes. The Knights of St John were the most famous invaders, arriving in 1309 and staying until they were defeated by Suleiman the Magnificent in 1522.

Ottoman architecture is most prominent on larger, wealthier islands, such as Kos and Rhodes. After centuries of Turkish rule, the Italians arrived in 1912 and began a regime of persecution. Mussolini built many imposing public buildings, notably in the town of Lakkí on Léros. After years of occupation, the islands were finally united with the Greek state in 1948.

Geographically, the Dodecanese vary dramatically in character: some are dry, stark and barren, such as Chálki and Kásos, while Tílos and volcanic Nísyros are fertile and green. Astypálaia and Pátmos, with their whitewashed houses, closely resemble Cycladic islands; the pale houses of Chóra, on Pátmos, are spectacularly overshadowed by the dark monastery of St John. Rhodes is the capital of the island group, and is one of the most popular holiday destinations due to its endless sandy beaches and many sights.

The climate of these islands stays hot well into the autumn, providing a long season in which to enjoy the beaches. These vary from black pebbles to silver sands, and deserted bays to shingle strips packed with sunbathers.

One monk's method of travelling around on the holy island of Pátmos

◀ Aerial view of Póthia, on the island of Kálymnos

Exploring the Dodecanese

The Dodecanese offer an unparalleled range of landscapes and activities. There are beautiful beaches with all kinds of watersports, safe yachting harbours, lush valleys and barren mountains, caves and fjords, and even the semi-active volcano on Nísyros. Historical sights in the group are just as diverse, including the 11th-century Monastery of St John on Pátmos, the Hellenistic Asklepieion of Kos, the medieval walled city of the Knights of Rhodes and the unique traditional village of Olympos on Kárpathos. This island group divides neatly into north and south. Kos in the north and Rhodes, the group's capital, in the south make good bases for air and ferry travel.

The domed entrance to the New Market in Rhodes town

Getting Around

Kos, Rhodes and Kárpathos have international airports; those at Léros, Astypálaia and Kásos are domestic. Travelling by sea, it is wise to plan where you want to go, as some islands do not share direct connections even when quite close. Also journeys can be long – it takes 9 hours from Rhodes to Pátmos. If possible allow time for changes in the weather. The cooling *meltémi* wind is welcome in the high summer but, if strong, can mean ferries will not operate and even leave you stranded. Bus services are good, especially on the larger islands, and there are always cars and bikes for hire or taxis available, though the standard of roads can vary.

Kastellórizo
inset map

Locator Map

An aerial view of Sými town with its Neo-Classical houses

Sými Town
Agía Marína
Sými
Panormítis

Rhodes
Triánda
Koskinoú
Kalavárda
Faliráki
Skála
Petaloúdes
Kameírou
Profítis Ilías
800m
Alimiá
Nímporió
Emponas
Atáviros
1210m
Archángelos
Láerma
Charáki
Monólithos
Líndos
Apolakkiá
Rhodes
Gennádio
Kattaviá

Islands at a Glance

Astypálaia *p178*
Chálki *pp202–3*
Kálymnos *pp172–3*
Kárpathos *pp206–7*
Kastellórizo *p203*
Kos *pp174–7*
Léros *pp170–71*
Lipsí *p170*
Nísyros *pp178–9*
Pátmos *pp166–9*
Rhodes *pp184–201*
Sými *pp182–3*
Tílos *p181*

0 kilometres   25
0 miles   15

Key

— Main road

===== Minor road

— Scenic route

--- High-season, direct ferry route

△ Summit

Ro
Kastellórizo
Kastellórizo
Strongylí
Rhodes

For keys to symbols *see back flap*

Pátmos
Πάτμος

Known as the Jerusalem of the Aegean, Pátmos's religous significance dates from St John's arrival in AD 95 and the founding of the Monastery of St John *(see pp168–9)* in 1088. Monastic control declined as the islanders grew rich through shipbuilding and trade, and in 1720 the laymen and monks divided the land. Today Pátmos tries to maintain itself as a centre for both pilgrims and tourists.

Lámpi

Christós

Vagiá

Livádi Kalc

Kámpos

Léfkes

Kámpos beach

PATMOS

Agathonisi

Lipsí
Léros

Melói

Skála

Kastélli

Holy Cave of the Apocalypse

Monastery of St John

Pirae
Sám
My
Asty

Chóra

Grikos

Tragonis

Diakóft

Mount Prás
775 m (2,54

Psili Ammos

Cape Genoúpas

0 kilometres 2
0 miles 1

Skála

Ferries, yachts and cruise ships dock at Skála, the island's port and main town, which stretches around a wide sheltered bay. As there are many exclusive gift shops and boutiques, Skála has a smart, upmarket feel. There are several travel and shipping agencies along the harbourfront.

Skála's social life centres on the café-bar Aríon, a Neo-Classical building that doubles as a meeting place and waiting point for ferries. From the harbourfront caïques and small cruise boats leave daily for the island's main beaches.

Environs

The sandy town beach can get very crowded. To the north, around the bay, lies the shingly, shaded beach at **Melói**. There is an excellent camp site and taverna, and taxi boats also run back to Skála. Above Skála lie the ruins of the ancient acropolis at **Kastélli**. The remains include a

Hellenistic wall. The little chapel of **Agios Konstantínos** is perched on the summit, where the wonderful views at sunset make the hike up from Mérichas Bay well worthwhile.

Chóra

From Skála, an old cobbled pathway leads up to the Monastery of St John *(see pp168–9)*. The panoramic views to Sámos and Ikaría are ample reward for the long trek. A maze of white narrow lanes with over

40 monasteries and chapels, Chóra is a gem of Byzantine architecture. Many of the buildings have distinctive window mouldings, or *mantómata*, decorated with a Byzantine cross. Along the twisting alleys, some doorways lead into vast sea captains' mansions, or *archontiká*, that were built to keep marauding pirates at bay. Down the path to Skála is the church of **Agía Anna**. Steps decked with flowers lead down from the path to the

View of Skála from the Monastery of St John

Souvenirs on sale on the pathway to the Monastery of St John

church (1090), which is dedicated to the mother of the Virgin Mary. Inside the church is the **Holy Cave of the Apocalypse**, where St John saw the vision of fire and brimstone and dictated the *Book of Revelation* to his disciple, Próchoros. On view is the rock where the *Book of Revelation* was written, and the indentation where the saint is said to have rested his head. There are 12th-century wall paintings and icons from 1596 of St John and the Blessed Christodoulos *(see p168)* by the Cretan painter Thomás Vathás. St John is said to have heard the voice of God coming from the cleft in the rock, still visible today. The rock is divided into three, symbolizing the Trinity.

Votive offerings from pilgrims to Pátmos

Near Plateía Xánthou is an *archontikó*, **Simantíris House**, preserved as a Folk Museum. Built in 1625 by Aglaïnós Mousodákis, a wealthy merchant, it still has the original furnishings and contains objects from Mousodákis's travels, such as Russian samovars.

Nearby, the tranquil convent of **Zoödóchou Pigís**, built in 1607, has some fine frescoes and icons and is set in peaceful gardens.

🏛 **Holy Cave of the Apocalypse**
Between Skála and Chóra. **Tel** 22470 1276. **Open** Tue, Thu & Sun.

🏛 **Simantíris House**
Chóra. **Open** daily.

Around the Island

Pátmos has some unspoiled beaches and a rugged interior with fertile valleys. Excursion boats run to most beaches and buses from Skála serve Kámpos, Gríkos and Chóra.

The island's main resort is **Gríkos**, set in a magnificent bay east of Chóra. It has a shingly beach with fishing boats, watersports facilities and a handful of tavernas. From here the bay curves past the uninhabited Tragonísi islet south to the bizarre Kallikatsoús rock, perched on a sand spit, which looks like the cormorant it is named after. The rock has been hollowed out to make rooms, possibly by 4th-century monks, or it could have been the 11th-century hermitage mentioned in the writings of Christodoulos.

On the southwestern coast is the island's best beach, **Psilí Ammos**, with its stretch of fine sand and sweeping dunes. It is the unofficial nudist beach and is also popular with campers. Across the bay, the Rock of Genoúpas is marked by a red buoy. This is where, according to legend, the evil magician Genoúpas challenged St John to a duel of miracles. Genoúpas plunged into the sea to bring back effigies of the dead, but God then turned him to stone. Cape Genoúpas has a grotto that is said to be where the wizard lived.

Situated in the more fertile farming region in the north of the island, **Kámpos** beach, reached via the little hill-village of Kámpos, is another popular beach with watersports and a few tavernas. From Kámpos a track leads eastwards to the good pebble beaches at **Vagiá, Geranoú** and **Livádi**.

Windy **Lámpi** on the north coast is famous for its coloured and multipatterned pebbles. There are two garden tavernas and a little chapel set back from the reedbeds. You can walk here from the hamlet of Christós above Kámpos.

Holy Cave of the Apocalypse where St John lived and worked

Pátmos: Monastery of St John
Μονή του Αγίου Ιωάννου του Θεολόγου

The 11th-century Monastery of St John is one of the most important places of worship among Orthodox and Western Christian faithful alike. It was founded in 1088 by a monk, the Blessed Christodoulos, in honour of St John the Divine, author of the *Book of Revelation*. One of the richest and most influential monasteries in Greece, its towers and buttresses make it look like a fairy-tale castle, but were built to protect its religious treasures, which are now the star attraction for the thousands of pilgrims and tourists.

Monastery of St John above Chóra

The Hospitality of Abraham
This is one of the most important of the 12th-century frescoes that were found in the chapel of the Panagía. They had been painted over but were revealed after an earthquake in 1956.

KEY

① **The Chapel of Christodoulos** contains the tomb and silver reliquary of the Blessed Christodoulos.

② **The monks' refectory** has two tables made of marble taken from the Temple of Artemis, which originally occupied the site.

③ Kitchens

④ Inner courtyard

⑤ Chapel of John the Baptist

⑥ **The treasury** houses over 200 icons, 300 pieces of silverware and a dazzling collection of jewels.

⑦ **The Chapel of the Holy Apostles** lies just outside the gate of the monastery.

⑧ **The main entrance** has slits for pouring boiling oil over marauders. This 17th-century gateway leads up to the cobbled main courtyard.

★ **Icon of St John**
This 12th-century icon is the most revered in the monastery and is housed in the *katholikón*, the monastery's main church.

Chapel of the Holy Cross

This is one of the monastery's 10 chapels built because church law forbade Mass being heard more than once a day in the same chapel.

VISITORS' CHECKLIST

Practical Information
Chóra, 4 km (2.5 miles) S of Skála.
Tel 22470 31234. **Open** 8am–1:30pm Mon–Sat, 10am–1pm Sun;
4–6pm Tue, Thu & Sun. 🗝 treasury. ✝ 🅦 patmosmonastery.gr

Transport
🚌 Monastery & Treasury

Chrysobull

This scroll of 1088 in the treasury is the monastery's foundation deed, sealed in gold by the Byzantine Emperor Alexios I Comnenos.

★ Main Courtyard

Frescoes of St John from the 18th century adorn the outer narthex of the *katholikón*, whose arcades form an integral part of the courtyard.

Niptír Ceremony

The Orthodox Easter celebrations on Pátmos are some of the most important in Greece. Hundreds of people pack Chóra to watch the *Niptír* (washing) ceremony on Maundy Thursday. The abbot of the Monastery of St John publicly washes the feet of 12 monks, re-enacting Christ's washing of his disciples' feet before the Last Supper. The rite was once performed by the Byzantine emperors as an act of humility.

Embroidery of Christ washing the disciples' feet

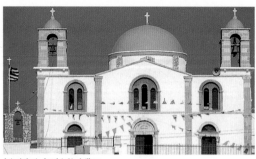

Agios Ioánnis church in Lipsí village

Lipsí
Λειψοί

🏔 700. 🚌 Lipsí town. 🛈 Town hall, Lipsí (22470 41209). 🚤 Platýs Gialós 4 km (2.5 miles) N of Lipsí town.
🌐 lipsi-island.com

Little Lipsí is a magical island characterized by green hills dotted with blue and white chapels, and colourfully painted village houses. It is one of many islands claiming to be the enchanted place where Calypso beguiled Odysseus. Officially owned by the monastery at Pátmos since Byzantine times, Lipsí has excellent beaches, and is popular for day excursions from Pátmos and Kálymnos.

The island is only 10 sq km (4 sq miles) and remains a haven for traditional Greek island life, producing some good local wines and cheeses.

The main settlement, **Lipsí town** is based around the harbour. Here the blue-domed church of **Agios Ioánnis** holds a famous icon of the Panagía. Ancient lilies within the frame miraculously spring into bloom on 23 August, the feast of the Yielding of the Annunciation. In the town hall the **Nikofóreion Ecclesiastical Museum** features an odd collection of finds, from neatly labelled bottles of holy water to traditional costumes.

These sights are all signposted from the harbour, and there are informal taxi services to the more distant bays and beaches of **Platýs Gialós, Monodéntri** and the string of sandy coves at **Katsadiás**.

🏛 **Nikofóreion Ecclesiastical Museum**
Open May–Sep: am only.

Léros
Λέρος

🏔 8,000. ✈ Parthéni. 🚌 Lakkí, Agía Marína (hydrofoils). 🚤 Plateía Plátanos, Plátanos. 🛈 Harbourfront, Lakkí (22470 22109).

Once famous as the island of Artemis, Léros's more recent history, as the home of Greece's prison camps and later mental hospitals, has kept tourism low-key. The hospitals still provide the main source of employment for the locals. However, life here is traditional, and the people are very welcoming and friendly.

The island was occupied by the Knights of St John in 1309, by the Turks from 1522 to 1831, and by the Italians in 1912, when they built naval

Neo-Classical façade of Maliamate villa, Agía Marína

bases in Lakkí bay. Under German rule from 1943 until the Allied liberation, Léros was eventually united with Greece in 1948. When the military Junta took power in 1967 they exiled political dissidents to Léros's prison camps.

Today, Léros is keen to emphasise its strong cultural and educational heritage. Famous for its musicians and poets, the island has preserved traditional folk dance and music through Artemis, the youth cultural society.

Lakkí
Lakkí, the main port and former capital, has one of the best natural harbours in the Aegean, and served as an anchorage point in turn for the Italian, German and then the British fleets. Today it resembles a disused film set full of derelict Art Deco buildings, the remains of Mussolini's vision of a Fascist dream town. Lakkí is a ghost town during the day, but the seafront cafés come to life in the evening.

Around the bay at Lépida, the former Italian naval base now houses the State Therapeutical Hospital and within the complex is a mansion once used as Mussolini's

The Art Deco Architecture of Lakkí

Mussolini's vision of a new Roman Empire took shape here in 1923, when Italian architects and town planners turned their energies to building the new town. A quite remarkable example of Art Deco architecture, Lakkí was built around wide boulevards by the engineers Sardeli and Caesar Lois, an Austrian. The model town was all curves and featured a saucer-shaped market building with a clock tower, completed in 1936; a cylindrical town hall and Fascist centre, dating to 1933–4; and the vast Albergo Romana, later the Léros Palace Hotel. The Albergo, with the cinema and theatre complex, was completed in 1937 for visiting Italian performers. These days the majority of the buildings are crumbling and neglected.

Lakkí's Art Deco cinema building

airport at Parthéni in the north. There are a few carved blocks of stone and fragments of pillars. The goddess still has some influence in Léros, however, as property passes down the female family line.

Early Christian basilicas have been found in the area, and south of the airport the 11th-century church of **Agios Geórgios**, built by the monk Christodoulos *(see p161)* using temple columns, has a fresco of the saint.

Agios Isídoros, on the west coast above sandy Goúrnas Bay, has a white chapel on an islet that can be reached by means of a narrow causeway.

At Drymónas, with its coves and oleander gorge, is the church of the **Panagía Gourlomáta**, which translates as the "goggle-eyed Virgin". Reconstructed in 1327 from an 11th-century chapel, the church takes its name from the wide-eyed expression of the Madonna seen in one of its frescoes.

The resort of **Xirókampos**, lying in a bay to the south of the island, is overlooked by ancient Palaiókastro, the former site of the 3rd-century castle of Lépida. The huge Cyclopean walls remain, and within them is the church of Panagía, that is home to some fine mosaics.

 **Historic and Folk Museum**
Belénis Castle, Alínda.
Open May–Sep: daily.

tree-fringed beach and harbour. The road north leads to Agía Marína, and is lined with impressive Neo-Classical mansions built between 1880 and 1920. **Agía Marína** is the principal port for hydrofoils. Following the coastal road north to Krithóni, the **British War Cemetery** is a site of pilgrimage for those who lost relatives in the 1943 Battle of Léros. Beaches line the road leading further north to **Álinda**, the island's main resort, which has a long beach with watersports and seafront cafés. Alínta's **Historic and Folk Museum** is housed in the twin-towered Belénis Castle, built by an expatriate benefactor, Paríssis Belénis. Little remains of the once-powerful Temple of Artemis, now overlooking the

summer residence. Also in Lépida is the 11th-century church of **Agios Ioánnis Theológos** (St John the Divine), built over the remains of a Byzantine church by the monk Christodoulos *(see p168)*.

Around the Island
Léros is a pretty, green island with an indented coastline sweeping into vast gulfs, the "four seas" of Léros. With craggy hills and fertile valleys, it is good walking country.

To defy the Italians, the Lerians abandoned Lakkí and made the village of **Plátanos** the capital. Straddling a hilltop, its houses spill down to the little port of Pantéli and to the fishing village of Agía Marína.

Perched above Plátanos, the Byzantine kástro offers fine views. Renovated by the Venetians and the Knights of St John, it houses the church of **Megalóchari or Kyrá tou Kástrou** (the Madonna of the Castle) famous for its miraculous icon. Nearby Pantéli is a fishing village with a

Plátanos village with the kástro in the background

Kálymnos
Κάλυμνος

Famous today as the sponge-fishing island, Kálymnos's history can be traced back to a Neolithic settlement in Vothýnoi, near Póthia; it was colonized after the 1450 BC devastation of Crete. The people have been known for their resilience since the 11th-century massacre by the Seljuk Turks, which a few survived in fortified Kastélli.

Emporeiós
Palaiónissos
Kolonóstilo
Kastélli
Télendos
Armeós
Arginónta
Masoúri
Drasónia
Myrtiés
KALYMNOS
Armiés
Kamári
Metóchi
Plátanos
Pánormos
Péra
Kástro
Rína
Daskalió
Cave
Chorió
Cave of
Seven Virgins
Castle of
the Knights
Póthia
Vothýnoi

Kos,
Nisyros,
Psérimos

Piraeus
Nerá
Léros
Astypálaia

0 kilometres   5
0 miles   3

Póthia

The capital and main port of the island is a busy working harbour. Wedged between two mountains, the town's brightly painted houses curve around the bay.

Póthia is home to Greece's last sponge fleet and there is a sponge-diving school on the eastern side of the harbour. The waterfront is lined with cafés and the main landmarks are the pink, domed Italianate buildings, including the old **Governor's Palace**, which now houses the market, and the silver-domed cathedral of **Agios Christós** (Holy Christ).

This 19th-century cathedral has a reredos (screen) behind the altar by Gian-noúlis Chalepás *(see p48).* The *Mermaid* at the harbour is one of 43 works that were donated to the island by local sculptors Irene and Michális Kókkinos.

The **Archaeological Museum**, housed in a Neo-Classical mansion, has been lavishly

reconstructed and there is a collection of Neolithic and Bronze Age finds from the island plus local memorabilia. The **Sponge Factory**, just off Plateía Eleftherías, has a complete history of sponges.

🏛 Archaeological Museum
Near Plateía Kýprou. **Tel** 22430 23113. **Open** 9am–4pm Tue–Sat (winter: call for times). **Closed** main public hols.

🏛 Sponge Factory
Off Plateía Eleftherías. **Tel** 22430 28501. **Open** daily.

Around the Island

Kálymnos is easy to get around with a good bus service to the villages and numerous taxis. This rocky island has three mountain ranges, the peaks offset by deep fjord-like inlets.

Northwest of Póthia the suburb of Mýloi, with its three derelict windmills, blends into **Chorió**, the pretty white town and former capital. On the way, standing to the left, is the ruined **Castle of the Knights**, and above, via steps from Chorió, is the citadel of **Péra Kástro**. Following a Turkish attack, this fortified village was inhabited from the 11th to the 18th century. It has good views and nine white chapels stand on the crags. The **Cave of Seven Virgins** (Eptá Parthénon) shows traces of nymph worship. Legend has it that the seven virgins hid here from pirates, but disappeared in the bottomless

The *Mermaid* at
Póthia harbour

The pretty waterfront of Póthia

The deep Vathý inlet with the settlement of Rína at its head

VISITORS' CHECKLIST

Practical Information
16,000. Plateía Taxi, Póthia
(22430 59141). Póthia: Mon–
Sat. Easter celebrations
around island: Easter Sat; Sponge
week at Póthia: week following
Greek Easter. **kalymnos-isl.gr**

Transport
Póthia. behind
marketplace, Póthia.

channel below. The main resorts on the island are strung out along the west coast. The sunset over the islet of Télendos from **Myrtiés** is one of Kálymnos's most famous sights. Although Myrtiés and neighbouring Masoúri have now grown into noisy tourist centres, the Armeós end of Masoúri is less frenetic. To the north is the fortified **Kastélli**, the refuge of survivors from the 11th-century Turkish massacre. The coast road from here is spectacular, passing fish farms, inlets and the fjord-like beach at **Arginónta**. A visit to the northernmost fishing hamlet, **Emporeiós** makes a good day out and is in craggy walking country.

You can walk to **Kolonóstilo** (the Cyclops Cave), which is named after its massive stalactites.

In the southeast is the most beautiful area of Kálymnos: the lush Vathý valley, which has three small villages at the head of a stunning blue inlet. Backed by citrus groves, **Rína**, named after St Irene, is a pretty hamlet with a working boatyard. **Plátanos**, the next village, has a huge plane tree and the remains of Cyclopean walls. There is a 3-hour trail from here via **Metóchi**, the third Vathý village, across the island to Arginónta. Caïques from Rína visit the **Daskalió Cave** in the side

of the sheer inlet, and Armiés, Drasónia and Palaiónissos beaches on the east coast.

Outlying Islands
Excursion boats leave Póthia daily for **Psérimos** and the islet of **Nerá** with its Moní Stavroú. Psérimos has an often busy, sandy beach and a popular festival of the Assumption on 15 August.

Télendos, reached from Myrtiés, is perfect for a hideaway holiday, with a few rooms to rent and a handful of tavernas, plus shingly beaches. There are Roman ruins, a derelict fort and the ruined Moní Agíou Vasileíou, dating from the Middle Ages. The Byzantine castle of Agios Konstantínos also stands here.

Sponge Fishing around Kalymnos

Kálymnos has been a sponge-fishing centre from ancient times, although fishing restrictions and sponge blight in the 1980s have threatened the trade. Once in great demand, sponges were used for the Sultan's harem, for padding in armour

Sea sponge and later for cosmetic and industrial purposes. Divers were weighed down with rocks or used crude air apparatus, and many men were drowned or died of the bends. The week before Kálymnos's fleet sets out to fish is the *Ipogros* or Sponge Week Festival. Divers are given a celebratory sendoff with food, drink and dancing in traditional costume.

A stone was used to weigh divers to keep them near to the seabed.

Diving equipment varied greatly over the years. Early diving suits were made from rubber and canvas with huge helmets. You can see some on display in the sponge factory at Póthia and on stalls where divers sell their wares.

This black-figure Greek vase depicts an early sponge-diving scene. The diver, pictured standing at the front of the boat, is preparing to enter the sea to search for sponges. The vase dates back to around 500 BC.

Kos
Κως

The second largest of the Dodecanese, Kos has a pleasant climate and fertile land, famous for producing the kos lettuce. Kos has attracted settlers since 3000 BC, and Hippocrates' teachings *(see p176)* increased the island's

renown. By the 4th century BC Kos was a strong trading power, though it declined after the Romans arrived in 130 BC. The Knights of St John ruled from 1315, and the Turks governed from 1522 to 1912. Italian and German occupation followed until unification with Greece in 1948.

Yachts moored in the harbour at Kos town

❶ Kos Town
Κως

🏙 15,000. 🚢 🚌 Aktí Koudouríotou. 🛈 Vasiléos Georgíou 1 (22420 28420). 🏪 daily. 🏙 Kos town.

Dominated by its Castle of the Knights, old Kos town was destroyed in the 1933 earthquake. This revealed many ancient ruins, which the Italians excavated and restored.

The harbour bristles with boats, and pavement cafés heave with tourists during the high season. There are palm trees, pines and gardens full of jasmine. Ancient and modern sit oddly side by side: Nafklírou, the "street of bars", runs beside the ancient agora, at night lit up by strobes and lasers.

Hippocrates' ancient plane tree, in Plateía Platánou, is said to have been planted by him 2,400 years ago. Despite its 14 m (46 ft) diameter, the present tree is only about 560 years old and is probably a descendant

The water fountain near Hippocrates' plane tree

of the original. The nearby fountain was built in 1792 by the Turkish governor Hadji Hassan, to serve the Mosque of the Loggia. The water gushed into an ancient marble sarcophagus.

🏰 Castle of Knights
Platánou. **Tel** 22420 27927. **Open** Jun–Sep: 8am–3pm Tue–Sun; Oct–May: 8:30am–3pm Tue–Sun. 🎫

The 16th-century castle gateway is carved with gargoyles and an earlier coat of arms of Fernández de Heredia, the Grand Master from 1376 to 1396. The outer keep and battlements were built between 1450 and 1478 from stone and marble, including blocks from the Asklepieíon *(see p176)*. The fortress was an important defence for the Knights of Rhodes against Ottoman attack.

🏛 Ancient Agora
South of Plateía Platánou.
This site is made up of a series of ruins; from the original Hellenistic city to Byzantine buildings. Built over by the Knights, the ancient remains were revealed in the 1933 earthquake. Highlights include the 3rd-century-BC stoa Kamára tou Fórou (Arcade of the

Sights at a Glance

1. Kos Town
2. Asklepieíon
3. Asfendíou Villages
4. Tigkáki
5. Palaió Pylí
6. Kardámaina
7. Antimácheia
8. Kamári

VISITORS' CHECKLIST

Practical Information

31,000. Kos town (22420 28420). Hippocrates Cultural Festival: Jul–Sep; Panagía at Kardámaina: 8 Sep; Agios Geórgios Festival at Palaió Pylí: 23 Apr.

Transport

27 km (16 miles) W of Kos town. Aktí Koudouriótou, Kos town. Kos town.

Forum), the 3rd-century-BC Temple of Herakles, mosaic floors depicting Orpheus and Herakles, and ruins of the Temple of Pándemos Aphrodite. A 5th-century Christian basilica was also discovered, along with the Roman Agora.

🏛 Archaeological Museum

Plateía Eleftherías. **Tel** 22420 28326. **Closed** for renovation; phone to confirm opening times before visiting.

The museum has an excellent collection of the island's Hellenistic and Roman finds, including a 4th-century-BC marble statue of Hippocrates.

The main hall displays a 3rd-century-AD mosaic of Asklepios and 2nd-century statues of Dionysos with Pan and a satyr. The east wing exhibits Roman statues and north Hellenistic finds; the west room has later huge statuary.

🏛 Roman Remains

Grigoríou E. **Tel** 22420 28326. **Closed** for renovation; phone ahead.

The most impressive of these ruins is the Casa Romana, built in the Pompeiian style. It had 26 rooms and three pools surrounded by shady courtyards lined with Ionian and Corinthian columns. There are mosaics of dolphins, lions and leopards. The dining room has decorated marble walls and several rooms are painted. In the grounds are the excavated thermal baths and part of the main Roman road, covered with ancient capitals and Hellenistic fragments. Set back off the road down an avenue of cypresses

Kos lettuce on a market stall in Plateía Eleftherías

is the ancient *Odeion* (theatre). It has rows of marble benches (first-class seats) and limestone blocks for the plebeians.

The western excavations opposite reveal a mix of historical periods. There are Mycenaean remains, a tomb dating from the Geometric period and Roman houses with some fine mosaics. One of the most impressive sights is the gym or *xystó* with its 17 restored Doric pillars.

Rows of marble benches for the Roman audiences that came to the ancient Odeion

Around Kos Island

Mainly flat and fertile, Kos is known as the "Floating Garden". It has a wealth of archaeological sites and antiquities, Hellenistic and Roman ruins, and Byzantine and Venetian castles. Most visitors, however, come for Kos's sandy beaches. Those on the southwest shore are some of the finest in the Dodecanese, while the northwest bays are ideal for watersports. Much of the coast has been developed, but inland you can still see remnants of Kos's traditional lifestyle.

The seven restored columns of the Temple of Apollo at the Asklepieíon

❷ Asklepieíon

Ασκληπιείο

4 km (2.5 miles) NW of Kos town. 🚌 **Tel** 22420 28326. **Open** 8am–8pm Tue–Fri, 8am–3pm Sat–Mon & pub hols (Nov–Mar: call to check times). 🖼

With its white marble terraces cut into a pine-clad hill, the Asklepieíon site was chosen in the 4th century BC for rest and recuperation and still exudes an air of tranquillity. The views from the sanctuary are breathtaking and it is one of Greece's most important Classical sites.

Temple, school and medical centre combined, it was built after the death of Hippocrates and was the most famous of ancient Greece's 300 *asklepieia* dedicated to Asklepios, god of healing. The doctors, priests of Asklepiados, became practitioners of Hippocrates' methods. The cult's symbol was the snake, once used to seek healing herbs, and is the emblem of modern Western medicine. There are three levels: the lowest has a 3rd-century-BC porch and 1st-century-AD Roman baths; the second has a 4th-century-BC

Altar of Apollo and a 2nd- to 3rd-century-AD Temple of Apollo; on the third level is the Doric Temple of Asklepios from the 2nd century BC.

❸ Asfendíou Villages

Χωριά Ασφενδίου

14 km (9 miles) W of Kos town. 🚌

The Asfendíou villages of Zía, Asómatos, Lagoúdi, Evangelístria and Agios Dimítrios are a cluster of picturesque hamlets on the wooded slopes of Mount Dikaíos. These mountain villages have managed to retain their traditional character, with whitewashed houses and attractive Byzantine churches. The highest village, **Zía**, has become the epitome of a traditional Greek village, at least to the organizers of the many coach tours that regularly descend upon it. The more adventurous traveller can take the very rough track from the Asklepieíon via tiny Asómatos to Zía. The lowest village, **Lagoúdi**, is less commercialized and a road leads from here to Palaió Pylí.

❹ Tigkáki

Τιγκάκι

12 km (7 miles) W of Kos town. 🚌 🚘 Tigkáki.

The popular resorts of Tigkáki and neighbouring Marmári have long white sand beaches ideal for windsurfing and other watersports. Boat trips are available from Tigkáki to the island of Psérimos opposite.

The nearby **Alykés Saltpans** are a perfect place for bird-watching. The many wetland species here include small waders like the avocet, and the black-winged stilt with its long pink legs.

Hippocrates

The first holistic healer and "father of modern medicine", Hippocrates was born on Kos in 460 BC and died in Thessaly in about 375 BC. He supposedly came from a line of healing demigods and he learned medicine from his father and grandfather: his father was a direct descendant of Asklepios, the god of healing, his mother of Herakles. He was the first physician to classify diseases and introduced new methods of diagnosis and treatment. He taught on Kos before the Asklepieíon was established, and wrote the Hippocratic Oath, to cure rather than harm, still sworn by medical practitioners worldwide.

Palaió Pylí castle perched precariously on a cliff's edge

❺ Palaió Pylí

Παλαιό Πυλί

15 km (9 miles) W of Kos town.
to Pylí. **Tel** 22470 24776.

The deserted Byzantine town of Palaió Pylí is perched on a crag 4 km (2 miles) above the farming village of Pylí, with the remains of its castle walls built into the rock. Here the Blessed Christodoulos built the 11th-century church of the Ypapandís (Presentation of Jesus), before he went to Pátmos *(see p167)*. In Pylí lies the Classical *thólos* tomb of the mythical hero-king Chármylos It has 12 underground crypts, which are now surmounted by the church of Stavrós.

❻ Kardámaina

Καρδάμαινα

26 km (16 miles) SW of Kos town.
Kardámaina.

Once a quiet fishing village noted for its ceramics, Kardámaina is the island's biggest resort – brash, loud and packed with young British and Scandinavian tourists. It has miles of crowded golden sands and a swinging nightlife. It is quieter further south, with some exclusive developments. Sights include a Byzantine church and the remains of a Hellenistic theatre.

❼ Antimácheia

Αντιμάχεια

25 km (16 miles) W of Kos town.

The village of Antimácheia is dominated by its Venetian castle and windmills. The castle, located near the airport, was built by the Knights of Rhodes *(see pp192–3)* as a prison in the 14th century, and was constantly bombarded by pirates. Its massive crenellated battlements and squat tower now overlook an army base, and there are good views towards Kardámaina. The inner gateway still bears the coat of arms of the Grand Master Pierre d'Aubusson (1476–1503) and there are two small chapels within the walls.

Antimácheia castle battlements

Environs

The road north from Antimácheia leads to the charming port of **Mastichári**. There are good fish tavernas here and a long sandy beach that sweeps into dunes at the western end. On the way to the dunes, the ruins of an early Christian basilica, with good mosaics, can be seen.

❽ Kamári

Καμάρι

15 km (9 miles) SW of Kos town.
Paradise 7 km (4 miles) E.

Kamári is a good base for exploring the southwest coast, where the island's best beaches can be found. Mostly reached via steep tracks from the main road, the most famous is Paradise beach with fine white sands. Kamári beach leads to the 5th-century-AD Christian basilica of Agios Stéfanos which has mosaics and Ionic columns.

Environs

Kéfalos, on the mountainous peninsula inland from Kamári, is known for its thyme, honey and cheeses. Sights include the ruined Castle of the Knights, said to be the lair of a dragon. According to legend, Hippocrates' daughter was transformed into a dragon by Artemis, and awaits the kiss of a knight to resume human form. Above Kéfalos is the windmill of Papavasílis, and nearby at Palátia are the remains of **Astypálaia**, the birthplace of Hippocrates. Neighbouring Aspri Pétra cave has yielded remains. The journey to **Moní Agíou Ioánni**, 6 km (4 miles) south of Kéfalos, passes through dramatic scenery, and a track leads to the beach of Agios Ioánnis Theológos.

The traditional Greek windmill at Kéfalos, just outside Kamári

Chóra overlooking Astypálaia's main harbour, Skála

Astypálaia

Αστυπάλαια

📷 1,200. ✈ 11 km (7 miles) E of Astypálaia town. 🚢 🚌 Astypálaia town. ℹ️ near Kástro, Astypálaia town (22430 61778).

With its dazzling white fortified town of Chóra and its scenic coastline, the island of Astypálaia retains an exquisite charm. A backwater in Classical times, Astypálaia flourished in the Middle Ages when the Venetian Quirini family ruled from 1207 to 1522.

The most westerly of the Dodecanese, it is a remote island with high cliffs and a hilly interior. There are many coves and sandy bays along the coast, which was once the lair of Maltese pirates.

Astypálaia town incorporates the island's original capital, Chóra, which forms its maze-like upper town. The splendid Venetian kástro of the Quirini family is on the site of the ancient acropolis. Houses were built into the kástro's walls for protection, and the Quirini coat of arms can still be seen on the gateway. Within its walls are two churches: the silver-domed, 14th-century Panagía Portaïtissa (Madonna of the Castle Gates), and the 14th-century Agios Geórgios (St George), built on the site of an ancient temple.

A 2-hour hike westwards from the derelict windmills above Chóra leads to **Agios Ioánnis** and its gushing waterfall. **Livádi**, the main resort, lies south of

Chóra in a fertile valley with citrus groves and cornfields. It has a long beach. The nudist haunt of **Tzanáki** lies a short distance to the south. From Livádi a dirt track leads north to **Agios Andréas**, a remote and attractive cove, a 90-minute trek away.

North of Chóra, on the narrow land bridge between the two sides of the island, lies **Maltezána** (also known as Análipsi), the fastest-growing resort on the island. Named after the marauding pirates who once frequented it, Maltezána was where the French Captain Bigot set fire to his ship in 1827 to prevent it being captured.

On the northeastern peninsula is the "lost lagoon", a deep inlet at the hamlet of **Vathý**. From here you can visit the caves of Drákou and Negrí by boat, or the Italian Kastellano fortress, built in 1912, 3 km (2 miles) to the south.

A typical housefront in Mandráki on Nísyros

Nísyros

Νίσυρος

📷 1,000. ✈ 🚢 🚌 Mandráki harbour. ℹ️ 22420 31204. 🚖 Gialiskári 2 km (1 mile) E of Mandráki; Páloi 4 km (2 miles) E of Mandráki. 🌐 nisyros.gr

Almost circular, Nísyros is on a volcanic line which passes through Aígina, Póros, Mílos and Santoríni. In 1422 there was a violent eruption and its 1,400 m- (4,593 ft-) high peak exploded, leaving a huge caldera (*see p180*). Everything flourishes in the volcanic soil and there is some unique flora and fauna.

According to mythology, Nísyros was formed when the enraged Poseidon threw a chunk of Kos on the warring giant, Polyvotis, who was submerged beneath it, fiery and fuming. In ancient times, it was famous for its millstones, often known as the "stones of Nísyros". Now the island prospers from pumice mining on the islet of Gyalí to the north.

Mandráki

Boats dock at Mandráki, the capital, with quayside tavernas, ticket agencies and buses shuttling visitors to the volcano. Mandráki's narrow two-storey houses have brightly painted wooden balconies, often hung with strings of drying tomatoes and onions. A maze of lanes congregates at Plateía Iróön, with its war memorial. Other roads weave south, away from the sea, past the *kípos* (public

Kos

Tilos,
Rhodes

Páloi

Mandráki
Gialiskári
Kolkáki
Loutrá
aiókastro
Emporeiós
Liés

Pachiá
Ammos

NISYROS

Profítis Ilias
698 m (2,290 ft)

Stéfanos
crater

Agios Ioánnis
Theológos
Nikiá

Argos

Avláki

0 kilometres   3
0 miles   2

orchard) to the main square, Plateía Ilikioménon. At night, the area is bustling: shops that resemble houses are open, with traditional painted signs depicting their wares. The lanes become narrow and more winding as you approach the medieval Chóra district. In the nearby Langádi area, the balconies on the houses almost touch across the street.

The major attractions in Mandráki are the 14th-century kástro and the monastery. The former is the castle of the Knights of St John (see pp192–3), built in 1325 high up the cliff face. The monastery, **Moní Panagías Spilianís**, lies within the kástro and dates from around 1600. Inside, a finely carved iconostasis holds a Russian-style icon, decked in gold and silver offerings, of the Virgin and Child. The fame of the church grew after Saracens failed to find its treasure of silver, hidden by being worked into the Byzantine icons. The library holds rare editions and a number of ecclesiastical treasures.

The main square in Nikiá with its *choklákia* mosaic

The **Historical and Folk Museum**, on the way up to the kástro, has a reconstructed traditional island kitchen, embroideries and a small collection of local photographs.

Excursion boats offer trips from Mandráki to **Gyalí** and the tiny **Agios Antónios** islet beyond. Both destinations have white sandy beaches.

Historical and Folk Museum
Kástro. **Open** May–Sep: daily.

Around the Island

Nísyros is lush and green with terraces of olives, figs and almond trees contrasting with the strange grey and yellow moonscape of the craters. No visit would be complete without an excursion to the volcano, and by day the island is swamped with visitors from Kos. However, it is quiet when the excursion boats have left.

Above Mandráki lies the **Palaiókastro**, the acropolis of ancient Nísyros, dating back 2,600 years. Remains include Cyclopean walls made from massive blocks carved from the volcanic rock, and Doric columns.

Nísyros is pleasant for walking. Visits to the volcano must include the pretty village of **Nikiá** (see p180), with its *choklákia* mosaic in the round "square", and abandoned **Emporeiós** which clings to the rim of the crater.

To the east of Mandráki, **Páloi** is a pretty fishing village with good tavernas and a string of dark volcanic sand beaches. Two kilometres (1 mile) west of the village, at **Loutrá**, an abandoned spa can be found.

The *meltémi* wind blows fiercely on Nísyros in high season, and the beaches east of Páloi can often be littered with debris.

Mandráki, the capital of Nísyros as viewed from the sea

For keys to symbols see back flap

The Geology of Nísyros

Fuming and smelling of rotten eggs, the centre of Nísyros is a semi-active caldera – a crater formed by an imploded mountain. Its eruption, around 24,000 years ago, was accompanied by an outpouring of pumice, forming a blanket 100 m (328 ft) thick on the upper slopes of the island. When formed, the caldera was 3 km (2 miles) in diameter. It is now occupied by two craters and five solidified lava domes, forced upwards in the last few thousand years, including Profítis Ilías, the largest in Europe. Further eruptions in 1873 built cones of ash 100 m (328 ft) high.

Steep paths descend to the crater floor, where the surface is hot enough to melt rubber-soled shoes. Gas vents let off steam, at 98° C (208° F), which bubbles away beneath the earth's crust.

Paths lead visitors around the caldera.

Profítis Ilías dome is almost 700 m (2,300 ft) high.

Ash cones have been produced in the recent life of the caldera.

Original caldera wall

Lava dome

The Stéfanos crater, which is 300 m (985 ft) wide and 25 m (82 ft) deep, was created by an explosion of pressurized water and superheated steam.

Nisyros Caldera

This huge caldera contains several water-filled mini craters. The largest is the still-active Stéfanos crater, which has a number of hot springs, boiling mud pots and gas vents. There is a stench of sulphur and numerous pure sulphur crystals are eagerly snapped up by would-be geologists.

Nikiá is the more appealing of Nísyros's two rim villages with its brightly painted houses and *choklákia* pebble mosaics. There are good views from Nikiá of the crater, and a path down to the caldera.

The oldest volcanic minerals found on Nísyros date back 200,000 years. There are vast amounts of pumice around the caldera and rich deposits of sulphur and kaoline.

Sulphur

Kaoline

Pumice

Tílos
Τήλος

🏘 500. 🚢 🚌 Livádia. ℹ️ Megálo Chorió (22460 44222). 🚕 Eristós 10 km (6 miles) NW of Livádia.

Remote Tílos is a tranquil island, with good walking, and, as a resting stop on migration paths, it offers rich rewards for bird-watchers. Away from the barren beaches, Tílos has a lush heartland, with small farms growing everything from tobacco to almonds. Its hills are scattered with chapels and ruins of Crusader castles, outposts of the Knights of St John, who ruled from 1309 until 1522.

There is a strong tradition of music and poetry on the island – the poet Erinna, famous for the *Distaff*, was born here in the 4th century BC. In the 18th and 19th centuries Tílos was known for weaving cloth for women's costumes, still worn by some islanders today.

Livádia

Livádia, the main settlement, has a tree-fringed pebble beach sweeping round its bay. The blue and white church of **Agios Nikólaos** dominates the waterfront, and has an iconostasis carved in 1953 by Katasáris from Rhodes. On the beach road, the tiny, early Christian basilica of **Agios Panteleïmon kai Polýkarpos** has an attractive mosaic floor.

The pebble beach at Livádia

Around the Island

Buses run from Livádia to Megálo Chorió and Erystos, and mopeds can be hired; otherwise you are on foot.

Built on the site of the ancient city of Telos, **Megálo Chorió** is 8 km (5 miles) uphill from Livádia. The kástro was built by the Venetians who incorporated a Classical gateway and stone from the ancient acropolis. The **Palaeontological Museum** has midget fossilized mastodon (elephant) bones from the Misariá region, and a gold treasure trove, found in a Hellenistic tomb in the Kená region of the island.

ΠΕΣΟΝΤΕΣ
Detail of the War Memorial at Livádia

The church of Archángelos Michaíl (1827) was built against the kástro walls. It has silver icons from the original Taxiárchis church, a gilded 19th-century iconostasis and the remains of 16th-century frescoes.

South of Megálo Chorió lies **Erystos**, a long sandy beach. **Agios Antónis** beach to the west of Megálo Chorió has the petrified remains of human skeletons. These "beach rocks" are thought to be of sailors caught in the lava when Nísyros erupted in 600 BC. Perched on a cliff

on the west coast, the Byzantine **Moní Agíou Panteleïmonos** is the island's main sight. In a cluster of trees, this fortified monastery with red pantiled roofs is famous for its sunset views. Built in 1470 it has circular chapels, a mosaic courtyard and medieval monks' cells. The dome of the church has a vision of *Christ Pantokrátor* (1776) by Gregory of Sými. Other important artifacts include 15th-century paintings of Paradise and the apostles, and a carved iconostasis that dates from 1714. The fossilized bones of mini masto-dons from 7000 BC were discovered in the **Charkadió Grotto**, a ravine in the Misariá area. The ruined fortress of Misariá marks the spot.

Mikró Chorió, below Misariá, has about 220 roofless, abandoned houses. Those residents who had stone roofs took them with them to Livádia when the population abandoned the village in the 1950s. Quiet during the day, at night the ruins are illuminated, and one house has been restored as a bar. There is also the mid-17th-century church of **Timía Zóní**, which has 18th-century frescoes, and the chapels of **Sotíros, Eleoúsas** and **Prodrómou**, with 15th-century paintings.

🏛 **Palaeontological Museum**
Megálo Chorió. **Open** 9am–2pm daily; request key at town hall.

One of many almond orchards on Tílos

Sými
Σύμη

Ever since classical times, rocky, barren Sými has thrived on the success of its sponge-diving fleet and boat-building industry, which once launched 500 ships a year. By the 17th century it was the third-richest island in the Dodecanese. The Italian occupation in 1912 and the arrival of artificial sponges and steam power ended Sými's good fortunes. Its population had fallen from 23,000 to 6,000 by World War II, and the mansions built in its heyday crumbled.

Sými Town

The harbour area, Gialós, is one of the most beautiful in Greece, surrounded by Neo-Classical houses and elaborate churches built on the hillside. Gialós is often busy with day trippers, particularly late morning and early afternoon.

A clock tower (1884) stands on the western side of the harbour where the ferries dock; beyond is the shingle bay of Nos beach. Next door to the town hall, the **Maritime Museum** has an interesting record of Sými's seafaring past.

Gialós is linked to the upper town, Chorió, by a road and also by 375 marble steps. Chorió comprises a maze of lanes and distinctive houses, often with traditional interiors. The late 19th-century church of **Agios Geórgios** has an unusual pebble mosaic of fierce mermaids who, in Greek folklore, are responsible for storms that sink ships. The **Sými Museum**, high up in Chorió, has a small but interesting collection of costumes and traditional items. Beyond the museum is the ruined Byzantine **kástro** and medieval walls. Megáli Panagía church, the jewel of the kástro, has an important post-Byzantine icon of the *Last Judgment*, from the late 16th century, by the painter Geórgios Klontzás.

🏛 **Maritime Museum**
Plateía Ogdóis Maḯou. **Tel** 22460 72363. **Open** Apr–Oct: daily. 🅿

🏛 **Sými Museum**
Chorió. **Tel** 22460 71114. **Open** 8:30am–3pm daily. **Closed** main public hols.

Map labels: Tilos, Nimos, Rhodes, Agía Marína, Emporeiós, Nos, Moní Agíou Michaḯl Roukounióti, Sými Town, Noúlia 250 m (820 ft), Pédi, Agios Nikó, Agios Aimilianós, Cape Kefála, SYMI, Agios Géorgios Dissálona, Agios Vasílios, Nanoú, Pidima, Gialesino, Megalonisi, Marathoún, Panormítis, Moní Taxiárchi Michaḯl Panormíti, Teftloysa

0 kilometres 4
0 miles 2

The pastel-coloured houses of Chorió on the ancient acropolis overlooking Sými's harbour

For keys to symbols *see back flap*

The traditional craft of boat building in Sými town

Environs

The road from Gialós to Chorió passes the hill of **Noúlia**, also known as Pontikókastro. On the hill are the remains of 20 windmills and an ancient tomb monument believed to have been erected by the Spartans in 412–411 BC.

Around the Island

Sými's road network is limited but there are plenty of tracks over its rocky terrain. East of Sými town, an avenue of eucalyptus trees leads down through farmland to **Pédi** bay, a beach popular with local families. From here taxi boats run to **Agios Nikólaos** beach and there are paths to Agios Nikólaos and **Agía Marína**.

The 18th-century church of **Moní Agíou Michaïl Roukou-nióti**, 3 km (2 miles) west of Sými town, is built like a desert fortress in Gothic and folk architecture. It houses 14th-century frescoes and a rare 15th-century, semicircular icon of the *Hospitality of Abraham* by Cretan artist Stylianós Génis.

Sými's most popular sight is **Moní Taxiárchi Michaïl Panormíti** in Panormítis bay, a place of pilgrimage for Greek sailors worldwide. Its white buildings, spanning the 18th to 20th centuries, line the water's edge. The pleasant horseshoe-shaped harbour is dominated by the elaborate mock-Baroque bell tower, a 1905 copy of the famous bell tower of Agía Foteiní in Izmir.

The monastery is famous for its icon of the Archangel Michael, Sými's patron saint and guardian of seafarers. Despite being removed to Gialós, it mysteriously kept returning to Panormítis so the monastery was founded

here. The single-nave *katholikón* was built in 1783 on the remains of an early Byzantine chapel also dedicated to the saint.

According to tradition, if you ask a favour of St Michael, you must vow to give something in return. As a result, the interior is a dazzling array of votive offerings, or *támata*, from pilgrims, including small model ships in silver and gold.

The mock-Baroque bell tower of Moní Taxiárchi Michaïl Panormíti

The intricate Baroque iconostasis by Mastrodiákis Taliadoúros is a remarkable piece of wood-carving. The walls and ceiling are covered in smoke-blackened 18th-century frescoes by the two Sýmiot brothers Nikítas and Michaïl Karakostís.

The sacristy museum is full of treasures, including a post-Byzantine painting of the 10 saints, Agioi Déka, by the Cretan Theó-doros Poulákis. There are prayers in bottles, which have floated miraculously into Pan-ormítis, containing money for the mon-astery from faithful sailors. The cloister has a *choklákia* court-yard of zigzag pebble mosaics *(see p202)* and an arcaded balcony.

West of the monastery, past the taverna, is a memorial to the former abbot, two monks and two teachers executed by the Germans in 1944 for running a spy radio for British commandos. Small Panormítis beach is here and there are woodland walks to **Marathoúnta.**

Moní Taxiárchi Michaïl Panormíti
Panormítis bay. **Open** Tue–Sun.

Rhodes

Ρόδος

Rhodes, the capital of the Dodecanese, was an important centre in the 5th to 3rd centuries BC. It was part of both the Roman and Byzantine empires, before being conquered by the Knights of St John. They occupied Rhodes from 1306 to 1522, and their medieval walled city still dominates Rhodes town. Ottoman and Italian rulers followed. Fringed by sandy beaches, and with good hiking and lively nightlife, Rhodes attracts thousands of tourists each year.

5 Ancient Kámeiros
The stunning ruins of this once-thriving Doric city include a 6th-century-BC Temple of Athena Polias.

6 Skála Kameírou
A pleasant place to relax, Skála Kameírou is an attractive harbour that once served the ancient city of Kámeiros.

Kritinía castle built by the Knights of Rhodes, was one of their larger strongholds (see p197).

Siána is a pretty traditional hill-village, known for its locally distilled spirit, soúma (see p197).

7 Emponas
The slopes around this traditional town have been cultivated with vines by the Emery winery since the 1920s.

8 Monólithos
The village is dominated by the 15th-century castle, perched high on a massive rock. It was built by the Knights of Rhodes.

9 Moní Skiádi
This monastery was built in the 18th and 19th centuries and is famous for its icon of the Panagía, or the Blessed Virgin.

10 Moní Thárri
Hidden away in the countryside, this monastery has a domed church that is home to several frescoes, some dating to the 12th century.

For hotels and restaurants in this region see pp309–10 and pp326–8

4 Petaloúdes
Called butterfly valley, this tranquil place is, in fact, home to thousands of moths during the summer.

3 Moní Filerímou
The monastery is set on the beautiful hillsides of Mount Filérimos. The main church dates back to the 14th century.

VISITORS' CHECKLIST

Practical Information
115,000. Rhodes town (22410 27423). Rodíni Park Wine Festival, outskirts of Rhodes town: end Aug.

Transport
25 km (16 miles) SW of Rhodes town. Commercial harbour, Rhodes town.

Sými, Kos
Chálki, Piraeus, Astypálaia
Kastellórizo

Rhodes Town
Triánda
Paradísi
Ancient Ialyssós
Réni Koskinoú
Moní Filerímou
Koskinoú **15**
Thérmes Kalithéas
Kalithéa
Kalythiés
Faliráki **14**
Petaloúdes
Psínthos
Ladikó Bay
Afántou
ODES
Loútan
Kolympia
Eptá Pigés **13**
Moní Tsampíkas
Tsampíka
Archángelos **12**
Stégna
Charáki
Líndos **11**
Péfkoi

2 Ancient Ialyssós
Set on a plateau with commanding views, this ancient site dates back to 2500 BC. The ruins include remains of a 3rd-century-BC acropolis.

1 ★ Rhodes Town
Mandráki harbour is at the centre of Rhodes town, which is one of Greece's most popular tourist destinations.

15 Koskinoú
This small village offers visitors the opportunity to see traditional Rhodian houses and *choklákia* pebble mosaics (see p202).

Faraklós was once used by the Knights of Rhodes as a prison. Today it overlooks Charáki village (see p198).

13 Eptá Pigés
This is an enchanting beauty spot that takes its name from the "seven springs" that are the source for the area's central reservoir.

14 Faliráki
This fun-packed resort offers all sorts of nightlife and watersports, and is particularly popular with the young.

12 Archángelos
A popular place to visit, Archángelos is set in attractive countryside, and maintains a tradition of handicraft production.

0 kilometres 10
0 miles 6

11 ★ Líndos
One of the island's most visited sites, the acropolis at Líndos towers over the town from its clifftop position.

For keys to symbols *see back flap*

❶ Street-by-Street: Rhodes Old Town
Παλιά Πόλη Ρόδου

The town of Rhodes has been inhabited for more than 2,400 years. A city was first built here in 408 BC, and when the Knights of St John arrived in 1309 they built their citadel over these ancient remains. The Knights' medieval citadel, dominated by the towers of the Palace of the Grand Masters, forms the centre of the old town. The new town *(see pp194–5)* lies beyond the original walls. Of the walls' 11 gates, Koskinoú (St John's) gate, which leads into the Bourg quarter *(see p189)*, has the best view of the city's defences.

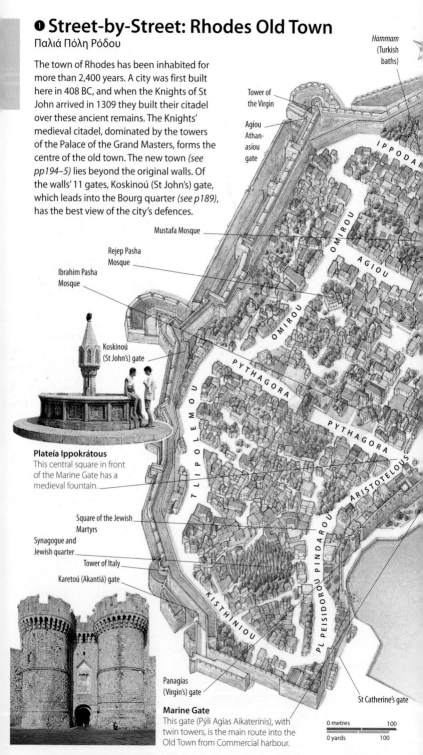

Hammam (Turkish baths)

Tower of the Virgin

Agíou Athanasíou gate

IPPODA...

OMIROU

AGIOU

Mustafa Mosque

Rejep Pasha Mosque

Ibrahim Pasha Mosque

OMIROU

PYTHAGORA

PYTHAGORA

Koskinoú (St John's) gate

T'LIPOLEMOU

ARISTOTELOUS

Plateía Ippokrátous
This central square in front of the Marine Gate has a medieval fountain.

Square of the Jewish Martyrs

Synagogue and Jewish quarter

Tower of Italy

Karetoú (Akantiá) gate

PL PEISIDOROU PINDAROU

KISTHINIOU

Panagías (Virgin's) gate

St Catherine's gate

Marine Gate
This gate (Pýli Agías Aikaterínis), with twin towers, is the main route into the Old Town from Commercial harbour.

0 metres 100
0 yards 100

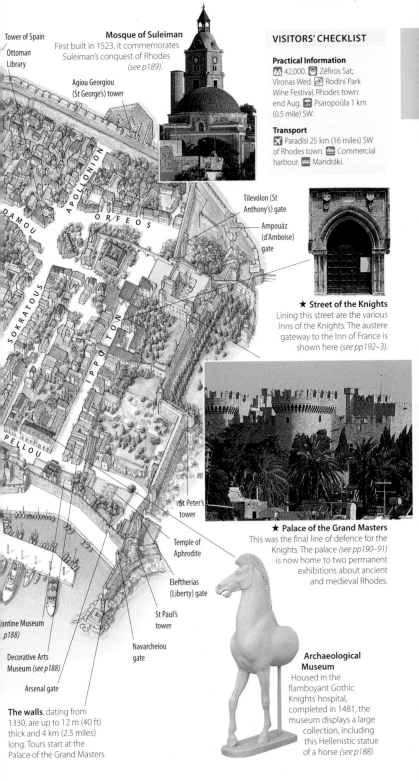

Tower of Spain

Ottoman Library

Mosque of Suleiman
First built in 1523, it commemorates Suleiman's conquest of Rhodes
(see p189).

Agíou Georgíou (St George's) tower

APOLLONION

DAMOU

ORFEOS

SOKRATOUS

IPPOTON

PELLOU

VISITORS' CHECKLIST

Practical Information

42,000. Zéfiros Sat; Víronas Wed. Rodíni Park Wine Festival, Rhodes town: end Aug. Psaropoúla 1 km (0.5 mile) SW.

Transport

Paradísi 25 km (16 miles) SW of Rhodes town. Commercial harbour. Mandráki.

Tilevólon (St Anthony's) gate

Ampouáz (d'Amboise) gate

★ **Street of the Knights**
Lining this street are the various Inns of the Knights. The austere gateway to the Inn of France is shown here *(see pp192–3)*.

St Peter's tower

★ **Palace of the Grand Masters**
This was the final line of defence for the Knights. The palace *(see pp190–91)* is now home to two permanent exhibitions about ancient and medieval Rhodes.

Temple of Aphrodite

Eleftherías (Liberty) gate

St Paul's tower

antine Museum p188)

Decorative Arts Museum *(see p188)*

Navarcheíou gate

Arsenal gate

The walls, dating from 1330, are up to 12 m (40 ft) thick and 4 km (2.5 miles) long. Tours start at the Palace of the Grand Masters.

Archaeological Museum
Housed in the flamboyant Gothic Knights' hospital, completed in 1481, the museum displays a large collection, including this Hellenistic statue of a horse *(see p188)*.

Exploring Rhodes Old Town

Dominated by the Palace of the Grand Masters, this medieval citadel is surrounded by moats and 4 km (2.5 miles) of walls. Eleven gates give access to the old town, which is divided into the Collachium and the Bourg. The Collachium was the Knights' quarter, and dates from 1309. The Bourg housed the rest of the population, which included Jews and Turks as well as Greeks. As one of the finest walled cities in existence, the Old Town is now a World Heritage Site.

An arched street in the old town

The imposing 16th-century d'Amboise gate

The Collachium

This area includes the Street of the Knights (see pp192–3) and the Palace of the Grand Masters (see pp190–91). The main gates of entry from the new town are d'Amboise gate and Eleftherías (Liberty) gate. The former was built in 1512 by Grand Master d'Amboise, leading from Dimokratías to the palace. The Eleftherías gate was built by the Italians and leads from Eleftherías to Plateía Sýmis. An archway leads from here into Apelloú.

🏛 Archaeological Museum

Plateía Mouseíou. **Tel** 22413 65270/ 75674. **Open** Apr–Oct: 8am–7:40pm daily (from 1:30pm Mon); Nov–Mar: 8am–2:40pm Tue–Sun. **Closed** main public hols. 🚫

The museum is housed in the Gothic Hospital of the Knights, built in 1440–81. Most famous of the exhibits is the 1st- century-BC marble *Aphrodite of Rhodes*. Other gems include a 2nd-century-BC head of Helios the Sun God, discovered at the Temple of Helios on the nearby hill of Monte Smith. The grave stelae from the necropolis of Kámeiros give a good insight into 5th-century-BC life. Exhibits also include coins, jewellery and ceramics from the Mycenaean graves at nearby Ialyssós.

Aphrodite of Rhodes, Archaeological Museum

🏛 Decorative Arts Museum

Plateía Argyrokástrou. **Tel** 22413 65246. **Closed** for renovation; call to check opening times. 🚫 ♿

This museum features Lindian tiles, costumes and a reconstructed traditional Rhodian house.

🏛 Medieval Rhodes and Ancient Rhodes Exhibitions

Palace of the Grand Masters. **Tel** 22413 65270. **Open** Apr–Oct: 8am–7:40pm daily; Nov–Mar: 8am–2:40pm Tue–Sun. **Closed** main public hols. 🚫 ♿

Both of these permanent exhibitions can be seen as part of a tour of the Palace of the Grand Masters (see pp190–91). The Medieval Rhodes exhibition is titled: Rhodes from the 4th century AD to the Turkish Conquest (1522). It gives an insight into trade and everyday life in Byzantine and medieval times, with Byzantine icons, Italian and Spanish ceramics, armour and militaria.

The exhibition Ancient Rhodes: 2,400 Years is situated off the inner court. It displays finds from 45 years of archaeological investigations on the island.

🏛 Byzantine Museum

Apéllou. **Tel** 22410 27657. **Closed** for renovation; exhibits currently in the Palace of the Grand Masters (see pp190–91). 🚫

Dating from the 11th century, this Byzantine church became the Knights' cathedral, but was converted under Turkish rule into the Mosque of Enderum, known locally as the Red Mosque. Now a museum, it houses a fine collection of icons and frescoes. Among the exhibits are striking examples of 12th-century paintings in the dynamic Comnenian style from Moní Thárri (see p198) and late 14th-century frescoes from Chálki's abandoned church of Agios Zacharías.

Courtyard at the Knights' Hospital, now the Archaeological Museum

For hotels and restaurants in this region see pp309–10 and pp326–8

🄝 Medieval City Walls

A masterpiece of medieval military architecture, the huge walls run for 4 km (2.5 miles) and display 151 escutcheons of Grand Masters and Knights.

The Bourg

Close to d'Amboise gate is the restored clock tower, which has excellent views.

The Bourg's clock tower

It was built in 1852 on the site of a Byzantine tower and marks the end of the Collachium. The Bourg's labyrinth of streets begins at Sokrátous, the Golden Mile of bazaar-style shops, off which lie shady squares with pavement cafés and tavernas. The architecture is a mix of medieval, Neo-Classical and Levantine. Between the houses, with rickety wooden balconies, Ottoman mosques can be found.

Other than the major sights listed below, the Hospice of the Tongue of Italy (1392) on Kisthiníou is worth a visit, as is the Panagía tis Níkis (Our Lady of Victory). It stands near St Catherine's gate, and was built by the Knights in 1480 after the Virgin had appeared to them, inspiring victory over the Turks.

🄒 Mosque of Suleiman the Magnificent

Orféos Sokrátous. **Closed** for renovation.

The pink mosque was constructed in 1522 to commemorate the Sultan's victory over the Knights. Rebuilt in 1808, using material from the original mosque, it remains one of the town's major landmarks. Its superb, but unsafe, minaret had to be removed in 1989, and the once-mighty mosque is now crumbling. It is sadly closed to the public.

🄛 Library of Ahmet Havuz

44 Orféos. *Tel* 22410 74090. **Open** 9:30am–4pm Mon–Sat. **Closed** main public hols.

The Library of Ahmet Havuz (1793) houses the chronicle of the siege of Rhodes in 1522. This is a collection of very rare Arabic and Persian manuscripts, including beautifully illuminated 15th- and 16th-century Korans, which were restored to the library in the early 1990s, having been stolen, then rediscovered in London.

🄜 Museum of Modern Greek Art

Plateía Symi 2. *Tel* 22410 23766. **Open** 8am–2pm Tue–Sat.
🅆 mgamuseum.gr

This non-profit museum displays engravings and paintings by local artists, sculptures and historical documents providing an insight into 19th- and 20th-century Greek art.

🄜 Hammam

Plateía Aríonos. *Tel* 22410 27739. **Open** 10am–5pm Mon–Fri, 8am–5pm Sat. 🅿

The *hammam* was built by Mustapha Pasha in 1765. For decades a famous place of rest and relaxation for Eastern nobility, it is now used by Greeks, tourists and the Turkish minority. Your own soap and towels are essential, and sexes are segregated.

🄒 Mosque of Ibrahim Pasha

Plátanos. *Tel* 22410 73410. **Open** daily. 🅿 donation.

Situated off Sofokléous, the Mosque of Ibrahim Pasha was built in 1531 and refurbished in 1928. The mosque has an exquisite interior.

🄒 Mosque of Rejep Pasha

Ekátonos. **Closed** for renovation.

Built in 1588, Rejep Pasha is one of the most striking of the 14 or so mosques in the Old Town. The mosque, which has a fountain made from Byzantine and medieval church columns, contains the sarcophagus of the Pasha. The tiny Byzantine church of Agios Fanoúrios is close by.

The Jewish Quarter

East from Hippocrates Square, the Bourg embraces Ovriakí. This was the Jewish Quarter from the 1st century AD until German occupation in 1944, when the Jewish population was transported to Auschwitz.

East along Aristotélous is **Plateía Evraíon Mart'yron** (Square of the Jewish Martyrs), named for all those who perished in the concentration camps. There is a bronze sea-horse fountain in the centre, and to the north is Admiralty House, an imposing medieval building. The Synagogue is on Simíou.

The dome of the Mosque of Suleiman the Magnificent

Rhodes: Palace of the Grand Masters
Παλάτι του Μεγάλου Μαγίστρου

A fortress within a fortress, this was the seat of 19 Grand Masters, the nerve centre of the Collachium, or Knights' Quarter, and last refuge for the population in times of danger. Built in the 14th century, it survived earthquake and siege, but was blown up by an accidental explosion in 1856. It was restored by the Italians in the 1930s for Mussolini and King Victor Emmanuel III. The palace has some priceless mosaics from sites in Kos, after which some of the rooms are named. It also houses two exhibitions: Medieval Rhodes, and Ancient Rhodes *(see p188)*.

Chamber with Colonnades
Two elegant colonnades support the roof and there is a 5th-century-AD early Christian mosaic.

★ Medusa Chamber
The mythical Gorgon Medusa, with hair of writhing serpents, forms the centrepiece of this important late Hellenistic mosaic. The chamber also features Chinese and Islamic vases.

The First Grand Master

The first Grand Master, or Magnus Magister, of the Knights was Foulkes de Villaret (1305–19), a French knight. He negotiated to buy Rhodes from the Lord of the Dodecanese, Admiral Vignolo de Vignoli. This left the Knights with the task of conquering the island's inhabitants. The Knights of Rhodes *(see pp192–3)*, as they became, remained here until their expulsion in 1522. The Villaret name lives on in Villaré, one of the island's white wines.

Foulkes de Villaret

MAGNUS FRATER FULCUS 1305 MAGISTER DE VILLARET 1319

Laocoön Chamber
A copy of the sculpture of the death of the Trojan, Laocoön, and his sons dominates the hall. The 1st-century-BC original by Rhodian masters Athenodoros, Agesandros and Polydoros is in the Vatican.

VISITORS' CHECKLIST

Practical Information
Ippotón. **Tel** 22410 23359.
Open Apr–Oct: 8am–7:40pm
daily; Nov–Mar: 8am–2:40pm
Tue–Sun. **Closed** main
public holidays.
limited.

★ **Central Courtyard**
The palace is built around a courtyard paved
with geometric marble tiles. The north side is
lined with Hellenistic statues taken from the
odeion in Kos *(see p175).*

★ **Main Gate**
This imposing entrance, built by
the Knights, has twin horseshoe-
shaped towers with swallowtail
turrets. The coat of arms is that of
Grand Master del Villeneuve, who
ruled from 1319 to 1346.

Entrance

Street of the Knights
(see pp192–3)

KEY

① **First Cross-Vaulted
Chamber**

② **The Second Cross-Vaulted
Chamber**, once used as the
governor's office, is paved with
an intricately decorated, early
Christian mosaic of the 5th
century AD from Kos.

③ **Thyrsus Chamber**

④ **Chamber of the Sea Horse
and Nymph**

⑤ **The battlements** and heavy
fortifications of the palace were
to be the last line of defence in
the event of the city walls being
breached.

⑥ **Entrance to Ancient Rhodes
exhibition** *(see p188)*

⑦ **The Chamber of the Nine
Muses** has a late Hellenistic
mosaic featuring busts of the
Nine Muses of Greek myth.

⑧ **The First Chamber**, with
its 16th-century choir stalls,
features a late Hellenistic
mosaic.

⑨ **Entrance to Medieval
Rhodes exhibition** *(see p188)*

⑩ **Grand staircase**

⑪ **The Second Chamber** has
a late Hellenistic mosaic and
carved choir stalls.

Rhodes: Street of the Knights

One of the old town's most famous sights, the medieval Street of the Knights (Odós Ippotón) is situated between the harbour and the Palace of the Grand Masters *(see pp190–91)*. It is lined by the Inns of the Tongues, or nationalities, of the Order of St John. Begun in the 14th century in Gothic style, the Inns were used as meeting places for the Knights. The site of the German Inn is unknown, but the others were largely restored by the Italians in the early 20th century.

This residence was built for the head of the Tongue of Aragon, Diomede de Vilaragut.

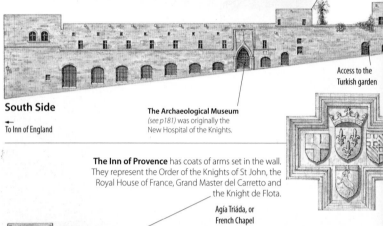

Access to the Turkish garden

South Side

To Inn of England

The Archaeological Museum *(see p181)* was originally the New Hospital of the Knights.

The Inn of Provence has coats of arms set in the wall. They represent the Order of the Knights of St John, the Royal House of France, Grand Master del Carretto and the Knight de Flota.

Agía Triáda, or French Chapel

North Side

Palace of the Grand Masters

Arched bridge connecting Inn of Spain and Inn of Provence

The Knights of Rhodes

Founded in the 11th century by merchants from Amalfi, the Order of Hospitallers of the Knights of St John guarded the Holy Sepulchre and tended Christian pilgrims in Jerusalem. They became a military order after the First Crusade (1096–9), but had to take refuge in Cyprus when Jerusalem fell in 1291. They then bought Rhodes from the Genoese pirate Admiral Vignoli in 1306, and eventually conquered the Rhodians in 1309. A Grand Master was elected for life to govern the Order, which was divided into seven Tongues, or nationalities: France, Italy, England, Germany, Provence, Spain and Auvergne. Each Tongue protected an area of city wall known as a Curtain. The Knights fortified the Dodecanese with around 30 castles, and their defences are some of the finest examples of medieval military architecture.

The Knights were drawn from noble Roman Catholic families. Those who entered the Order of the Knights of St John swore vows of chastity, obedience and poverty. Although Knights held all the major offices, there were also lay brothers.

Odós Ippotón, the Street of the Knights, lies along a section of ancient road that led all the way down to the harbour. It was here that the Knights would muster in times of attack.

Archway to Ippárchou

Archway to Láchitos

Arched bridge connecting Inn of Spain and Inn of Provence

Palace of the Grand Masters →

The Inn of Spain is one of the largest inns. Its assembly hall was over 150 sq m (1,600 sq ft). On the exterior there is a small and simple coat of arms of the Spanish Tongue.

The Inn of France's armorial bearings are the French royal fleur-de-lys, and those of Grand Master Petrus d'Amboise.

The Inn of Italy has a marble escutcheon bearing the arms of the Grand Master Fabricius del Carretto.

Palace of Grand Master Villiers de l'Isle Adam (1521–34)

Inn of Auvergne →

The Great Siege of Rhodes in 1522 resulted in the Knights being defeated by the Turks. From a garrison of 650 Knights, only 180 survived. They negotiated a safe departure, although the Rhodians who fought with them were slaughtered. Seven years later, the Knights found sanctuary on the island of Malta. Their final defeat came in 1798, when Malta was annexed by Napoleon.

Pierre d'Aubusson, Grand Master from 1476 to 1503, is featured in this market scene. He oversaw a highly productive time in terms of building in Rhodes, including completion of the Hospital (now the Archaeological Museum).

Exploring Rhodes New Town

The New Town grew steadily over the last century, and became firmly established during the Italian Fascist occupation of the 1920s with the construction of the grandiose public buildings by the harbour. The New Town is made up of a number of areas, including Néa Agora and Mandráki harbour in the eastern half of town. The Italian influence remains in these areas, with everything from pizzerias to Gucci shops. The town's west coast is a busy tourist centre, with lively streets and a crammed beach.

Government House, previously the Italian Governor's Palace

Mandráki harbour with the two statues of deer at its entrance

Mandráki Harbour

The harbour is the hub of life, the link between the Old and New towns where locals go for their evening stroll, or *vólta*. It is lined with yachts and excursion boats for which you can book trips in advance.

A bronze doe and stag guard the harbour entrance, where the Colossus was believed to have stood. The harbour sweeps round to the ruined 15th-century fortress of **Agios Nikólaos**, now a lighthouse, on the promontory past the three medieval windmills.

Elegant public buildings, built by the Italians in the 1920s, line Mandráki harbour: the post office, law courts, town hall, police station and the National Theatre all stand in a row. The **National Theatre** often shows Rhodian character plays based on folk customs.

Nearby, on Plateía Eleftherías, is the splendid church of the **Evangelismós** (Annunciation), a 1925 replica of the Knight's Church of St John, which has a lavishly decorated interior. The Archbishop's Palace is next door beside a giant fountain, which is a copy of the Fontana Grande in Viterbo, Italy. Further along, the mock-Venetian-Gothic **Government House** (Nomarchía) is ornately decorated and surrounded by fine vaulted arcades. Unfortunately there is no access for tourists or the general public.

At the north end of Plateía Eleftherías is the **Mosque of Murad Reis**, with its graceful minaret. It was named after a Turkish admiral serving under Suleiman who was killed in the 1522 siege of Rhodes. Situated in the grounds is Villa Kleoboulos, which was the home of the British writer Lawrence Durrell between 1945 and 1947. Also in the grounds is a cemetery for Ottoman notables.

Heading north from the area around Mandráki harbour, a pleasant stroll along the waterfront via the crowded Elli beach leads to the northern tip of the New Town.

The minaret o the Mosque o Murad Reis

The Hydrobiological Institute is situated on the coastal tip, housing the **Aquarium**. Set in a subterranean grotto, this is the only major aquarium in Greece, displaying nearly 40 tanks of fish. Opposite, on the north point of the island is Aquarium Beach, which is good for windsurfing and paragliding.

🔲 **Aquarium**
Hydrobiological Institute, Kássou. **Tel** 22410 27308. **Open** Apr–Oct: 9am–8:30pm daily; Nov–Mar: 9am–4:30pm daily. **Closed** main public hols. 🐾 ♿

The Colossus of Rhodes

One of the Seven Wonders of the Ancient World, the Colossus was a huge statue of Helios, the sun god, standing at 32–40 m (105–130 ft). Built in 305 BC to celebrate Rhodian victory over Demetrius, the Macedonian besieger, it was sculpted by Chares of Líndos. It took 12 years to build, using bronze from the battle weapons, and cost 9 tons (10 imperial tons) of silver. Traditionally pictured straddling Mandráki harbour, it probably stood at the Temple of Apollo, now the site of the Palace of the Grand Masters in the Old Town *(see pp186–7)*. An earthquake in 227 BC caused it to topple over.

Painting of the Colossus by Fischer von Erlach, 1700

Rhodes New Town

① Mosque of Murad Reis
② National Theatre
③ Government House
④ Evangelismós
⑤ Agios Nikólaos
⑥ Mandráki Harbour
⑦ Néa Agora

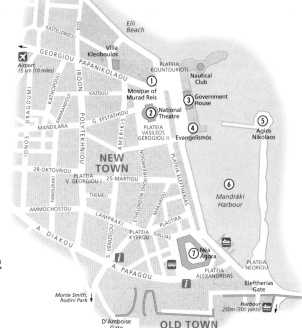

0 metres 250
0 yards 250

Néa Agora
Mandráki is backed by the New Market or Néa Agora, with its Moorish domes and lively cafés. Inside the market are food stalls, gift shops, small *souvláki* bars and cafés. It is popular as a meeting place for people coming from outlying villages

The domed centrepiece of the New Market from Mandráki Harbour

and islands. Behind the Néa Agora, in the grounds of the Palace of the Grand Masters, a sound and light show is held. This takes place daily in one of four languages and tells the story of the overthrow of the Knights by Suleiman the Magnificent in 1522.

Monte Smith
Monte Smith, a hill to the west of town, offers panoramic views over Rhodes town and the coast. It is named Monte Smith after the English Admiral Sir Sidney Smith, who kept watch from there for Napoleon's fleet in 1802. It is also known as Agios Stéphanos.
The hill is the site of a 3rd-century-BC Hellenistic city which was excavated by the Italians. They restored the 3rd-century-BC stadium, the 2nd-century-BC acropolis and a small theatre or odeion. This was built in

an unusual square shape and is used for performances of ancient drama in the summer. Only three columns remain of the once-mighty Temple of Pythian Apollo, and there are other ruins of the temples of Athena Polias and Zeus. Nearby, on Voreíou Ipeírou, are the remains of the Asklepieíon, a temple dedicated to the god of healing, Asklepios.

Rodíni Park
The beautiful Rodíni Park, 3 km (2 miles) to the south of Rhodes town, is now home to the Rhodian deer sanctuary, and perfect for a break away from the crowded centre. It is the site where the orator Aeschines built the School of Rhetoric in 330 BC, attended by both Julius Caesar and Cassius, although there are no remains to visit. Sights include a 3rd-century-BC necropolis with Doric rock tombs and several Ptolemaic, rock-cut tombs. In medieval times the Knights grew their herbs at Rodíni.

For keys to symbols *see back flap*

Exploring Western Rhodes

The windswept west coast is a busy strip of hotels, bars and restaurants, along shingly beaches from Rhodes town to the airport at Paradísi. But head south and the landscape becomes green and fertile, with vineyards and wooded mountain slopes, dotted with traditional farming villages. The attractions include Moní Filerímou, Ancient Kámeiros, the wine-making village of Emponas, and the enchanting valley of Petaloúdes, the place that gives Rhodes its name as the "Island of Butterflies". Further south is a dramatic mix of scenery with castle-topped crags and sea views to the islands of Chálki.

❷ Ancient Ialyssós

Αρχαία Ιαλυσός

15 km (9 miles) SW of Rhodes town. 🚌 to Triánda. **Open** 8am–7:40pm Tue–Sat, 8:30am–2:40pm Sun (Nov–Mar: 8:30am–2:40pm Tue–Sun). **Closed** main public hols.

Ialyssós fused with two other Doric city-states, Líndos and Kámeiros, to create one capital, Rhodes, in 408 BC. As this new centre grew, Ialyssós, Líndos and Kámeiros lost their former importance. However, Ialyssós proved a much-fought-over site: the Byzantines were besieged by the Genoese there in 1248; the Knights *(see pp192–3)* used it as a base before taking Rhodes in 1309; and it was Suleiman's headquarters before his assault on the Knights in 1522. The Italians used it again for gun positions during World War II.

The only remnant of the acropolis is the 3rd-century-BC **Temple of Athena Polias and Zeus Poliefs** by the church of Agios Geórgios. The restored lion-head fountain, to the south, is 4th century BC.

❸ Moní Filerímou

Μονή Φιλερήμου

15 km (9 miles) SW of Rhodes town. **Tel** 22410 92202. 🚌 to Triánda. **Open** 8am–7:40pm Tue–Sat, 8:30am–2:40pm Sun. 🅿️

One of Rhodes' beauty spots, the hillsides of Filérimos are home to cypresses and pines. Among the trees sits Moní Filerímou, its domed chapels decorated with the cross of the Knights and the coat of arms of Grand Master Pierre d'Aubusson. A place of worship for 2,000 years, layers of history and traditions can be seen, from Phoenician to Byzantine, Orthodox and Catholic.

The main attraction is Our Lady of Filérimos, the Italian reconstruction of the Knights' 14th-century church of the Virgin Mary. It is a complex of four chapels: the main one, built in 1306, leads to three others. The innermost chapel has a Byzantine floor decorated with a red mosaic fish.

The Italians erected a Calvary, from the entrance of the

❹ Petaloúdes

Πεταλούδες

26 km (16 miles) SW of Rhodes town. 🚌 **Tel** 22410 91998

Petaloúdes, or Butterfly Valley, is a narrow leafy valley with a stream crisscrossed by wooden bridges. It teems, not with butterflies, but with Jersey tiger moths from June to September. Thousands are attracted by the golden resin of the storax trees, which exude vanilla-scented gum used for incense. Cool and pleasant, Petaloúdes attracts walkers as well as lepidopterists, and is at its most peaceful in the early morning before all the tour buses arrive.

There is a walk along the valley to the **Moní Panagías Kalópetras**. This rural church, built in 1782, is a tranquil resting place, and the fine views are well worth the climb.

Jersey tiger moth

❺ Ancient Kámeiros

Αρχαία Κάμειρος

36 km (22 miles) SW of Rhodes town. **Tel** 22410 40037. 🚌 **Open** Apr–Oct: 8am–7:40pm Tue–Sat, 8am–2:40pm Sun (Nov–Mar: 8am–2:40pm Tue–Sun). **Closed** main public hols. 🅿️ 🅰️ to lower sections only.

Discovered in 1859, this Doric city thrived in the 5th century BC. Founded by Althaemenes of Crete, it was probably destroyed in a large earthquake in 142 BC, yet it remains one of the best-preserved Classical Greek cities.

There are remains of a 3rd-century-BC Doric temple, an altar to Helios, public baths and a 6th-century-BC cistern, which supplied 400 families. The 6th-century-BC Temple of Athena Polias is on the top terrace, below which are remains of the Doric stoa, 206 m- (675 ft-) long.

Moní Filerímou in its woodland setting

Monólithos castle in its precarious position overlooking the sea

⑥ Skála Kameírou

Σκάλα Καμείρου

50 km (30 miles) SW of Rhodes town. 🏔 100. 🚌

The fishing harbour of Skála Kameírou makes a good place for lunch. It was the Doric city of an ancient port, and the outline of a Lycian tomb remains on the cliff side. Nearby, **Kritiniá castle** is one of the Knights' more impressive ruins. Its three levels are attributed to different Grand Masters. Clinging to the hillside, a cluster of white houses form the picturesque village of **Kritiniá.**

⑦ Emponas

Έμπωνας

55 km (34 miles) SW of Rhodes town. 🏔 1,500. 🚌

Situated in the wild foothills of **Mount Attávyros**, the atmospheric village of Emponas has been home

to the Cair winery since the 1920s and is also famous for its folk dancing and festivals. Although the village is popular for organized Greek nights, Emponas has maintained its traditional ways.

⑧ Monólithos

Μονόλιθος

80 km (50 miles) SW of Rhodes town. 🏔 250. 🚌 🚌 Foúrni 5 km (3 miles) SW.

Named after its monolith, a crag with a dramatic 235 m- (770ft-) drop to the sea, Monólithos is the most important village in the southwest.

Situated at the foot of Mount Akramýtis, the village is 2 km (1 mile) from **Monólithos castle**. This impregnable 15th-century fortress, built by Grand Master d'Aubusson, is perched spectacularly on the vast grey rock. Its massive walls enclose two small 15th-century chapels, Agios Panteleïmon and Agios

Geórgios, both decorated with frescoes. Views from the top are impressive.

Down a rough road south from the castle is the sheltered sandy beach of **Foúrni**, which has a seasonal taverna.

Environs
Between Emponas and Monólithos, the pretty hill village of **Siána** is famous for its honey and fiery *soúma* – a kind of grape spirit, like the Cretan raki. The villagers were granted a licence by the Italians to make the spirit, and you can sample both the firewater and honey at the roadside cafés. The village houses have traditional clay roofs, and the domed church of **Agios Panteleïmon** has restored 18th-century frescoes.

⑨ Moní Skiádi

Μονή Σκιάδι

8 km (5 miles) S of Apolakkiá. 🚌 to Apolakkiá. **Open** daily. ♿

Moní Skiádi is famous for its miraculous icon of the Panagía or the Blessed Virgin. When a 15th-century heretic stabbed the Virgin's cheek, it was supposed to have bled, and the brown stains are still visible. The present monastery was built during the 18th and 19th centuries around the 13th-century church of Agios Stavrós, or the Holy Cross. At Easter the holy icon is carried from village to village until finally coming to rest for a month on the island of Chálki.

Sunset over the village of Emponas and Mount Attávyros

For hotels and restaurants in this region see pp309–10 and pp326–8

Exploring Eastern Rhodes

The sheltered east coast has miles of beaches and rocky coves, the crowded holiday playgrounds of Faliráki and Líndos contrasting with the deserted sands in the southeast. For sightseeing purposes the way east divides into two sections: from the southern tip of the island at Prasonísi up to Péfkoi, and then from Líndos up to Rhodes town. The landscape is a rich patchwork, from the oasis of Eptá Pigés and the orange groves near Archángelos, to the stretches of rugged coastline and sandy bays.

⑩ Moní Thárri

Μονή Θάρρι

40 km (25 miles) S of Rhodes town.
🚌 to Laérma. **Open** daily.

From the inland resort of Lárdos follow signs to Láerma, which is just north of Moní Thárri, famous for its 12th-century frescoes. Reached through a forest, the domed church was hidden from view in order to escape the attention of marauding pirates.

According to legend, it was built in the 9th century by a mortally ill Byzantine princess, who miraculously recovered when it was completed.

The 12th-century north and south walls remain, and there are vestiges of the 9th-century building in the grounds. The nave, apse and dome are covered with frescoes. Some walls have four layers of paintings, the earliest dating as far back as 1100, while there are three layers in the apse dating from the 12th–16th

A friendly resident in the pretty village of Asklipieío

centuries. These are more distinct, and depict a group of prophets and a horse's head. The monastery has been extended and has basic accommodation for visitors.

About 8 km (5 miles) south along a rough track is the pleasant village of **Asklipieío**, with the frescoed church of Kímisis tis Theotókou.

⑪ Líndos

See pp200–01.

The rooftops of Archángelos village in the Valley of Aíthona

⑫ Archángelos

Αρχάγγελος

33 km (20 miles) S of Rhodes town.
🚍 3,000. 🚌 🚐 Stégna 3 km (2 miles) E.

The island's largest village, Archángelos lies in the Valley of Aíthona, which is renowned for its oranges. The town itself is famous for its colourful pottery, which is produced using time-honoured methods and features traditional Rhodian motifs. Pottery has been one of the main sources of income for generations of villagers.

The townspeople have their own dialect and are fiercely patriotic – some graves are even painted blue and white.

In the centre, the church of **Archángeloi Michaïl and Gavriíl**, the village's patron saints, is distinguished by a tiered bell tower and pebble-mosaic courtyard.

Above the town are ruins of the **Crusader castle**, built by Grand Master Orsini in 1467 as part of the Knights' defences against the Turks. Inside, the chapel of Agios Geórgios has a modern fresco of the saint in action against the dragon. To the east of the town lies the bay of **Stégna**, a quiet and sheltered stretch of sand.

Environs
South past Malónas is the castle of **Faraklós**. It was a pirate stronghold before the Knights saw them off and turned it into a prison. The fortress overlooks Charáki, a pleasant fishing hamlet, now growing into a holiday resort, with a pebble beach that is lined with fish tavernas.

⑬ Eptá Pigés

Επτά Πηγές

26 km (16 miles) S of Rhodes town.
🚌 to Kolýmpia. 🚐 Tsampíka 5 km (3 miles) SE.

Eptá Pigés, or Seven Springs, is one of the island's leading woodland beauty spots. Peacocks strut beside streams

The sandy beach at Tsampíka

and waterfalls, where the seven springs feed a central reservoir. The springs were harnessed to irrigate the orange groves of Kolýmpia to the east. The lake can be reached either by a woodland trail, or you can shuffle ankle-deep in water through a 185 m (605 ft) tunnel. This quiet spot is home to a restaurant serving traditional Greek fare. Diners sit at wooden tables in the shadow of plane and pine trees.

Environs
Further east along the coast, the Byzantine **Moní Tsampíkas** sits on a mountain-top at 300 m (985 ft). Legend has it that the 11th-century icon in the chapel was found by an infertile couple, who later conceived a child. The chapel hence became a place of pilgrimage for childless women to come to pray to the icon of the Virgin. They also pledge to name their child Tsampíka or Tsampíkos, names unique to the Dodecanese.

Below the monastery lies **Tsampíka** beach, a superb stretch of sand that becomes very crowded in the tourist season. Various watersports, such as jet- and water-skiing, along with eateries, are also available here.

⓮ Faliráki
Φαληράκι

15 km (9 miles) S of Rhodes town.
🏠 400. 🚌

Faliráki, one of the island's most popular resorts, consists of long sandy beaches surrounded by whitewashed hotels, holiday apartments and restaurants. Also a good base for families who like a lively holiday with plenty of activities, it is a brash and loud resort that caters mostly for a younger crowd. As well as a huge waterside complex, **Faliráki Water Park**, there are all types of watersports to enjoy. There are bars and discos, and numerous places to eat, from fish and chips to Chinese. Other diversions include bungee jumping.

Peacock at Eptá Pigés

🎡 Faliráki Water Park
Faliráki. **Tel** 22410 84403.
Open May–Oct: 9:30am–6pm daily (Jun–Aug: to 7pm). 🏊

Environs
Slightly inland, the village of **Kalythiés** offers a more traditional break. Its attractive Byzantine church, **Agía Eleoúsa**, contains some interesting frescoes. Further southeast, rocky **Ladikó Bay** is worth a visit. It was used as a location for filming *The Guns of Navarone*.

Golfers can visit the 18-hole course at **Afántou** village, with its pebbly coves and beaches, popular for boat trips from Rhodes town. Set in apricot orchards, Afántou means the "hidden village", and it is noted for its hand-woven carpets.

⓯ Koskinoú
Κοσκινού

10 km (6 miles) S of Rhodes town.
🏠 1,200. 🚌 🚉 Réni Koskinoú 2 km (1 mile) NE.

The old village of Koskinoú is characterized by its traditional Rhodian houses featuring the *choklákia* pebble mosaic floors and courtyards. There is an attractive church of **Eisódia tis Theotókou**, which has a multi-tiered bell tower. Nearby, **Réni Koskinoú** has good hotels, restaurants and beaches.

Environs
South of Koskinoú lies **Thérmes Kalithéas**, Kalithea Spa, once frequented for its healing waters. Though no longer in use, the site is used in films and offers visitors a unique combination of nature, architecture and history. The spa is set in lovely gardens, reached through pinewoods. There is now a busy lido here, and the rocky coves are popular for scuba-diving and snorkelling.

A church with a tiered bell tower in Koskinoú village

⓫ Líndos
Λίνδος

Líndos was first inhabited around 3000 BC. Its twin harbours gave it a head start over Rhodes' other ancient cities of Kámeiros and Ialyssós as a naval power. In the 6th century BC, under the benevolent tyrant Kleoboulos, Líndos thrived and grew rich from its many foreign colonies. With its dazzling white houses, Crusader castle and acropolis dramatically overlooking the sea, Líndos is a magnet for tourists. Second only to Rhodes town as a holiday resort, it is now a National Historic Landmark, with development strictly controlled.

A traditional Líndian doorway

Exploring Líndos Village

Líndos is the most popular excursion from Rhodes town, and the best way to arrive is by boat. The narrow cobbled streets can be shoulder to shoulder with tourists in high summer, so spring or autumn are more relaxed times to visit. Líndos is a sun trap, and is known for consistently recording the highest temperatures on the island. In winter, the town is almost completely deserted.

Traffic is banned, so the village retains much of its charm, and donkeys carry people up to the acropolis (be warned that they proceed rather quickly downhill). It is very busy, with a bazaar of gift shops and fast-food outlets. Happily there are also several

Líndos lace seller on the steps to the acropolis

good tavernas and, at the other end of the scale, there are a number of stylish restaurants offering international cuisine. Some quiet, romantic little places can be found, with views of the bay and the sea.

The village's winding lanes are fronted by imposing doorways which lead into the flower-filled courtyards of the unique Líndian houses. Mainly built by rich sea-captains between the 15th and 18th centuries, these traditional houses are called *archontiká*. They have distinctive carvings on the stonework, like ship's cables or chains (the number of chains supposedly corresponds to the number of ships owned), and are built round *choklákia* pebble mosaic courtyards *(see p202)*. A few of them are open to the public for viewing. The older houses mix Byzantine and Arabic styles and a few have small captain's rooms built over the doorway. Some of the *archontiká* have been converted into apartments and restaurants.

In the centre of the village lies the Byzantine church of the **Panagía**, complete with its graceful bell tower and pantiled domes. Originally a 10th-century basilica, it was rebuilt beween 1489 and 1490. The frescoes inside were painted by Gregory of Sými in 1779.

On the path leading to the acropolis, are a number of women selling the lace for which Líndos is renowned. Lindian stitchwork is sought after by

Líndos Stoa
This colonnade or stoa was built in the Hellenistic period around 200 BC.

The battlements were built in the 13th century by the Knights of Rhodes.

A trireme warship is carved into the rock.

The Acropolis at Lindos

Perched on a sheer precipice 125 m (410 ft) above the village, the acropolis is crowned by the 4th-century-BC Temple of Lindian Athena, its remaining columns etched against the skyline. The temple was among the most sacred sites in the ancient world, visited by Alexander the Great and supposedly by Helen of Troy and Herakles. In the 13th century, the Knights Hospitallers of St John fortified the city with battlements much higher than the original walls.

The acropolis overlooking Líndos town and bay

museums throughout the world; it is said that even Alexander the Great wore a cloak stitched by Lindian women. The main beach at Líndos, **Megálos Gialós**, is where the Líndian fleet once anchored, and it sweeps north of the village round Líndos bay. It is a popular beach and it tends to get very crowded in summer, but a wide selection of watersports are available. It is also safe for children, and several tavernas can be found along the beachfront.

Environs
Tiny, trendy **Pallás** beach is linked to Líndos's main beach by a walkway. Nudists make for the headland, around which is the more exclusive **St Paul's Bay**, where the apostle landed in AD 43, bringing Christianity to Rhodes. An idyllic, almost enclosed cove, it has azure waters and a white chapel dedicated to St Paul, with a festival on 28 June.

Although called the **Tomb of Kleoboulos**, the stone monument on the promontory north of the main beach at Líndos bay had nothing to do with the great Rhodian tyrant. The circular mausoleum was constructed around the 1st century BC, several centuries after his death. In early Christian times the tomb was converted into the church of Agios Aimilianós, though who was originally buried here still remains a mystery.

Péfkos, 3 km (2 miles) south of Líndos, has small sandy beaches fringed by pine trees, and is fast developing as a popular resort.

Lárdos is a quiet inland village, 7 km (4 miles) west of Líndos. **Lárdos Bay**, 1 km (0.5 mile) south of the village, has sand dunes bordered by reeds, and is being developed with upmarket village-style hotels designed to blend in with the landscape.

Vaulted structures support the terrace.

The Doric stoa was built in the 3rd century BC.

Temple of Lindian Athena, 4th century BC

Agios Ioánnis, the church of St John, was built in the 13th century.

Medieval entrance to the acropolis

The palace of the commander of the fortress was added in the period of the Knights.

Roman temple of Diocletian, 3rd century AD

Reconstruction of the Acropolis (c. AD 300)

Temple of Lindian Athena

Propylaia

Doric stoa

Nimporió with Agios Nikólaos church towering above the surrounding buildings

Chálki
Χάλκη

📇 280. 🚢 Nimporió. 🛈 Piátsa, Nimporió (22460 45207). 🎪 Chorió: Panagía 15 Aug. 🚍 Nimporió. 🌐 chalki.gr

Chálki was once a thriving sponge-fishing island, but was virtually abandoned when its sponge divers emigrated to Florida in search of work in the early 1900s. Tourism has grown steadily as the island has been smartened up. Once fertile, Chálki's water table was infiltrated by sea water and the island is now barren with fresh water shipped in by tanker. Sheep and goats roam the rocky hillside, there is little cultivation and produce is imported from Rhodes.

Nimporió

Chálki's harbour and only settlement, Nimporió is a quiet and picturesque village with a Neo-Classical flavour. The main

A goat farmer in Chálki on his journey home

sight in Nimporió is the church of **Agios Nikólaos** with its elegant bell tower, the highest in the Dodecanese, tiered like a wedding cake. The church is also known for its magnificent black and white *choklákia* pebble mosaic courtyard depicting birds and the tree of life. The watchful eye painted over the main door is to ward off evil spirits.

A row of ruined windmills stands above the harbour, which also boasts an Italianate

town hall and post office plus a fine stone clock tower. Nearby is sandy **Póntamos** beach, which is quiet and shallow and suitable for children.

Around the Island

The island is almost traffic-free, so it is ideal for walkers. An hour's walk uphill from Nimporió is the abandoned former capital of **Chorió**. Its Crusader castle perches high on a crag, worth a visit for the coat of arms and Byzantine frescoes

Choklákia Mosaics

A distinctive characteristic of the Dodecanese, these decorative mosaics were used for floors from Byzantine times onwards. An exquisite art form as well as a functional piece of architecture, they were made from small sea pebbles, usually black and white but occasionally reddish, wedged together to form a kaleidoscope of raised patterns. Kept wet, the mosaics also helped to keep houses cool in the heat.

Early examples featured abstract, formal and mainly geometric designs such as circles. Later on the decorations became more flamboyant, with floral patterns and symbols depicting the lives of the householders with ships, fish and trees. Aside from Chálki, the houses of Líndos also have fine mosaics (*see pp200–201*). On Sými the church of Agios Geórgios (*see p182*) depicts a furious mermaid about to dash a ship beneath the waves.

A *choklákia* mosaic outside Moní Taxiárchi in Sými

Circular *choklákia* mosaic in Chálki

in the ruined chapel. On a clear day you can see Crete. The Knights of St John (see pp192–3) built it on an ancient acropolis, using much of the earlier stone.

The Byzantine church of the **Panagía** below the castle has some interesting frescoes and is the centre for a giant festival on 15 August. Clinging to the mountainside opposite is the church of **Stavrós** (the Cross).

From Chorió you can follow the road west to the Byzantine **Moní Agíou Ioánnou Prodrómou** (St John the Baptist). The walk takes about 3–5 hours, or it is a 1-hour drive. The monastery has an attractive shaded courtyard. It is best to visit in the early morning or to stay overnight: the caretakers will offer you a cell. You can walk from Nimporió to the pebbly beaches of **Kánia** and **Dyó Gialí** or take a taxi boat.

Outlying Islands

Excursions run east from Nimporió to deserted **Alimiá** island, where Italy berthed some submarines in World War II. There are several small chapels and a ruined castle.

The interior of Moní Agíou Ioánnou Prodrómou

Kastellórizo
Καστελλόριζο

275. ✈ 2.5 km (1.5 miles) S of Kastellórizo town. ⛴ Kastellórizo town. ℹ 500 m (1,640 ft) N of port (22460 49269).

Remote Kastellórizo is the most far-flung Greek island, just 2.5 km (1.5 miles) from Turkey but 120 km (75 miles) from Rhodes. It was very isolated until the

airport opened up tourism in 1987. Kastellórizo has no beaches, but clear seas full of marine life, including monk seals, and it is excellent for snorkelling. Known locally as Megísti (the biggest), it is the largest of 14 islets.

The island's population has declined from 15,000 in the 19th century to nearly 300 today. From 1920 it was severely oppressed by the Italians who occupied the Dodecanese, and in World War II it was evacuated and looted.

Despite hardships, the waterside bustles with tavernas and sometimes impromptu music and dancing. It is a strange backwater, but the indomitable character of the islanders is famous through-out Greece.

Kastellórizo town is the island's only settlement, with reputedly the best natural harbour between Piraeus and Beirut. Above the town is the ruined fort or **kástro** with spectacular views over the islands and the coast of Turkey. It was named the Red Castle (Kastello Rosso) by the Knights of St John due to its red stone, and this name was adopted by the islanders. The **Castle Museum** contains costumes, frescoes and photographs. Nearby, cut into the rock, is Greece's only **Lycian Tomb**, from the ancient Lycian civilization of Asia Minor. It is noted for its Doric columns.

Most of the old Neo-Classical houses stand in ruins, blown up during World War II or destroyed by earthquakes.

A traditional housefront in Kastellórizo town

However, many buildings have been restored thanks to the rise in tourism. The Italian film *Mediterraneo* was set here and since then the island has attracted many Italian tourists.

Highlights worth seeing include the elegant cathedral of **Agioi Konstantínos kai Eléni**, incorporating granite columns from the Temple of Apollo in Patara, Anatolia.

From town a path leads up to four white churches and the **Palaiókastro**. This Doric fortress and acropolis has a 3rd-century-BC inscription on the gate referring to Megísti.

A boat trip southeast from Kastellórizo town to the spec-tacular **Parastá Cave** should not be missed; it is famed for its stalactites and the strange light effects on the vivid blue waters.

🏛 Castle Museum
Kastellórizo town. **Tel** 22460 49269. **Open** 8:30am–3pm Tue–Sun.

Kastellórizo town with Turkey in the background

Kárpathos
Κάρπαθος

Wild, rugged Kárpathos is the third-largest island in the Dodecanese. Dramatically beautiful, it has remained largely unspoiled despite increasing tourism. Like most of the Dodecanese, it has had a chequered history, including periods of domination by both the Romans and Byzantines. Once known as Porfiris, after the red dye that is manufactured locally, the island's name today is thought to derive from the word *arpaktós* (robbery), as the island was a popular pirate lair in medieval times.

Kárpathos Town
Kárpathos town, also known as Pigádia, is the island's main port and capital, sheltered in the southeast of Vróntis bay. Once an ordinary working town, it now has hotels strung out all around the bay. The waterfront is bustling with cafés and restaurants that serve international fare. Opposite the Italianate town hall, **Kárpathos park** has an open-air display of ancient objects. Exhibits include an early Christian marble font and objects discovered in 5th-century-BC Mycenaean tombs on the island.

Environs
South of Kárpathos town there is a pretty walk through olive groves to the main resort of **Amoopí**, 7 km (4 miles) away, with its white-washed houses, coves and a string of sandy beaches. Above Amoopí, the village of **Menetés**, nestling at 350 m (1,150 ft) on the slopes of Mount Profítis Ilías, has quaint vine-covered streets. The traditional pastel-coloured houses have attractive courtyards and gardens. The village boasts two churches, the imposing church of the Assumption of the Virgin, and the Byzantine church of Agíos Mamás.

Around the Island
A mountainous spine divides the wild north from the softer, fertile south. On the west coast, 8 km (5 miles) from Menetés, the village of **Arkása** has been transformed into a resort. In 1923, the 4th-century church of Agía Anastasía was discovered. It contained some fine early Byzantine mosaics, the best of which depicts two deer gazing into a water jug, now in the Rhodes Archaeological Museum *(see p188)*.

Apéri, 8 km (5 miles) north of Kárpathos town, was the island's capital until 1892, and is said to be one of the richest villages in Greece. It sits 300 m (985 ft) up Mount Kalí Límni and has fountains and fine houses with exquisite gardens dating from the 1800s.

Sariá

Chálki, Rhodes

Vroukoúnda
Avlóna
Olympos
Diafáni

KARPATHOS

Apélla
Kalí Límni 685 m (2,250 ft)
Lefkós
Kyrá Panagiá
Apéri
Othos
Kárpathos Town
Menetés
Arkása
Amoopí
Profítis Ilías 510 m (1,670 ft)
Crete (Sitéia)

0 kilometres 5
0 miles 3

Armáthia
Agía Marína
Erý
Kásos
Crete (Sitéia, Irákleio)
Chélathros Bay

The white mansions of Apéri, clustered on the hillside

◀ Kastellórizo town, on the island of the same name

For keys to symbols *see back flap*

Windmills in the traditional village of Olympos

VISITORS' CHECKLIST

Practical Information
6,500. Kárpathos town (22450 22222). Panagía at Olympos: 15 Aug.

Transport
17 km (11 miles) S of Kárpathos town. Kárpathos town, Diafáni. corner of 28th Oktovríou & Dimokratías, Kárpathos town.

Othos, just to the west of Apéri, is the highest village on the island, at 450 m (1,500 ft) above sea level. It is also one of the oldest, with traditional Karpathian houses. One of the houses is a **Folk Museum** with textiles and pottery on show. There is also a family loom and tools for traditional crafts.

The west coast resort of **Lefkós** is considered to be the jewel of the island by the Karpathians, with its three horseshoe bays of white sand. On the east coast, **Kyrá Panagiá**, with its pink-domed church, is another beautiful cove of fine white sand. **Apélla**, the next beach along, is a stunning crescent of sand with azure water.

Diafáni, a small, colourful village on the northeast coast, has a handful of tavernas and hotels and both sand and shingle beaches. A 20-minute bus-ride away is the village of **Olympos**, which spills down from a bleak ridge 600 m (1,950 ft) up. Founded in 1420, and virtually cut off from the rest of the island for centuries by its remote location, this village is now a strange mix of medieval and modern. The painted houses huddle together in a maze of steps and alleys just wide enough for mules. One traditional house, with just a single room containing many embroideries and bric-a-brac, is open to visitors. Customs and village life are carefully preserved and traditional dress is daily wear for the older women who still bake their bread in outdoor ovens.

From Olympos a rough track leads north to **Avlóna**, inhabited only in the harvest season by local farmers. From here, **Vrouk-oúnda**, the site of a 6th-century-BC city, is a short walk away. Remains of the protective city walls can be seen, as can burial chambers cut into the cliffs.

Folk Museum
Othos village. **Tel** 22460 49283.
Open Apr –Oct: Tue–Sun.
Closed Nov–Mar.

Outlying Islands
North of Avlóna is the island of **Sariá**, site of ancient Nísyros,

where the ruins of the ancient city can be seen. Excursion boats go there from Diafáni.

Barely touched by tourism, **Kásos**, off the south coast of Kárpathos, was the site of a massacre by the Turks in 1824, commemorated annually on 7 June in the capital, Frý. Near the village of Agía Marína are two fine caves, Ellinokamára and Sellái, both with stalactites and stalagmites. Chélathros Bay is ideal for sun-lovers, as are the quiet beaches of the tiny offshore islet of **Armáthia**.

The Traditions of Olympos

The traditional costume of Olympos women consists of white pantaloons with an embroidered tunic or a dark skirt with a long patterned apron. Fabrics are heavily embroidered in lime green, silver and bright pinks. Daughters wear a collar of gold coins and chains to indicate their status. The society was once strictly matriarchal. Today the mother passes on her property to the first-born daughter and the father to his son, ensuring that the personal fortunes of each parent are preserved through the generations.

Matriarch at the Olympos windmills

Traditional houses in Olympos often have decorative balconies and the initials of the owners sculpted above the entrance. Consisting of one room built around a central pillar with fold-away bedding, they are full of photographs and souvenirs. People flock to Olympos from all over the world for the Festival of the Assumption of the Virgin Mary, from 15 August, one of the most important festivals in the Orthodox church. The village celebrations of music and dance last three days. Traditional instruments are played, including the *lýra*, which stems from the ancient lyre, the bagpipe-like goatskin *tsampoúrás*, and the *laoúto*, which is similar to a mandolin.

Interior of an Olympos house

THE CYCLADES

ANDROS · TINOS · MYKONOS · DELOS · SYROS · KEA · KYTHNOS
· SERIFOS · SIFNOS · PAROS · NAXOS · AMORGOSIOS · SIKINOS
· FOLEGANDROS · MILOS · SANTORINI

Deriving their name from the word *kyklos*, meaning "circle", because they surround the sacred island of Delos, the Cyclades are the most visited island group. They are everyone's Greek island ideal, with their dazzling white houses, twisting cobbled alleyways, blue-domed churches, hilltop windmills and stunning beaches.

The islands were the cradle of the Cycladic civilization (3000–1000 BC). The early Cycladic culture developed in the Bronze Age and has inspired artists ever since with its white marble figurines. The Minoans from Crete colonized the islands during the middle Cycladic era, making Akrotíri on Santoríni a major trading centre. During the late Cycladic period the Mycenaeans dominated, and Delos became their religious capital. The Dorians invaded the islands in the 11th century BC, a calamity that marked the start of the Dark Ages.

Venetian rule (1204–1453) had a strong influence, evident today in the medieval kástra seen on many islands and the Catholic communities on Tínos, Náxos and Sýros.

There are 56 islands in the group, 24 inhabited, some tiny and undisturbed, others famous holiday playgrounds. They are the ultimate islands for sun, sea and sand holidays, with good nightlife on Mýkonos and Ios. Sýros, the regional and commercial capital, is one of the few islands in the group where tourism is not the mainstay. Cycladic life is generally centred on the village, which is typically divided between the harbour and the upper village, or chóra, often topped with a kástro.

Most of the Cyclades are rocky and arid, with the exceptions of wooded and lush-valleyed Andros, Kéa and Náxos. This variety ensures the islands are popular with artists, walkers and those seeking quiet relaxation.

The sandy cove of Kolympíthres beach, Páros

◄ Main walkway between the two neighbourhoods of Chóra, Sérifos

Exploring the Cyclades

The Cyclades are best known for their beaches and whitewashed clifftop villages with stunning views; most famously, Firá on Santoríni. Mýkonos and Íos are well-established beach destinations, while more remote islands such as Mílos and Amorgós also have beautiful stretches of sand. Packed in July and August, these usually arid islands are beautiful in spring, when they are carpeted with wild flowers. Varying in character, some of the islands, such as Síkinos, are quiet and traditional, whereas others, such as Íos, are more nightlife-orientated. The Cyclades also offer a rich ancient history, evident in the ruins of ancient Delos.

Getting Around

Páros and Sýros are the travel hub of the Cyclades. Ferries serve most of the islands from here and link to Crete and the Dodecanese. The islands are buffeted by the strong *meltémi* wind from July to September. It provides relief from the heat, but can play havoc with ferry timetables. Mýkonos and Santoríni have international airports, and islands with domestic airports include Sýros, Mílos, Páros and Náxos.

Andr

Gávrio

Skiáthos, Rafína

Mpatsí   Arnás

Palaiópoloi   M

Lávrio

Korissía

Ioulis

Kéa   *Gyáros*

Poliessa

Piraeus

Sýros

Ermoúpol

Loutrá

Chóra

Mérichas   *Kýthnos*

Poseidonía

Kanála

Piraeus

Galaní

Sérifos   Chóra

Livádi

Piraeus

Kamáres   Kástro

Apollonía   *Desp*

Sífnos

Platýs Gialós

Kímolos

Antímilos   Psathi

Pláka   Apollonía   *Polýaigos*

Mílos   Adámas

Zefyría

Provatás

Crete

Folégandros

Karavostá

Folégand

Key

===== Minor road

——— Scenic route

----- High-season, direct ferry route

Boat garages in Mandrákia, Mílos

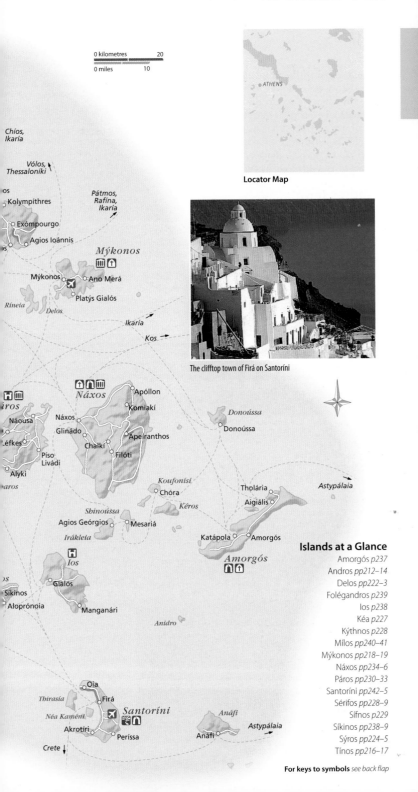

0 kilometres 20
0 miles 10

Chios, Ikaría

Vólos, Thessaloníki

Kolympíthres

Exómpourgo

Agios Ioánnis

Pátmos, Rafína, Ikaría

Locator Map

ATHENS

Mýkonos

Mýkonos

Ano Merá

Platýs Gialós

Ríneia

Delos

Ikaría

Kos

The clifftop town of Firá on Santoríni

Náxos

Apóllon

Komiakí

Náxos

Glinádo

Apeíranthos

Chalkí

Filóti

Donoússa

Donoússa

Náousa

Léfkes

Píso Livádi

Alykí

Koufonísi

Chóra

Kéros

Tholária

Aigiális

Astypálaia

Schinoússa

Agios Geórgios

Mesariá

Irákleia

Katápola

Amorgós

Amorgós

Ios

Gialós

Síkinos

Aloprónoia

Manganári

Anidro

Oía

Firá

Thirasía

Santoríni

Anáfi

Néa Kaméni

Akrotíri

Períssa

Anáfi

Astypálaia

Crete

Islands at a Glance

For keys to symbols *see back flap*

Andros
'Ανδρος

The northernmost of the Cyclades, Andros is lush and green in the south, scorched and barren in the north. The fields are divided by distinctive dry-stone walls. The island was first colonized by the Ionians in 1000 BC. In the 5th century BC, Andros sided with Sparta during the Peloponnesian War *(see p36)*. After Venetian rule, the Turks took power in 1566 until the War of Independence. Andros has long been the holiday haunt of wealthy Athenian shipping families.

❶ Andros Town
Χώρα

🏠 1,680. 🚌 Plateía Agías Olgas. 🛈 22820 22275.

The capital, Andros town, or Chóra, is located on the east coast of the island 20 km (12 miles) from the island's main port at Gávrio.

An elegant town with magnificent Neo-Classical buildings, it is the home of some of Greece's wealthiest shipowners. The pedestrianized main street is paved with marble slabs and lined with old mansions converted into public offices among the *kafeneía* and small shops.

The Hermes of Andros, in the Archaeological Museum

include the *Matron of Herculaneum*, which was found with the *Hermes*, and finds from the 10th-century-BC city at Zagorá. There are also finds from Ancient Palaiópoli *(see p214)* near Mpatsí, architectural illustrations and a large collection of ceramics.

The **Museum of Modern Art**, which was endowed by the Goulandrís family, has an excellent collection of paintings by 20th-century artists such as Picasso and Braque and leading Greek artists such as Alékos Fasianós. The sculpture garden has works by Michális Tómpros (1889–1974).

Plateía Kaïri
This is the main square in the town's Ríva district and is home to the **Archaeological Museum**, built in 1981. The museum's most famous exhibit is the 2nd-century-BC *Hermes of Andros*, a fine marble copy of the 4th-century-BC bronze original. Other exhibits

🏛 **Archaeological Museum**
Plateía Kaïri. **Tel** 22820 22820. **Open** Tue–Sun. **Closed** main public hols; Tue–Thu in winter. ♿

🏛 **Museum of Modern Art**
Plateía Kaïri. **Tel** 22820 22444. **Open** 8:30am–3pm daily. **Closed** main public hols. ♿ ♿

Káto Kástro and Plateía Ríva
From Plateía Kaïri an archway leads into the maze of streets that form the medieval city, Káto Kástro, wedged between Parapórti and Nimorió bays. The narrow lanes lead to wind-swept Plateía Ríva at the end of the peninsula, jutting into the sea and dominated by the heroic statue of the *Unknown Sailor* by Michális Tómpros. Just below, a precarious stone bridge leads to the islet opposite, with the Venetian castle, **Mésa Kástro**, built in 1207–33.

The **Maritime Museum** has model ships, photographs and a collection of nautical instruments on display, is situated inside the town hall.

On the way back to the centre of the town is the church of **Panagía Theosképasti**, built in 1555 and dedicated to the Virgin Mary. Legend has it that the priest could not afford the wood for the church roof, so the ship delivering the wood set sail again. It ran into a storm and the crew prayed to the Virgin for help, promising to return the cargo to Andros. The seas were miraculously calmed and the church became known as Theosképasti, meaning "sheltered by God".

Statue of the *Unknown Sailor*

🏛 **Maritime Museum**
Plateía Ríva. **Tel** 22823 22275. **Open** 8:30am–3pm daily. **Closed** main public hols. ♿

Environs
Steniés, 6 km (4 miles) northwest of Andros town, is very beautiful and popular with wealthy shipping families. Fifteen minutes' walk southwest of Steniés, the 17th-century Mpístis-Mouvelás tower is a fine example of an Andriot house.

Below Steniés lies **Giália** beach, where there is a fish taverna and trees for shade. In **Apoíkia**, 3 km (2 miles) west, mineral water is bottled from the Sáriza spring. You can taste the waters at the spring.

Typical white houses and a small church in Káto Kástro

Around Andros Island

Prosperous, neat and dotted with many white dovecotes first built by the Venetians, Andros retains its traditional charm while playing host to international holiday-makers. There are a number of unspoiled sandy beaches, watersport facilities, wild mountains and a good network of footpaths However,unless you are a keen trekker, car or bike hire is essential, as the bus service is quite limited.

❷ Mesariá

Μεσαριά

8 km (5 miles) SW of Andros town.
🅜 850. 🚌

From Andros town the road passes through the medieval village of Mesariá with ruined tower-houses and the restored pantiled Byzantine church of the **Taxiárchis**, built by Emperor Emanuel Comnenus in 1158.

Springs gush from marble lion's-head fountains in the leafy village of **Ménites**, just above Mesariá on the slopes of the Petalo mountain. Ménites is known both for its nightingales and for the taverna overlooking a stream.

Steps lead up to the pretty restored church of **Panagía i Koúmoulos** (the Virgin of the Plentiful), thought to be built on the site of an ancient Temple of Dionysos.

Sights at a Glance

❶ Andros Town
❷ Mesariá
❸ Moní Panachrántou
❹ Palaiókastro
❺ Mpatsí
❻ Gávrio

❸ Moní Panachrántou

Μονή Παναχράντου

12 km (7 miles) SW of Andros town.
Tel 22820 51090. **Open** daily.

This spectacular monastery is perched 230 m (755 ft) above sea level in the mountains southwest of Andros town. It can be reached either by a 2-hour steep walk from Mesariá or a 3-hour trek from Andros town.

It was founded in 961 by Nikifóros Fokás, who later became Byzantine Emperor as reward for his help in the liberation of Crete from Arab occupation. The fortified monastery is built in Byzantine style and today houses just three monks. The church holds many treasures, including the skull of Agios Panteleïmon, believed to have healing powers. Visitors flock here to see the skull on the saint's annual festival day.

Moní Panachrántou overlooking the valley

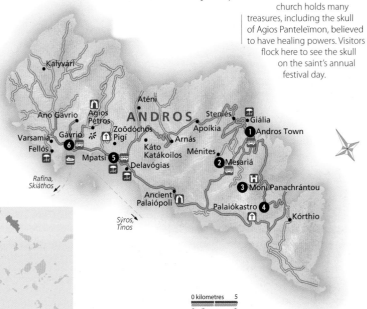

❹ Palaiókastro

Παλαιόκαστρο

18 km (11 miles) SW of Andros town.
Open unrestricted access.

High on a rocky plateau inland is the ruined Venetian Palaiókastro built between 1207 and 1233. Its alternative name, the Castle of the Old Woman, is after a woman who betrayed the Venetians to the Turks in the 16th century. After tricking her way inside the castle, she opened the gates for the Ottoman Turks. Appalled by the bloody massacre that followed, she hurled herself off the cliffs near Kórthio, 5 km (3 miles) to the southeast, in remorse. The rock from which she jumped is known as Tis Griás to Pídima, or Old Lady's Leap.

❺ Mpatsí

Μπατσί

8 km (5 miles) S of Gávrio. 🚌 200. 🚐

Built around a sweeping sandy bay, Mpatsí is a pretty resort. It has a small fishing harbour and a maze of narrow lanes reached by white steps from the café-lined seafront. Despite the lively nightlife Mpatsí has retained its village atmosphere. The main beach is popular with families, while **Delavógias** beach, south along the coastal track, is a favourite with naturists. Agía Marína, further along, has a friendly, family-run taverna.

Environs
South of Mpatsí the original capital of Andros, **Ancient Palaiópoli**, was inhabited until around AD 1000 when the people moved to Mesariá *(see p213)*. It was largely destroyed in the 4th century AD by an earthquake, but part of the acropolis is still visible, as are the remains of some of the temples under the sea.

Inland lies **Káto Katákoilos** village, known for its island music and dance festivals. A rough track leads north from here to remote **Aténi**, a hamlet at the head of a lush valley. Two beautiful beaches lie further to the windy northeast, in the bay of Aténi. The garden village of **Arnás**, high on the slopes of the Kouvára mountain range, has flowing springs and is one of the island's greenest spots. The area has many dry-stone walls and is spectacular walking country.

🏛 **Ancient Palaiópoli**
9 km (6 miles) S of Mpatsí.
Open unrestricted access. ♿ limited.

❻ Gávrio

Γαύριο

🚌 450. 🚐 🚌 🚢 Fellós 4 km (2.5 miles) NW.

Gávrio is a rather characterless port which, at weekends, becomes packed with Athenians heading for their holiday homes. There is a beach, a good campsite and plenty of tavernas. During the high season it can

Agios Pétros tower near Gávrio

be the only place with rooms available, as Mpatsí is often pre-booked by package companies.

Environs
From Gávrio, it takes an hour or so to walk up to the tower of **Agios Pétros**, the island's best-preserved ancient monument. Dating from the Hellenistic era, the tower stands 20 m (66 ft) high in an olive grove below the hamlet of Káto Agios Pétros The upper storeys of the tower were reached by footholds and an internal ladder, and its inner hall was once crowned by a corbelled dome. The purpose of the tower remains a mystery, although it may have been built to serve as a watchtower to guard the nearby mines from attack by marauding pirates.

North of Gávrio there are good beaches beyond the village of Varsamiá, which has two sandy coves. **Fellós** beach is the best, but is fast being developed with holiday villas.

A turnoff from the coastal road, 8 km (5 miles) south of Gávrio, leads to the 14th-century convent, **Zoödóchos Pigí**, the Spring of Life. Only a handful of nuns remain where there were 1,000 monks, but they are happy to show visitors their collection of icons and Byzantine tapestries.

The beach at Mpatsí Bay on Andros

Cycladic Art

With their simple geometric shapes and purity of line, Cycladic marble figurines are the legacy of the islands' Bronze Age civilization *(see pp32–3)* and the first real expression of Greek art. They all come from graves and are thought to represent, or be offerings to, an ancient deity. The earliest figures, from before 3000 BC, are slim and violin-shaped. By the time of the Keros-Sýros culture of 2700–2300 BC, the forms are recognizably human and usually female. They range from palm-sized up to life-size, the proportions remaining consistent. Obsidian blades, marble bowls prefiguring later Greek art, abstract jewellery and pottery, including the strange "frying pans", also survive. The examples of Cycladic art shown here are from the Museum of Cycladic Art in Athens *(see p295)*. Cycladic artefacts are also in many museums throughout the Cyclades.

"Frying pan" pottery vessels take their name from their shape, but their function is unknown. They may have been used in religious rituals. Decorated with spirals or suns, they belong to the mature phase of Cycladic art.

"Violin" figurines, such as this one, date from the early Cycladic period of 3300–2700 BC. Often no bigger than a hand, the purpose of these highly schematic representations of the human form is unknown. In some graves up to 14 of these figurines were found; other graves had none.

Collared vases, or kandelas, carved from marble, are one of the high points of Cycladic art. Probably used for food storage, the four lugs on the sides would have allowed them to be hung from a support.

This male figurine, found together with a female figurine, is one of the few male figures to have been found. He is also atypical, in having one arm raised and a band slung across his chest.

This female figurine with folded arms is typical of Cycladic sculpture. The head is slightly tipped back, with only minimal markings for arms, legs and features.

Influence on Modern Art

Considered crude and ugly when first discovered in the 19th century, the simplicity of both form and decoration of Cycladic art exerted a strong influence on 20th-century artists and sculptors such as Picasso, Modigliani, Henry Moore and Constantin Brancusi.

Henry Moore's *Three Standing Figures*

The Kiss by Brancusi

Tínos

Τήνος

A craggy yet green island, Tínos was first settled by Ionians in Archaic times. In the 4th century BC it became known for its Sanctuary of Poseidon and Amphitrite. Under Venetian rule from medieval times, Tínos became the Ottoman Empire's last conquest in 1715. Tínos has over 800 chapels, and in the 1960s the military Junta declared it a holy island. Many Greek Orthodox pilgrims come to the church of the Panagía Evangelístria (Annunciation) in Tínos town. The island is also known for its many dovecotes (peristeriónes), scattered across the landscape.

Archaeological Museum exhibit from Exómpourgo

have healing powers, and the church became a pilgrimage centre for Orthodox Christians. Tínos becomes very busy during the festivals of the Annunciation and the Assumption, when the icon is paraded through the streets (see pp52–3) and the devout often crawl to Panagía Evangelístria.

The church is a treasury of offerings, such as an orange tree made of gold and silver, from pilgrims whose prayers have been answered. The icon itself is so smothered in gold and jewels it is hard to see the painting.

The crypt where it was found is known as the chapel of Evresis, or Discovery. Where the icon lay is now lined with silver, and the holy spring here, Zoödóchos Pigí, is said to have healing powers.

The vestry has gold-threaded ecclesiastical robes, and valuable

Tínos town and the small harbourfront

Tínos Town

A typical island capital, Tínos town has narrow streets, whitewashed houses and a bustling port lined with restaurants and hotels.

🏛 Panagía Evangelístria

Church & museums **Open** daily.
Tel 22830 22256. ♿

Situated at the top of Megalóchari, the main street that runs up from the ferry, Panagía Evangelístria, the church of the Annunciation, dominates Tínos town. The pedestrianized Evangelistrías, which runs parallel to Megalóchari, is packed with stalls full

of icons and votive offerings. Built in 1830, the church houses the island's miraculous icon. In 1822, during the Greek War of Independence, Sister Pelagía, a nun at Moní Kechrovouníou, had visions of the Virgin Mary showing where an icon had been buried. In 1823, acting on the nun's directions, excavations revealed the icon of the Annunciation of the Archangel Gabriel, unscathed after 850 years underground. Known in Greece as the Megalóchari (the Great Joy), the icon was found to

Pilgrim crawling to the Panagía Evangelístria

Pýrgos • Pánormos

• Istérnia

Kolympíthres

Kallóni
Kómi

TINOS

Kámpos

Exómpourgo
• Falátados

Kiónia

Potamiá — Santa
Moní Margaríti
Kechrovouníou

Stavrós

Tínos Town

Agios Ioánnis

Agios Fokás

Mýkonos,
Vólos

Sýros,
Páros

Andros, Skiáthos,
Thessaloníki

0 kilometres   5

0 miles   3

For keys to symbols see back flap

The pretty village of Pýrgos in the north of the island

copies of the gospels. Also within the complex is a museum with items by local sculptors and painters, including works by sculptors Antónios Sóchos, Geórgios Vitális and Ioánnis Voúlgaris. The art gallery has works of the Ionian School, a Rubens, a Rembrandt and 19th-century works by international artists.

Archaeological Museum
Megalóchari **Tel** 22830 22670. **Open** 9am–4pm Tue–Sun. **Closed** main public hols.

On Megalóchari, near the church, is the Archaeological Museum, which has displays of sculptures of nereids (sea-nymphs) and dolphins found at the Sanctuary of Poseidon and Amphitrite. There is also a 1st-century-BC sundial by Andronikos Kyrrestes, who designed Athens' Tower of the Winds (*see p291*), and some huge 8th-century-BC storage jars from ancient Tínos on the rock of Exómpourgo.

Environs
East of town, the closest beach is shingly **Agios Fokás**. To the west is the popular beach at **Stavrós**, with a jetty that was built in Classical times. To the north near Kiónia are the foundations of the 4th-century-BC **Sanctuary of Poseidon and Amphitrite**, his sea-nymph bride. The excavations here have yielded many columns, or *kiónia*, after which the surrounding area is named.

Around the Island
Tínos is easy to explore, as there are plenty of taxis and a good bus service around the island.

North of Tínos town is the 12th-century walled **Moní Kechrovouníou**, one of the largest convents in Greece.

You can visit the cell where Sister Pelagía had her visions and the chest where her embalmed head is kept.

At 640 m (2,100 ft) high, the great rock of **Exómpourgo** was the site of the Archaic city of Tínos and later became home to the Venetian fortress of St Elena. Built by the Ghisi family after the Doge handed over the island to

The interior of the 12th-century Moní Kechrovouníou

them in 1207, the fortress was the toughest stronghold in the Cyclades, until it surrendered to the Turks in 1714. You can see remains of a few ancient walls on the crag, medieval houses, a fountain and three churches.

From Kómi, to the north, a valley runs down to the sea at **Kolympíthres**, with two sandy bays: one is deserted; the other has rooms and tavernas.

Overlooking the harbour of Pánormos in the northwest of the island, the pretty village of **Pýrgos** is famous for its sculpture school. The area is known for its green marble, and the stonework here is among the finest in the islands. Distinctive, carved marble fanlights and balconies decorate the island villages. There are examples at the **Museum of Marble Crafts**, one of seven museums dotted around the island. The old grammar school is now the School of Fine Arts, and a shop in the main square exhibits and sells the students' works.

Museum of Marble Crafts
Pýrgos. **Tel** 22830 31290. **Open** 10am–6pm daily (winter: to 5pm).

The Peristeriones (Dovecotes) of Tinos

The villages of Tínos are studded with around 1,300 beautiful white dovecotes (*peristeriónes*), all elaborately decorated. They have two storeys: the lower floor is for storage, the upper houses the doves and is usually topped with stylized winged finials or mock doves. The breeding of doves was introduced by the Venetians. Although also found on the islands of Andros and Sífnos, the *peristeriónes* of Tínos are considered the finest.

A dovecote in Kámpos with traditional elaborate patterns

Mýkonos
Μύκονος

Although Mýkonos is dry and barren, its sandy beaches and dynamic nightlife make this island one of the most popular in the Cyclades. Under Venetian rule from 1207, the islanders later set up the Community of Mykonians in 1615 and flourished as a self-sufficient society. Visited by intellectuals in the early days of tourism, today Mýkonos thrives on its reputation as the glitziest island in Greece.

Mýkonos harbour in the early morning

Mýkonos Town

Mýkonos town (or Chóra) is the supreme example of a Cycladic village – a tangle of dazzling white alleys and cube-shaped houses. Built in a maze of narrow lanes to defy the wind and pirate raids, the bustling port is one of the most photographed in Greece. Many visitors still get lost around the lanes today.

Taxi boats for the island of Delos (see pp222–3) leave from the quayside. The island's mascot, Pétros the Pelican, may be seen near the quay, hunting for fish.

Adjacent to the harbour is Plateía Mavrogénous, overlooked by the bust of revolutionary heroine Mantó Mavrogénous (1796–1848). She was awarded the rank of General for her victorious battle against the Turks on Mýkonos during the War of Independence in 1821.

The **Archaeological Museum**, housed in a Neo-Classical building south of the ferry port, has a large collection of Roman and Hellenistic carvings, 6th- and 7th-century-BC ceramics, jewellery and gravestones, as well as many finds from the ancient site on Delos.

Kástro, the oldest part of the town, sits high up above the waterside district. Built on part of the ancient castle wall is the excellent **Folk Museum**, one of the best in Greece. It is housed in an elegant seacaptain's mansion and has a fine collection of ceramics, embroidery and ancient and modern Mykonian textiles. Among the more unusual exhibits is the original Pétros the Pelican, now stuffed, who was the island's mascot for 29 years. The 16th-century Vonís Windmill is part of the Folk Museum and has been restored to full working order. It was one of the 30 windmills that were used by families all over the island to grind corn. There is also a small threshing floor and a dovecote in the grounds around the windmill.

The most famous church on the island, familiar from postcards, is the extraordinary **Panagía Paraportianí**, in the

Mantó Mavrogénous

Kástro. Built on the site of the postern gate *(parapórti)* of the medieval fortress, it is made up of four chapels at ground level with another above. Part of it dates from 1425, while the rest was built in the 16th and 17th centuries.

7th-century-BC amphora in the Archaeological Museum

From Kástro, the lanes run down into Venetía, or **Little Venice** (officially known as Alefkándra), the artists' quarter. The tall houses have painted balconies jutting out over the sea. The main square, Plateía Aléfkandras, is home to the large Orthodox cathedral of Panagía Pigadiótissa (Our Lady of the Wells).

The **Maritime Museum of the Aegean**, at the end of Matogiánni, features a collection of model ships from pre-Minoan times to the 19th century, maritime instruments, paintings and 5th-century-BC coins with nautical themes. The museum's restored traditional sailing boat is moored in the harbour for visits in the summer.

Next door, **Lena's House**, a 19th-century mansion, evokes the life of a Mykonian lady, Léna Skrivánou. Everything is preserved, even her needle-work to her chamberpot.

Works of Greek and international artists are on show at the **Municipal Art Gallery** on Matogiánni, and include an exhibition of works by local Mykonian painters.

Working 16th-century windmill, part of the Folk Museum

The famous Paraportianí church

🏛 Archaeological Museum
Harbourfront. **Tel** 22890 22325.
Open 8:30am–3pm Tue–Sun.
Closed main public hols. 🎫

🏛 Folk Museum
Harbourfront. **Tel** 22890 22591. **Open**
Apr–Oct: 4:30–8:30pm Mon–Sat. ♿

🏛 Maritime Museum of the Aegean
Enóplon Dynámeon. **Tel** 22890 22700.
Open Apr–Oct: 10:30am–1pm & 6:30–9pm daily; Nov–Mar: 8:30am–5pm daily. **Closed** main public hols. ♿ 🎫

🏛 Lena's House
Enóplon Dynámeon. **Tel** 22890 22390.
Open Apr–Oct: 6:30–9:30pm Mon–Sat. ♿ limited.

🏛 Municipal Art Gallery
Matogiánni. **Tel** 22890 27190.
Open Jun–Oct: daily. ♿

Around the Island
Mýkonos is popular primarily for its beaches. The best ones are along the south coast. At stylish **Platýs Gialós**, 3 km (2 miles) south of the town, regular taxi boats are available to ferry sun-worshippers from bay to bay. Backed by hotels and restaurants, this is the main family beach on the island, with watersports and a long sweep of sand. Serious sun-lovers head southeast to the famous nudist beaches. First is **Parágka**, or Agía Anna, a quiet spot

with a good taverna. Next is **Paradise**, with its neighbouring camp site, disco music and water-sports. The lovely cove of **Super Paradise** is gay and nudist. **Eliá**, at the end of the boat line, is also nudist.

In contrast to Mýkonos town, the inland village of **Ano Merá**, 7.5 km (4.5 miles) east, is traditional and largely unspoiled by tourism. The main attraction is the 16th-century **Panagía i Tourlianí**, dedicated to the island's protectress. Founded by two monks from Páros, the red-domed monastery was restored in 1767. The ornate marble tower was sculpted by Tíniot craftsmen. The monastery houses some fine 16th-century icons, vestments and embroideries. Northwest of the village is Palaiókastro hill,

VISITORS' CHECKLIST

Practical Information
🗺 9,400. ℹ Harbourfront, Mýkonos town (22890 22201); Plateía Karaóli & Dimitríou. 🎭 Fishermen's Festival, Mýkonos town: 30 Jun. 🌐 mykonos.gr

Transport
✈ 3 km (1.5 miles) SE of Mýkonos town. 🚌 Mýkonos town. 🚌 Polykandrióti, Mýkonos town (for north of island); on road to Ornós, Mýkonos town (for south of island).

once crowned by a Venetian castle. It is thought to be the site of one of the ancient cities of Mýkonos. Today it is home to the 17th-century working **Moní Palaiokástrou**. To the northwest, in the pretty village of **Maráthi**, is Moní Agíou Panteleïmona, founded in 1665. From here, the road leads to **Pánormos Bay** and **Fteliá**, a windsurfers' paradise.

Platýs Gialós beach, one of the best on Mýkonos

Delos
Δήλος

Tiny, uninhabited Delos is one of the most important archaeological sites in Greece and a UNESCO World Heritage Site. According to legend, Leto gave birth to Artemis and Apollo here. The Ionians arrived in about 1000 BC, bringing the worship of Apollo and founding the annual Delia Festival, during which games and music were played in his honour. By 700 BC, Delos was a major religious centre. First a place of pilgrimage, it later became a thriving commercial port particularly in the 3rd and 2nd centuries BC. It is now an open-air archaeological museum with mosaics and marble ruins covered in wild flowers in spring.

Archaeological Museum
This displays most of the finds from the island, including storage pots used for offerings and *koúroi* dating from the 7th century BC.

Stadium and Gymnasium

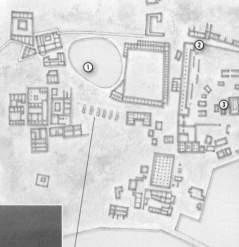

KEY

① **The Sacred Lake**, now dried up, was so called because it had witnessed Apollo's birth. A wall marks the lake's Hellenistic boundaries.

② **The Sanctuary of Dionysos** has remains of huge phallic monuments dating back to 300 BC.

③ **The Sanctuary of Apollo** has three temples: one dating from the 6th century BC, and two dating from the 5th century BC.

★ Lion Terrace
The famous lions (now replaced by replicas) were set up to overlook and protect the Sacred Lake. They were carved from Naxian marble at the end of the 7th century BC. Originally there were nine, but now only five remain.

2000 BC	1000 BC	750	500	250	AD 1
	1000 BC Ionians arrive on Delos and introduce Apollo worship	**422 BC** Athens exiles Delians to Asia Minor; Delians return the following year **426 BC** Second purification	**88 BC** Delos sacked by Mithridates		
		478 BC Athenians make Delos the centre of the first Athenian League	**166 BC** Romans return Delos to Athens; trade flourishes		
2000 BC Earliest settlement on Mount Kýnthos	**700 BC** Naxians in control of Sanctuary of Apollo	**550 BC** Polykrates, the tyrant of Sámos, conquers the Cyclades, but respects the sanctity of Delos	**314 BC** Delos declares independence from Athens **543 BC** First purification (removal of tombs) of Delos by Athenians	**250 BC** Romans settle in Delos **69 BC** Romans fortify Delos after sack by pirates	

House of the Dolphins
This house from the 2nd century BC contains a mosaic of two dolphins with an elaborate Greek key design and waved borders.

↑ *Mount Kýnthos*

House of the Masks
Probably a hostelry for actors, this house contains a 2nd-century-BC mosaic of Dionysos, god of theatre, riding a panther.

★ Theatre
Built in 300 BC to hold 5,500 spectators, the theatre was sited in a natural amphitheatre. On its west side, a huge, vaulted cistern collected rainwater draining from the theatre and supplied part of the town.

★ Theatre Quarter
In Hellenistic and Roman times the wealthy built houses near the theatre, many with opulent, colonnaded courtyards.

Key

 Theatre quarter

House of Dioscourides and Cleopatra
Two statues represent the couple Cleopatra and Dioscourides, who lived here in the 2nd century BC.

House of Dionysos
Inside the house is a mosaic depicting Dionysos riding a leopard. Twenty-nine *tesserae* are used just to make up the animal's eye.

Sýros
Σύρος

Rocky Sýros, or Sýra, is the commercial, administrative and cultural centre of the Cyclades. Archaeological digs have revealed finds of the Cycladic civilization dating from 2800 to 2300 BC. The inhabitants converted to Catholicism under the French Capuchins in the Middle Ages. The 19th century saw Sýros become a wealthy and powerful port in the eastern Mediterranean. Though Sýros does not live off tourism, more visitors arrive each year attracted by its traditional charm.

Town hall, designed by Ernst Ziller

The twin peaks of Ermoúpoli: Ano Sýros and Vrondádo

❶ Ermoúpoli
Ερμούπολη

🏠 13,000. 🚢 🚌 Aktí Ethnikís Antístassis. 🛈 Thymáton Sperchíon 11 (22813 61000).

Elegant Ermoúpoli, named after Hermes, the god of commerce, is the largest city in the Cyclades. In the 19th century it was Greece's leading port and a major coaling station with a huge natural harbour and thriving shipyard. Crowned by the twin peaks of Catholic Ano Sýros to the north, and the Orthodox Vrondádo to the south, the city is built like an amphitheatre around the harbour.

The Lower Town
The architectural glories of central **Plateía Miaoúli** have led to the town becoming a National Historical Landmark. Paved with marble and lined

with palm-shaded cafés and pizzerias, the grand square is the city's hub and meeting place, especially for the evening stroll, or *vólta*. There is also a marble bandstand and a statue

dedicated to the revolutionary hero Admiral Andréas Miaoúlis. The square is dominated by the vast Neo-Classical **town hall** (1876), designed by the German architect Ernst Ziller.

The **Archaeological Museum**, up the steps to the left of the town hall, houses bronze and marble utensils from the 3000 BC Cycladic settlement of Chalandrianí. Also on display are Cycladic statuettes and Roman finds. Left of the town hall is the **Historical Archives Office**.

Nearby, on Plateía Vardáka, is the **Apollo Theatre**, designed in 1864 by French architect Chabeau as a copy of La Scala, Milan. The first opera

Statue of Andréas Miaoúlis

Sights at a Glance

❶ Ermoúpoli
❷ Kíni
❸ Galissás
❹ Poseidonía
❺ Vári

0 kilometres   4
0 miles   2

Cape Diapór
Grámmata Bay
Sykamiá
Sýringas
Kastrí
Chalandri
Delfíni
Kíni ❷
Chartiana
Agía Varvára
Agía Paraskeví
Galissás
Armeós ❸
Agía Pákou
Profítis Ilías
❶ Ermoúpoli
Foínikas
Andre Tinos
Agathopés
❹ Poseidonía
Chroússa
SYROS
Mégas Gialós
Vári ❺
Mýkonos Náxos, Páros

Markos Vamvakaris

One of the greatest exponents of *rempétika*, the Greek blues, Márkos Vamvakáris (1905–72) was born in Ano Sýros. Synonymous with hash dens and the low life, *rempétika* was the music of the urban under-class. With strong Byzantine and Islamic influences, it is often played on the *baglama* or the bouzouki. Vamvakáris was a master of the bouzouki as well as a noted composer. Over 20 recordings have been made of his music, the earliest of which dates back to the 1930s. A bust of Vamvakáris looks out to sea from the small square named after him in Ano Sýros.

house in Greece, it is noted for its fine wall paintings of Mozart and Verdi and is still used for plays and concerts.

Across the street the 1871 **Velissarópoulos Mansion**, now housing the Labour Union, has an elaborate marble façade and splendid painted ceilings and murals. Beyond here is the church of **Agios Nikólaos** (1848) with a marble iconostasis by the 19th-century sculptor Vitális. Also by Vitális is the world's first monument of the unknown soldier, in front of the church.

The Upper Town

The twin bell towers and distinctive blue and gold dome of Agios Nikólaos mark the start of the **Vapória** district. Here Sýros's shipowners built their Neo-Classical mansions, with some of the finest plasterwork, frescoes and marble carvings in Greece. The houses cling to the coastline above the town's quays

Marble iconostasis by Vitális, in the church of Agios Nikólaos

and moorings at Tálira, Evangelístria and Agios Nikólaos.

The charming district of **Vrontádo**, on the eastern peak, has a number of excellent tavernas spread out on its slopes at night. The Byzantine church of the **Anástasis** on top of the hill has views to Tínos and Mýkonos.

A half-hour's climb along Omiroú, or a brief bus ride, is the fortified medieval quarter of **Ano Sýros**, on the western peak. It is also known as Apáno Chóra or Kástro. On the way is the Orthodox cemetery of **Agios Geórgios** with its elaborate marble mausoleums. Ano Sýros is a maze of white-washed passages, arches and steps forming a huddle of interlinking houses. The architecture is unique, making the most of minimal space with *stegádia* (slate or straw roofs) and tight corners. The main entrance into Ano Sýros is Kamára, an ancient passageway leading into the main road, or Piatsa. The **Vamvakáris Museum**, dedicated to the life and work of Márkos Vamva-káris, is situated just off this road. At the top of Ano Sýros, the Baroque **Aï-Giórgis**, known as the cathedral of St George, was built on the site of a 13th-century

A typical street in the Ano Sýros quarter

For hotels and restaurants in this region see pp310–11 and pp328–31

VISITORS' CHECKLIST

Practical Information
🗺 19,700. ℹ Ermoúpoli (22813 61000). ⛴ Ermoúpoli Maritime Festival: Jul; Agios Nikólaos processions at Ermoúpoli: 6 Dec.

Transport
✈ 1 km (0.5 miles) SE of Ermoúpoli. ⛴ Ermoúpoli. 🚌

church. The basilica contains fine icons. The Jesuit cloister was founded in 1744 around the church of Our Lady of Karmilou (1581), and has 6,000 books and manuscripts in its library. Below it, the Capuchin convent of **Agios Ioánnis** was a meeting place and a refuge from pirates. Its church was founded by Louis XIII of France as a poorhouse.

A ceiling in one of Ermoúpoli's mansions

🏛 **Archaeological Museum**
Plateía Miaoúli. **Tel** 22810 88487. **Open** Tue–Sun. **Closed** main public hols.

🏛 **Historical Archives Office**
Plateía Miaoúli. **Tel** 22810 86891. **Open** 8:30am–2:30pm Mon–Fri. **Closed** main public hols.

🏛 **Vamvakáris Museum**
Plateía Vamvakári, Ano Sýros. **Tel** 22813 60952. **Open** Jun–Sep: 10:30am–1:30pm & 7–10pm daily. **Closed** main public hols.

Around Sýros Island

Sýros has numerous attractive coves as well as popular resorts like Galissás and Kíni. The landscape is varied, with palm trees and terraced fields. In the northern region of Apáno Meriá the traditional farms built to house both families and animals are in total contrast to the Italianate mansions and holiday homes of the south. Sýros has good roads, especially in the south, and is easy to explore by car or bike. There is a regular bus from the harbour to Ano Sýros, the main resorts and outlying villages.

Kíni Bay and the town's harbour

❷ Kíni

Kíni

9 km (6 miles) NW of Ermoúpoli. 🏠 300. 🚌 🚤 Delfíni 3 km (2 miles) N.

The fishing village of Kíni is set in a horseshoe-shaped bay with two good sandy beaches. Kíni is a popular meeting place for watching the sunset over an oúzo, and it has some excellent fish tavernas.

North, over the headland, is the award-winning **Delfíni** beach – the largest on Sýros and popular with naturists.

Between Ermoúpoli and Kíni, set in pine-covered hills, is the red-domed convent of **Agía Varvára**. With spectacular views to the west, the Orthodox convent was once a girls'

The red-tiles roofs of Agía Varvára convent near Kíni

orphanage. The nuns run a weaving school and their knitwear and woven goods are on sale at the convent. The frescoes in the church depict the saint's martyrdom.

Environs

Boat services run from Kíni to some of the island's remote northern beaches. **Grámmata Bay** is one of the most spectacular, a deep sheltered inlet with golden sands where sea lilies grow in autumn. Some of the rocks here have a Hellenistic inscription carved on them, seeking protection for ships from sinking.

A boat trip around the tip of the island past Cape Diapóri to the east coast takes you to **Sykamiá** beach. Here there is a cave where the Syriot philosopher Pherekydes is thought to have lived during the summer months. A physicist and astronomer, Pherekydes pioneered philosophical thought in the mid-6th century BC, and was the inventor of the heliotrope, an early sundial. From Sykamiá you can see

the remains of the Bronze Age citadel of **Kastrí** with its six towers perched on a steep rock.

❸ Galissás

Γαλησσάς

7 km (4 miles) W of Ermoúpoli. 🏠 500. 🚌 🚤 Armeós beach 1 km (0.5 miles) N.

Lively Galissás has the most sheltered beach on the island, fringed by tamarisk trees and, across the headland to the north, **Armeós** beach is a haven for nudists. Galissás has both the island's campsites, making it popular with backpackers. In high season it can be a noisy place to stay, and is often full of bikers. To the south of the bay lies **Agía Pákou**, which is the site of the Classical city of Galissás.

Huge **Foínikas** bay, 3 km (2 miles) further south, was originally settled by the Phoenicians, and now houses more than 1,000 people. Foínikas is a popular resort, with a pier and moorings for yachts and fishing boats.

Sweeping Foínikas bay on the southwest coast of Sýros

❹ Poseidonía

Ποσειδωνία

12 km (7 miles) SW of Ermoúpoli. 🏠 700. 🚌 🚤 Agathopés 1 km (0.5 miles) S.

Poseidonía, or Dellagrázia, is one of the largest tourist sites on the island, with cosmopolitan hotels and restaurants. The island's first main road was

An Italianate mansion in Poseidonía

built in 1855 from Ermoúpoli through Poseidonía to Foínikas. The affluent village contains some Italianate mansions, which are the country retreats of wealthy islanders. A short walk to the southwest, quieter **Agathopés** is one of the island's best beaches, with safe waters protected by an islet opposite. **Mégas Gialós**, 3 km (2 miles) away on the west coast, is a pretty beach shaded by tamarisk trees.

❺ Vári

Βάρη

8 km (5 miles) S of Ermoúpoli.
🏠 1,150. 🚌 🚕 Vári.

Quaint, sheltered Vári has become a major resort, but it still has traditional houses. On the Chontrá peninsula, east of the beach, is the site of the island's oldest prehistoric settlement (4000–3000 BC).

Kéa

Kéa

🏠 2,400. 🚢 🚕 Korissía. 🛈 22883 60000. 🚕 Gialiskári 6 km (4 miles) NW of Ioulís. 🌐 kea.gr

Kéa was first inhabited in 3000 BC and later settled by Phoenicians and Cretans. In Classical times it had four cities: Ioulís, Korissía, Poiíessa and Karthaía. The remains of Karthaía can be seen on the headland opposite Kýthnos. It is a favourite spot for rich Athenians due to its proximity to Attica. Mountainous with fertile valleys, Kéa has been known since ancient times for its wine, honey and almonds.

Ioulís

The capital, Ioulís, or Ioulída, with its red terracotta-tiled roofs and winding alleyways, is perched on a hillside 5 km (3 miles) above Korissía. Ioulís has 26 windmills situated on the Mountain of the Mills. The town is a maze of tunnel-like alleys, and has a spectacular Neo-Classical **town hall** (1902) topped with statues of Apollo and Athena. On the west side are ancient bas-relief sculptures and in the entrance a sculpture of a woman and child found at ancient Karthaía.

The Kástro quarter is reached through a white archway, which stands on the site of the ancient acropolis. The Venetians, under the leadership of Domenico Micheli, built their castle in 1210 with stones from the ancient walls and original Temple of Apollo. There are panoramic views from here. The **Archaeological Museum** is inside a fine Neo-Classical house. Its displays include an interesting collection of Minoan finds from Agía Eiríni; artifacts from the four ancient cities; Cycladic figurines and ceramics; and a copy of the stunning, marble, 6th-century-BC *koúros* of Kéa. The smiling 6th-century-BC **Lion of Kéa** is carved into the rock 400 m (1,300 ft) north of the town.

🏛 **Archaeological Museum**
Tel 22880 22079. **Open** Apr–Oct: 8:30am–3:30pm daily; Nov–Mar: 8am–5pm Thu–Sun. **Closed** main public hols. 🖼

Around the Island

The port of **Korissía** can be packed with Greek families on holiday breaks; as can **Vourkári**, an attractive and popular resort further north on the island that is famous for its fish tavernas.

The archaeological site of **Agía Eiríni** is topped by the chapel of the same name. The Bronze-Age settlement was destroyed by an earthquake in 1450 BC, and was excavated from 1960 to 1968. First occupied at the end of the Neolithic period, around 3000 BC, the town was fortified twice in the Bronze Age and there are still remains of the great wall with a gate, a tower and traces of streets. Many of the finds are displayed in the Archaeological Museum in Ioulís. The most spectacular monument on Kéa is the Hellenistic tower at **Moní Agías Marínas**, 5 km (3 miles) southwest of Ioulís.

A Hellenistic tower at Moní Agías Marínas on Kéa

Kýthnos
Κύθνος

🏔 1,600. 🚢 🚌 Mérichas.
ℹ️ 22813 61100. 🌐 **kythnos.gr**

Barren Kýthnos attracts more
Greek visitors than foreign
tourists, although it is a
popular anchorage for flotilla
holidays. Its dramatic, rugged
interior and the sparsity of
visitors make it an ideal
location for walkers.

The local clay was tradi-
tionally used for pottery
and ceramics, but is also
used to make the red roofing
tiles that characterize all the
island's villages.

Known locally as Thermiá
because of the island's hot
springs, Kýthnos attracts visitors
to the thermal spa at Loutrá.
Since the closure of the iron
mines in the 1940s, the
islanders have lived off fishing,
farming and basket-weaving.
To celebrate festivals, such as
the major pre-Lenten carnival,
the islanders often wear
traditional costumes.

Chóra
Also known as Messariá, the
capital is a charming mix of
red roofs and Cycladic cube-
shaped houses. Also worth
visiting is the church of **Agios
Sávvas**, founded in 1613 by
the Venetian Cozzadini family
whose coat of arms it bears.
The oldest church is **Agía
Triáda** (Holy Trinity), a domed,
single-aisle basilica.

Interior of the church of Panagía Kanála in
Kanála town on Kýthnos

Around the Island
The road network is limited,
but buses connect the port of
Mérichas with Kanála in the
south and Loutrá in the north.
The remaining areas of the
island are mostly
within a walkable
distance of these
points. **Mérichas**, on
the west coast, has a
small marina and tree-
fringed beach, lined
with small hotels and
tavernas. Just to the
north, the sandy
beach of **Martinákia** is
popular with families.
Further along the coast are
the lovely beaches at **Episkopí**
and **Apókrisi**, overlooked by
Vryókastro, the Hellenistic ruins
of ancient Kýthnos. You can walk
to **Dryopída**, a good hour south

Potter at work
in Dryopída

of Chóra, down the ancient
cobbled way with dramatic
views. The town was named
after the ancient Dryopes tribe
whose king, Kýthnos, gave the
island its name. The charming
red-roofed village is divided into
two districts by the river valley:
Péra Roúga is lush with crops,
while Galatás was once a centre
for ceramics, but only one
pottery remains.

At **Kanála**, 5 km (3 miles) to
the south, holiday homes have
sprung up by the church of
Panagía Kanála, dedicated to
the Virgin Mary, the island's
patron saint. Set in attractive
shaded picnic grounds, the
church houses Kýthnos's most
venerated icon of the Virgin.
It is probably by master icono-
grapher, Skordílis, as Kýthnos
was a centre for icon-painting
in the 17th century. Kanála
beach has views to Sérifos and
Sýros and there are
good beaches nearby.

Loutrá is a straggling
resort on the northeast
coast with windswept
beaches. Its spa waters
are saturated with iron,
and since ancient times
the springs of Kákavos
and Agioi Anárgyroi
have been used as a
cure for ailments
ranging from gout, rheumatism
and eczema to gynaecological
problems. The Xenía Hotel,
situated next door to the
excellent Hydrotherapy Centre,
has late 19th-century marble
baths inside. A Mesolithic
settlement to the north, dating
from 7500–6000 BC, is the oldest
in the Cyclades.

Sérifos
Σέριφος

🏔 1,400. 🚢 🚌 Livádi.
ℹ️ 22810 51210. 🌐 **serifos.gr**

In mythology, the infant
Perseus and his mother Danae
were washed up on the shores
of rocky Sérifos, known as
"the barren one". Once rich
in iron and copper mines, the
island has bare hills with
small fertile valleys, and
long sandy beaches.

The red-roofed village of Dryopída on Kýthnos

The whitewashed village of Chóra on Sérifos

Ferries dock at **Livádi** on the southeast coast. The town is situated on a sandy, tree-fringed bay backed by hotels and tavernas. Follow the stone steps up from Livádi, or use the sporadic bus service to reach the dazzling white **Chóra** high above on the steep hillside. It is topped by the ruins of a 15th-century Venetian kástro. Many of its medieval cube-shaped houses, some incorporating stone from the castle, have been renovated as holiday homes by Greek artists and architects. It is an attractive town with chapels and windmills perched precariously, offering breathtaking views of the island.

Near to the northern inland village of Galaní, the fortified **Moní Taxiarchón** (Archangel), built in 1500, is run by a single monk. The monastery contains fine 18th-century frescoes by Skordílis and some valuable Byzantine manuscripts.

Sífnos
Σίφνος

🅰 2,400. 🚢 🚌 Kamáres.
ℹ 22843 60300. 🆆 sifnos.gr

Famous for its pottery, poets and chefs, Sífnos has become the most popular destination in the western Cyclades. Visitors in their thousands flock to the island in summer lured by its charming villages, terraced countryside dotted with ancient towers, Venetian dovecotes and

long sandy beaches. In ancient times Sífnos was renowned for its gold mines. The islanders paid yearly homage to the Delphic sanctuary of Apollo with a solid gold egg. One year they cheated and sent a gilded rock instead, incurring Apollo's curse. The gold mines were flooded, the island ruined and from then on was known as *sífnos*, meaning "empty".

Apollonia
The capital is set above Kamáres port and is a Cycladic labyrinth of white houses, flowers and belfries. It is named after the 7th-century-BC Temple of Apollo, which overlooked the town, now the site of the 18th-century church of the Panagía Ouranofóra. The **Museum of Popular Arts and Folklore** in the main square has a good collection of local pottery and embroideries.

🏛 **Museum of Popular Arts and Folklore**
Plateía Iróon. **Tel** 22840 33730. **Open** Apr–Oct: afternoons daily. 🎫

Around the Island
Sífnos is a small, hilly island, popular with walkers. Buses from Kamáres port connect it with Apollonía and Kástro, on the east coast. **Artemónas** is Apollonía's twin village, the second largest on Sífnos, with impressive Venetian houses sporting distinctive chimneys. The 17th-century church, Agios Geórgios tou Aféndi, contains several fine icons from the period. The church of Panagía Kónchi, with its cluster of domes, was built on the site of a temple of Artemis.

Kástro, 3 km (2 miles) east of Artemónas, overlooks the sea, the backs of its houses forming massive outer walls *(see pp26–7)*. Some buildings in the narrow, buttressed alleys bear Venetian coats of arms. There are ruins of a Classical acropolis in the village. The **Archaeological Museum** has a collection of Archaic and Hellenistic sculpture, and Geometric and Byzantine pottery.

A fountain in Kástro, Sífnos

The port of **Kamáres** is a straggling resort, with waterside cafés and tavernas. The north of the harbour was once lined with pottery shops making Sífnos's distinctive blue and brown ceramics, but only two remain. Taxi boats go from Kamáres to the pretty pottery hamlet of **Vathý**, in the south. An hour's walk to the east is the busy resort of **Platýs Gialós**, with its long sandy beach. This is also connected by bus to Apollonía and Kamáres.

🏛 **Archaeological Museum**
Kástro. **Tel** 22840 31022. **Open** Tue–Sun. **Closed** main public hols.

A chapel with steps leading down to a small quay at Platýs Gialós, Sífnos

Páros
Πάρος

Fertile, thyme-scented Páros is the third-largest Cycladic island. Since antiquity it has been famous for its white marble, which ensured the island's prosperity from the early Cycladic age through to Roman times. In the 13th century Páros was ruled by the Venetian Dukes of Náxos, then by the Turks from 1537 until the Greek War of Independence *(see pp46–7)*. Páros is the hub of the Cycladic ferry system and is busy in high season. Buffeted by strong winds in July and August, it is a windsurfer's paradise. There are several resorts, but it retains its charm with hill-villages, vineyards and olive groves.

An ornate chandelier in the interior of Ekatontapylianí

The famous windmill beside Paroikiá's busy port

❶ Paroikiá
Παροικιά

🏠 3,000. 🚢 🚌 harbour. 🛈 22840 24772. **Open** Apr–Oct. 🚌 Kriós 3 km (2 miles) N.

The port of Paroikiá, or Chóra, owes its foundations to the marble trade. Standing on the site of a leading early Cycladic city, it became a major Roman marble centre. Traces of Byzantine and Venetian rule remain, although earthquakes have caused much damage.

Today it prospers as a resort town, with its quayside wind-mill and commercialized waterfront crammed with ticket agencies, cafés and bars. The area behind the harbour is an enchanting Cycladic town, with narrow paved alleys, archways dating from medieval times and white houses overhung with cascading jasmine.

🏛 Ekatontapylianí
W Paroikiá. **Tel** 22840 21243.
Open daily.
The Ekatontapylianí (Church of a Hundred Doors) in the west of

town is the oldest in Greece in continuous use and a major Byzantine monument. Its official name is the Dormition of the Virgin.

According to legend, the church was founded by St Helen, mother of Constantine, the first Christian Byzantine emperor. After having a vision here showing the path to the True Cross, she vowed to build a church on the site but died before fulfilling her promise. In the 6th century AD the Emperor Justinian carried out her wish, commissioning the architect Ignatius to design a cathedral. He was the apprentice of Isidore of Miletus, master builder of Agía Sofía in Constantinople. The result was so impressive that Isidore, consumed with jealousy, pushed his pupil off the roof. Ignatius grabbed his master's foot and they both fell to their deaths. The pair are immortalized in

Theoktísti's footprint

stone in the north of the courtyard in front of the church.

Ekatontapylianí is made up of three interlocking buildings. It is meant to have 99 doors and windows. According to legend, when the 100th door is found, Constant-inople (Istanbul) will return to the Greeks. Many earthquakes have forced much reconstruction, and the main church building was restyled in the 10th century in the shape of a Greek cross. The sanctuary columns date from the pre-Christian era and the marble screen, capitals and iconostasis are of Byzantine origin.

Fishing boats, Paroikiá harbour

Náxos, Kos,
Kálymnos,
Ikaría

Lágeri

Náousa

Moní Longovárdas

Sýros, Tinos
Mýkonos
Kamínia Kriós Trís Ekklisíes

Piraeus
Paroikiá
Maráthi quarries

Moní Christoú
tou Dásous
Léfkes Kéfalos
Mólos
Pródromos
Petaloúdes Márpissa Moní Agíou
Antoníou
paros Town Pounta Píso Livádi
PAROS Pounta
Chrysí Aktí
Dryós

ntíparos
Alykí
↓ Santorini

0 kilometres   5
0 miles   2

Sights at a Glance

1 Paroikiá
2 Trís Ekklisíes
3 Náousa
4 Léfkes
5 Píso Livádi
6 Petaloúdes

hunting and a frieze of
Archilochus, the 7th-century-
BC poet and soldier from Páros.

Kástro
Built in 1260 on the site of the
ancient acropolis, the Venetian
kástro lies on a small hill at the
end of the main street of the
town. The Venetians used the
marble remains from the
Classical temples of Apollo
and Demeter to construct the
surviving eastern fortification of
the kástro. The ancient columns
have also been partially used to
form the walls of neighbouring
houses. Next to the site of the
Temple of Apollo stands the
300-year-old blue-domed
church of **Agía Eléni and
Agios Konstantínos**.

Environs
Taxi boats cross the bay from
Paroikiá to the popular sands of
Kamínia beach and Kriós, both
sheltered from the prevailing
wind. The ruins of an Archaic
sanctuary of Delian Apollo stand
on the hill above.

On the carved wooden iconostasis is an icon of the Virgin,
worshipped for its healing
virtues. Nearby a foot-print, set
in stone, is claimed to be that
of Agía Theoktísti, the island's
patron saint. The Greeks fit their
feet into the print to bring them
luck. Also displayed is her
severed hand.

From the back of the church a
door leads to the chapel of Agios
Nikólaos, an adapted 4th-
century-BC Roman building.
It has a double row of Doric
columns, a marble throne and a
17th-century iconostasis. Next
door, the 11th-century baptistry
has a marble font with a frieze of
Greek crosses. Ekatontapyliani
has no bell tower and instead
the bells are hung from a
tree outside.

Archaeological Museum
W Paroikiá. **Tel** 22840 21231. **Open**
9am–4pm Tue–Sun (Nov–May: 8am–
3pm). **Closed** main public hols.

The museum can be found
behind Ekatontapyliani. One
of its main exhibits is part of

the priceless Parian Chronicle,
a historical record of the artistic
achievements of ancient
Greece up to 264 BC. It is carved
on a marble tablet and was
discovered in the kástro walls
during the 17th century. Also
on display are finds from the
Temple of Apollo including the
5th-century-BC mosaic *Winged
Victory*, which depicts Herakles

The Legend of Agia Theoktisti

Páros's patron saint, Theoktísti, was a young woman captured by
pirates in the 9th century. She escaped to Páros and lived alone in
the woods for 35 years, leading a pious and frugal life. Found by a
hunter, she asked him to bring her some Communion bread. When
he returned with the bread, she lay down and died. Realizing she was
a saint, he cut off her hand to take as a relic, but found he could not
leave Páros until he reunited her hand with her body.

For keys to symbols *see back flap*

Around Páros Island

Páros is an easy island to explore, with an excellent bus service linking the three main towns: the capital Paroikía, the trendy fishing village resort of Náousa in the north and the central mountain town of Léfkes. There are plenty of cars and bikes for hire to get to the beaches and villages off the beaten track, and boat excursions and caïques to tour the remoter shores.

The mountain village of Léfkes, the medieval capital of Páros

❷ Trís Ekklisíes

Τρεις Εκκλησίες

3 km (2 miles) NE of Paroikiá. 🚌

North of Paroikiá the road to Náousa passes the remains of three 17th-century churches, Tris Ekklisíes, adapted from an original 7th-century basilica. That was in turn built from the marble of a 4th-century-BC *heróon*, or hero's shrine, tomb of the Parian poet Archilochus.

In the mountains further north, the remote, 17th-century **Moní Longovárdas** is a hive of activity. The monks make wine and books and work in the fields, and the abbot is famous for his icon-painting. Visitors are, however, discouraged and women are banned.

Main door at Moní Longovárdas

❸ Náousa

Νάουσα

12 km (7 miles) NE of Paroikiá. 🚗 2,100. 🛈 22840 53171. 🚌 🚏 Lageri 5 km (3 miles) NE.

With its brightly painted fishing boats and winding white alleyways, Náousa has become a cosmopolitan destination for the jet set, with expensive boutiques and relaxed bars. It is the island's second-largest town and the place to sit and watch the rich and the beautiful parade chic designer clothes along the waterfront.

The colourful harbour has a unique breakwater in the half-submerged ruin of a Venetian castle which has slowly been sinking with the coastline.

Every year, on the evening of 23 August, 100 torch-lit fishing caïques assemble to re-enact the battle of 1536 between the islanders and the pirate Barbarossa, ending with celebrations of music and dancing.

❹ Léfkes

Λεύκες

10 km (6 miles) SE of Paroikiá. 🚗 850 🛈 22840 41990. 🚌

The mountain road to Léfkes, the island's highest village, passes the abandoned marble quarries at Maráthi, last worked for Napoleon's tomb. It is possible to explore the ancient tunnels with a torch.

Léfkes, named after the local poplar trees, was the capital under Ottoman rule. It is a charming, unspoiled village with medieval houses covered in pink and red bougainvillea, a labyrinth of paved alleys, *kafeneía* in shaded squares and restaurants with terraces overlooking the green valley below. The only way to explore is on foot; parking is provided at the village entrance. Shops stock local weaving and ceramic handicrafts and the town has a tiny Folk Museum (key available from Town Hall).

Environs

From the windmills overlooking Léfkes, a Byzantine marble pathway leads 3 km (2 miles) southeast to **Pródromos**, an old fortified farming village. Walk a further 15 minutes past olive groves to reach **Mármara** village with its marble-paved streets. The pretty hamlet of **Márpissa** lies about 1.5 km (1 mile) south.

On Kéfalos hill, 2 km (1 mile) east of Márpissa, are the ruins of a 15th-century Venetian fortress and the 16th-century **Moní Agíou Antoníou**. The monastery is built from Classical remains and has a 17th-century fresco of the *Second Coming*.

Caïques at the attractive fishing harbour at Náousa

For hotels and restaurants in this region see pp310–11 and pp328–31

The convent of Moní Christoú tou Dásous near Petaloúdes

❺ Píso Livádi

Πίσω Λιβάδι

15 km (9 miles) SE of Paroikiá. 🚌 50. 🚐 to Márpissa. 🚢 Poúnta 1 km (0.5 mile) S.

Situated below Léfkes on the east coast of the island, the fishing village of Píso Livádi, with its sheltered sandy beach, has grown into a lively small resort. It was once the port for Páros's hill-villages and the island's marble quarries; today there are services operated over to nearby Agía Anna (see p234) on Náxos island. The small harbour has a wide range of bars and tavernas, with a disco nearby and occasional local activities and entertainments.

The beautiful and fashionable beach at Poúnta

Environs

Mólos, 6 km (4 miles) north, has a long sandy beach with dunes, tavernas and a windsurfing centre. Just to the south lies **Poúnta** (not to be confused with the village of Poúnta on the west coast), one of the best and most fashionable beaches in the Cyclades with a trendy laid-back beach bar. The island's most famous east-coast beach, 3 km (2 miles) south, is **Chrysí Aktí** (Golden Beach). With 700 m (2,300 ft) of golden sand it is perfect for families. It is also a well-known centre for watersports and has hosted the world windsurfing championships.

Dryós, 2 km (1 mile) further southwest, is an expanding resort but at its heart is a pretty village with a duck pond, tavernas, a small harbour with a pebbly beach and a string of sandy coves.

❻ Petaloúdes

Πεταλούδες

6 km (4 miles) SW of Paroikiá. 🚌 **Open** 1 Jun–20 Sep: daily. 🚫

Petaloúdes, or the Valley of the Butterflies, on the slopes of Psychopianá, is easily reached from Paroikiá. This lush green oasis is home to swarms of Jersey tiger moths, from May to August, which flutter from the foliage when disturbed. There are mule treks along the donkey paths that cross the valley. About 2 km (1 mile) north of Petaloúdes, the 18th-century convent of **Moní Christoú tou Dásous**, Christ of the Woods, is worth the walk, although women only are allowed into the sanctuary. Páros's second patron saint, Agios Arsénios, teacher and abbot, is also buried here.

Outlying Islands

The island of **Antíparos** used to be joined to Páros by a causeway. These days a small ferry links the two from the west-coast resort of Poúnta and there are also caïque trips from Paroikiá. Antíparos town has a relaxed and stylish café society, good for escaping from the Páros crowds. Activity centres around the quay and the Venetian **kástro** area. The kástro is a good example of a 15th-century fortress town, designed with inner courtyards and narrow streets to impede pirate attacks (see pp26–7). The village also has two 17th-century churches, Agios Nikólaos and Evangelismós.

The island has fine beaches, but the star attraction is the massive **Cave of Antíparos**, with a breathtaking array of stalactites and stalagmites, discovered during Alexander the Great's reign. In summer, boats run to the cave from Antíparos town and Poúnta on Páros. From where the boat docks, it is a half-hour walk up the hill of Agios Ioánnis to the cave mouth, then a dramatic 70 m (230 ft) descent into the cavern. Lord Byron and other visitors have carved their names on the walls. In 1673 the French ambassador, the Marquis de Nointel, held a Christmas Mass here for 500 friends. The church outside, Agios Ioánnis Spiliótis, was built in 1774.

Bougainvillea on a house in Antíparos town

Náxos
Νάξος

The largest of the Cyclades, Náxos was first settled in 3000 BC. A major centre of the Cycladic civilization (see pp32–3), it was one of the first islands to use marble. Náxos fell to the Venetians in 1207, and the numerous fortified towers (pýrgoi) were built, still evident across the island today. Its landscape is rich with citrus orchards and olive groves, and it is famous in myth as the place where Theseus abandoned the Cretan princess Ariadne.

Mosaic from the Archaeological Museum in Náxos town

The Portára gateway from the unfinished Temple of Apollo

❶ Náxos Town
Χώρα

🏛 15,000. 🚢 🚌 Harbourfront ℹ Harbourfront (22853 60100).

North of the port and reached by a causeway is the huge marble Portára gateway on the islet of Palátia, which dominates the harbour of Náxos town, or Chóra. Built in 522 BC, it was to be the entrance to the unfinished Temple of Apollo.

The town is made up of four distinct areas. The harbour bustles with its cafés and fishermen at work. To the south is Neá Chóra, or Agios Geórgios, a concrete mass of hotels, apartments and restaurants. Above the harbour, the old town divides into the Venetian kástro, once home of the Catholic nobility, and the medieval Bourg, where the Greeks lived.

The twisting alleys of the Bourg market area are lined with restaurants and gift shops. The Orthodox cathedral in the Bourg, the fine 18th-century **Mitrópoli**

Zoödóchou Pigís, has an iconostasis, painted by Dimítrios Válvis of the Cretan School in 1786.

Uphill lies the imposing medieval north gate of the fortified Kástro, built in 1207 by Marco Sanudo. Only two of the original seven gate-towers remain. Little is left of the 13th-century outer walls, but the inner walls still stand, protecting 19 impressive houses. These bear the coats of arms of the Venetian nobles who lived there, and many of the present-day residents are descended from these families. Their remains are housed in the 13th-century Catholic **cathedral**, in the Kástro, beneath marble slabs dating back to 1619.

During the Turkish occupation, Náxos was famous for its schools. The magnificent Palace of Sanoúdo, dating from 1627, which incorporates part of the Venetian fortifications, housed the French school. The most famous pupil was Cretan novelist Níkos Kazantzákis (see p280), who wrote Zorba the Greek.

Angel from the Roman Catholic cathedral

The building now houses the **Archaeological Museum**, with one of the best collections of Cycladic marble figurines (see p215) in the Greek islands, and some beautiful Roman mosaics.

🏛 Archaeological Museum
Palace of Sanoúdo. **Tel** 22850 22725. **Open** 8am–3pm Tue–Sun (daily Nov–Mar). **Closed** main public holidays. 🏛

Environs
The remains of the Temple of Dionýsos can be found in Iria, near Glinado, around 5 km (3 miles) from Náxos Town. A causeway leads to the **Grótta** area, north of Náxos town, named after its numerous sea caves. To the south the lagoon-like bay of **Agios Geórgios** is the main holiday resort, with golden sands and shallow water. The best beaches are out of town along the west coast. **Agía Anna** is a pleasant small resort with silver sands and watersports. For more solitude, head south 3 km (2 miles) over the dunes to **Pláka**, the best beach on the island. Further south the pure white sands of **Mikrí Vígla**, and **Kastráki**, named after a ruined Mycenaean fortress, are good for watersports.

The remote and beautiful Pláka beach south of Náxos town

Around Náxos Island

Inland, Náxos is a dramatic patchwork of rich gardens, vineyards, orchards and villages. These are backed by wild crags and dotted with Venetian watchtowers and a wealth of historical sites. Although there are organized tours from Náxos town and a good local bus service, a hired car is advisable to explore the island fully. The Tragaía region is, however, a walker's paradise.

Moní village in the Tragaía valley, surrounded by olive groves

❷ Mélanes Valley

Κοιλάδα Μελάνων

10 km (6 miles) S of Náxos town. to Kinídaros.

The road south of Náxos town passes through the Livádi valley, the heart of ancient marble country, to the Mélanes villages. In **Kournochóri**, the first village, is the Venetian Della Rocca tower. At **Mýloi**, near the ancient marble quarry at Flerió, lie two 6th-century-BC *koúroi*, huge marble statues. One, 8 m (26 ft) long, lies in a private garden,

open to visitors. The other, 5.5 m (18 ft) long, lies in a nearby field.

Environs
Southeast of Náxos town is **Glinádo**, home to the Venetian Bellonias tower, first of the fortified mansions on Náxos. The chapel of Agios Ioánnis Gýroulas in Ano Sagrí, south of Glinádo, is built over the ruins of a temple of Demeter.

❸ Tragaía Valley

Κοιλάδα Τραγαίας

15 km (9 miles) SE of Náxos town.

From Ano Sagrí the road twists to the Tragaía valley. The first village in the valley, **Chalkí**, is the most picturesque with its Venetian architecture and the old Byzantine Fragópoulos tower in its centre.

From Chalkí a road leads up to Moní, home of the most unusual church on Náxos, **Panagía Drosianí**. Dating from the 6th century, its domes are made from field stones.

Filóti is a traditional village, the largest in the region.

It sits on the slopes of Mount Zas, which, at 1,000 m (3,300 ft), is the highest in the Cyclades.

Sights at a Glance

❶ Náxos town
❷ Mélanes Valley
❸ Tragaía Valley
❹ Apeíranthos
❺ Komiakí
❻ Apóllon

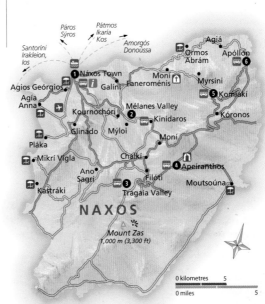

Terraced fields outside the village of Komiakí

❹ Apeíranthos

Απείρανθος

25 km (16 miles) SE of Náxos town.
🏛 1,500. 🚌

Apeíranthos was colonized in the 17th and 18th centuries by Cretan refugees fleeing Turkish oppression and coming to work in the nearby emery mine. It is the island's most atmospheric village, with marble-paved streets and 14th-century towers *(pýrgoi)* built by the Venetian Crispi family. Locals still wear traditional costume, women weave on looms and farmers sell their wares from donkeys.

The small **Archaeological Museum** has a collection of proto-Cycladic marble plaques depicting scenes from daily life as well as Neolithic finds. There is also a small **Geological Museum** on the second floor of the village school. Below the village is the port of **Moutsoúna,** where ships were once loaded with emery before the industry's decline. The fine beach is now lined with holiday villas.

🏛 **Archaeological Museum**
Off main road. **Tel** 22850 61725.
Open daily. **Closed** main
public hols. ♿

🏛 **Geological Museum**
Village school. **Tel** 22850 61724.
Open daily. **Closed** main
public hols. 📷

❺ Komiakí

Κωμιακή

42 km (26 miles) E of Náxos town.
🏛 500. 🚌

Approaching from Kóronos the road becomes a tortuous succession of hairpin bends before finally arriving in pretty Komiakí (also known as Koronída). This is the highest village on Náxos and a former home of the emery miners. It is covered with vines and is known for being the place where the local *kítro* liqueur originated more than two centuries ago. It is made from the leaves of citrus trees. There are wonderful views over the surrounding terraced vineyards. The village is the start of one of the finest walks on Náxos. The walk takes you down into the lush valley and the charming oasis hamlet of **Myrsíni**.

❻ Apóllon

Απόλλων

49 km (30 miles) NE of Náxos town.
🏛 100. 🚌

Originally a fishing village that is slowly turning into a resort, Apóllon gets busy in the summer with coach trips of people coming to visit the fish tavernas and the huge *koúros* found here. Steps lead up the hillside above the village to ancient marble quarries where the vast unfinished statue has lain abandoned since 600 BC. The bearded marble figure, which is believed to represent the god Apollo, is 10.5 m (35 ft) long and weighs 30 tonnes. There is also a lively festival in the village for St John the Baptist on 28 August.

Environs
At Agiá, 10 km (6 miles) west of Apóllon, stands the **Cocco Pýrgos**, built by the Venetian Cocco clan at the beginning of their rule of northern Náxos in 1770. *Pýrgoi* are fortified watchtowers that were built during the Venetian occupation of Náxos. Further along the north coast road lies the idyllic beach of **Ormos Abrám** with a good family-run taverna.

Dating from 1606, the abandoned **Moní Faneroménis** is 13 km (8 miles) south on the road winding down the west coast from Apóllon. Slightly further south towards Galíni, a road leads up to the most famous *Pýrgos*, the **High Tower** of the Cocco clan. It was built in 1660 in a commanding position overlooking a valley. During the 17th century a family feud between the Orthodox Cocco

The harbour at Moutsoúna, Náxos

The huge *koúros* in Apóllon's ancient quarries

and the Catholic Barozzi families broke out as a result of an insult. The feud led to the bombardment of the High Tower when a Barozzi woman persuaded her husband, who was a Maltese privateer, to besiege it. The Cocco clan managed to hold out, but the vendetta continued to rage for another 20 years until a marriage eventually united the two families.

A Venetian fortified watchtower, or *pýrgos*, west of Apóllon

Outlying Islands

Between Náxos and Amorgós lie **Donoússa, Koufoníssi, Irakliá** and **Schinoússa**, the "Back Islands". They all have rooms to rent, a post and tourist office.

Irakliá, the largest, boasts impressive stalactites in the Cave of Aï-Giánni as well as Cycladic remains. Koufoníssi consists of two islands, Ano (upper), the most developed of the Back Islands, with good sandy beaches, and the uninhabited Káto (lower). Schinoússa has wild beaches and great walking over cobbled mule tracks. Donoússa, the most northerly of the chain, is more isolated and food can be scarce. A settlement from the Geometric era was excavated on the island, but most of its visitors come for the fine sandy beaches at **Kéntros** and **Livádi**.

Amorgós
Αμοργός

🗺 1,800. 🚢 Katápola & Aigiáli. 🚌 Katápola & Aigiáli harbours. *i* Katápola quay (22853 60200). ✈ Ormos Aigiális 12 km (7 miles) NE of Amorgós town.

Dramatically rugged, the small island of Amorgós is narrow and long with a few beaches. Inhabited from as early as 3300 BC, its peak was during the Cycladic civilization, when there were three cities: Minoa, Arkesini and Egiali. In 1885 a find of ceramics and marble was taken to the Archaeological Museum in Athens *(see p290)*.

Chóra

The capital, Chóra, or Amorgós town, is a dazzling clutch of whitewashed houses with windmills standing nearby. Above the town is **Apáno Kástro**, a Venetian fortress, which was built by Geremia Ghisi in 1290. Chóra also boasts the smallest church in Greece, the tiny **Agios Fanoúrios**.

Environs

Star attraction on the island is the spectacular Byzantine **Moní Panagías Chozoviótissas**, below Chóra on the east coast. The stark white monastery clings to the 180 m (590 ft) cliffs. It is a huge fortress, built into the rock, housing the miraculous icon of the Virgin Mary.

Founded in 1088 by the Byzantine Emperor Alexios I Comnenos, the monastery has a library with a collection of ancient manuscripts.

Around the Island

The best way to get around the island is by boat or walking, although there is a limited bus service. The main port of **Katápola** in the southwest is set in a horseshoe-shaped bay with tavernas, *pensions*, fishing boats and a small shingly beach. The harbour area links three villages: Katápola in the middle where the ferries dock, quieter **Xylokeratídi** to the north and **Rachídi** on the hillside above. A track leads from Katápola to the hilltop ruins of the ancient city of **Minoa**. All that remains are the Cyclopean walls, the gymnasium and the foundations of the Temple of Apollo.

The northern port of **Ormos Aigiális** is the island's main resort, popular for its sandy beach. It is worth following the mule paths north to the hill-villages of **Tholária**, which has vaulted Roman *tholos* tombs, and **Lagáda**, one of the prettiest villages on the island, with a stepped main street painted with daisies.

The cliff-top Moní Chozoviótissas

The white walls and blue-domed churches of Ios town

Ios
Ιος

🏔 1,800. ⛴ Gialós. 🚌 Ios town. ℹ️
Ano Chóra, Ios town (22860 91936).
🚌 Mylopótas 2 km (1 mile) E of Ios
town. 🌐 **ios.gr**

In ancient times Ios was
covered in oak woods, later
used for shipbuilding. The
Ionians built cities at the port of
Gialós and at Ios town, later to
be used as Venetian
strongholds. Ios is
also known as the
burial place of
Homer, and 15 May is
the Omíria, or Homer
festival. A local
speciality is its
cheese, *myzíthra*,
similar to a soft
cream cheese.

Ios is renowned for
its nightlife and as a
result is a magnet for the young.
However, it remains a beautiful
island. Its mountainous
coastline has over 400 chapels
and some of the finest sands in
the Cyclades.

Windmill above Ios town

Ios town, also known as the
Village, is a dazzling mix of
white houses and blue-domed
churches fast being swamped
by discos and bars. There are
ruins of the Venetian fortress,
built in 1400 by Marco Crispi,
remains of ancient walls, and
12 windmills above the town.

The port of **Gialós**, or Ormos,
has a busy harbour, with yachts
and fishing boats, good fish
tavernas and quieter accommodation than Ios town. The
beach here is windy, although
a 20-minute walk west leads
to the sandy cove at Koumpará.

A bus service runs from here
to Ios town and the superb
Mylopótas beach which has
two camp sites. Excursion boats
run from Gialós to the beach
at **Manganári** bay, in the south
and **Psáthi** bay in the east.

On the northeast coast the
beach at **Agía Theodóti** is
overlooked by the medieval
ruins of Palaiókastro fortress.
A festival is held at nearby
Moní Agías Theodótis on
8 September to
mark the islanders'
victory over
medieval pirates.
You can see the
door the pirates
broke through only
to be scalded to
death by boiling oil.

Homer's tomb is
supposedly in the
north at **Plakotós**, an
ancient Ionian town which has
slipped down the cliffs over the
ages. Homer died on the island
after his ship was forced to dock
en route to Athens. The tomb
entrance, ruined houses and
the remains of the Hellenistic
Psarópyrgos tower can be
seen today.

Síkinos
Σίκινος

🏔 300. ⛴ Aloprónoia. 🚌 Síkinos
town. ℹ️ Kástro, Síkinos town (22860
51228). 🚌 Agios Geórgios 7 km
(3 miles) NE of Síkinos town.

Síkinos is quiet, very Greek
and one of the most ruggedly
beautiful islands in the Cyclades.
Known in Classical Greece as
Oinoe (wine island), it has
remained a traditional
backwater throughout history.
Fishing and farming are the
main occupations of the 300
or so islanders and, although
there are some holiday homes,
there is little mass tourism.

Síkinos town is divided into
twin villages: Kástro and the
pretty and unspoiled Chóra
perched high up on a ridge
overlooking the sea. Kástro is a
maze of lanes and *kafeneía*. At
the entrance to the village is
Plateía Kástrou, where the walls
of 18th-century stone mansions
formed a bastion of defence.
The church of the Pantánassa
forms the focal point, and
among the ruined houses
is a huge marble portico.

The partly ruined Moní
Zoödóchou Pigís, fortified
against pirate raids, looms down
from the crag above Chóra and
has icons by the 18th-century
master Skordílis.

In medieval Chóra there is
a private **Folk Museum**, which
is in the family home of an
American expatriate. It has
an olive press and a wide
range of local domestic
and agricultural artifacts.

From Chóra a path leads past
the ruined ancient Cyclopean
walls southwest to **Moní
Episkopís**, a good hour's trek.

The golden sands of Mylopótas beach, Ios

With Doric columns and inscriptions it is thought to be a 3rd-century-AD mausoleum, converted in the 7th century to the Byzantine church of Koímisis Theotókou. A monastery was added in the 17th century, but is now disused.

On the east coast 3 km (2 miles) southeast of Síkinos town, the port of **Aloprónoia**, also known as Skála, has a few small cafés that double as shops, a modern hotel complex and a wide sandy beach that is safe for children.

Folk Museum
Ano Chorió, Síkinos town. **Tel** 22860 51228. **Open** May–Sep: daily.

The sleepy port of Aloprónoia

Folégandros
Φολέγανδρος

650. Karavostásis. Chóra (22860 41285). Agáli 2 km (1 mile) W of Folégandros town.
folegandros.gr

Bleak and arid, Folégandros is one of the smallest inhabited islands in the Cyclades. It aptly takes its name from the Phoenician for "rocky". Traditionally a place of exile, this remote island passed quietly under the Aegean's various rulers, suffering only from the threat of pirate attack. Popular with photographers and artists for its sheer cliffs, terraced fields and striking chóra, it can be busy in peak season, but is still a good place for walkers, with a wild beauty and unspoiled

Koímisis tis Theotókou in Folégandros town

beaches. **Folégandros town** or Chóra, perched 300 m (985 ft) above the sea to avoid pirates, is spectacular. It divides into the fortified Kástro quarter *(see p26)* and Chóra, or main village. Kástro, built in the 13th century by Marco Sanudo, Duke of Náxos, is reached through an arcade. The tall stone houses back on to the sea, forming a stronghold along the ridge of the cliff with a sheer drop below. Within its maze of crazy-paved alleys full of geraniums are the distinctive two-storey cube houses with brightly painted wooden balconies.

In Chóra village life centres on four squares with craft shops and lively tavernas and bars. The path from the central bus stop leads to the church of Koímisis tis Theotókou (Assumption of the Virgin Mary). It was built after a silver icon was miraculously saved by an islander from medieval pirates who drowned in a storm. Forming part of the ancient town walls, it is thought to have once been the site of a Classical temple of Artemis.

Ferries dock at **Karavostási** on the east coast, a tiny harbour with a tree-fringed pebble beach, restaurants, hotels and rooms. There is a bus to Chóra, and **Livádi** beach is a short walk from the port. In season there are excursions available to the western beaches at **Agáli, Agios Nikólaos** and **Latináki**, as well as to

the island's most popular sight, the **Chrysospiliá** or Golden Cave. Named after the golden shade of its stalactites and stalagmites, the grotto lies just below sea level in the northeast cliffs.

Ano Meriá, 5 km (3 miles) to the west of Folégandros town, is a string of farming hamlets on either side of the road, surrounded by terraced fields. There are wonderful sunset views from here and on a clear day it is possible to see Crete in the distance. There is a good **Ecology and Folk Museum** with a display of farming implements, and reconstructions of traditional peasant life. On 27 July a major local festival is held for Agios Panteleïmon.

From Ano Meriá steep paths weave down to the remote beaches at **Agios Geórgios** bay and **Vígla**.

Ecology and Folk Museum
Ano Meriá. **Tel** 22860 41069. **Open** 8:30am–3pm daily.

Traditional houses in Kástro, Folégandros town

Mílos

Μήλος

Volcanic Mílos is the most dramatic of the Cyclades with its extraordinary rock formations, hot springs and white villages perched on multicoloured cliffs. Under the Minoans and Mycenaeans the island became rich from trading obsidian. However, the Athenians brutally captured and colonized Mílos in the 4th century BC. Festooned with pirates, the island was ruled by the Crispi dynasty during the Middle Ages and was claimed by the Turks in 1580. Minerals are now the main source of the island's wealth, although tourism is growing.

View across the houses of Pláka in the mid-morning sun

Pláka

On a clifftop 4 km (2.5 miles) above the port of Adámas, Pláka is a pretty mix of churches and white cube houses. These blend into the suburb of Trypití, which is topped by windmills.

It is believed that Pláka is sited on the acropolis of ancient Mílos, built by the Dorians between 1100 and 800 BC. The town was then destroyed by the Athenians and later settled by the Romans.

The principal sight is the **Archaeological Museum**, its entrance hall dominated by a plaster copy of the *Venus de Milo*, found on Mílos. The collection includes Neolithic finds, particularly obsidian, Mycenaean pottery, painted ceramics, and terracotta

animals from 3500 BC, found at the ancient city of Philakopí. The most famous of the ceramics is the *Lady of Phylakopi*, an early Cycladic goddess decorated in Minoan style. However, the Hellenistic 4th-century-BC statue of Poseidon and the *koúros* of Mílos (560 BC) are now in the National Archaeological

Museum in Athens *(see p290)*. There are also finds from the neighbouring island of Kímolos. The **History and Folk Museum** is housed in a 19th-century mansion in the centre of Pláka. It has costumes, four-poster beds and handicrafts.

The *Lady of Phylakopi* in t Archaeologic Museum

Steps lead to the ruined **kástro**, which was built by the Venetians on a volcanic plug 280 m (920 ft) above sea level. Only the houses that formed the outer walls of the fortress remain.

Above the kástro, the church of Mésa Panagía was bombed during World War II. It was rebuilt and renamed **Panagía Schiniótissa** (Our Lady of the Bushes) after an icon of the Virgin Mary appeared in a bush where the old church used to stand.

Just below, the church of **Panagía Thalassítra** (Our Lady of the Sea), built in 1728, has icons of Christ, the Virgin Mary and Agios Eleuthérios.

The massive stone blocks of the Cyclopean walls that formed the city's East Gate in 450 BC remain, while 15 m (50 ft) west there are marble

The twin rocks, known as The Bears, on the approach to Adámas

relics and a Christian baptismal font from a Byzantine basilica. A Roman amphitheatre nearby is still used for performances.

🏛 **Archaeological Museum**
Main square. **Tel** 22870 28026.
Open 8am–3pm Tue–Sun.
Closed 1 May. 🔲

🏛 **History and Folk Museum**
Pláka. **Tel** 22870 21292.
Open 10am–2pm Tue–Sat.
Closed main public hols. ♿ 🔲

Inside the Christian Catacombs

Environs
In the nearby town of Trypití are well-preserved 1st-century-AD **Christian Catacombs**. Carved into the hillside, the massive complex of galleries has tombs in arched niches, each one containing up to seven bodies. The catacomb network is 184 m (605 ft) long, with 291 tombs. Archaeologists believe that as many as 8,000 bodies were interred here.

From the catacombs, a track leads to the place where the *Venus de Milo* was discovered, now marked by a plaque. It was found on 8 April 1820, by a farmer, Geórgios Kentrótas. He uncovered a cave in the corner of his field with half of the ancient marble statue inside. The other half was found by a visiting French officer and both

halves were bought as a gift for Louis XVIII, on 1 March 1821. The statue is now on show in the Louvre, Paris. The missing arms are thought to have been lost in the struggle for possession.

🔲 **Christian Catacombs**
Trypití, 2 km (1 mile) SE of Pláka.
Tel 22870 21625. **Open** 8:30am–6:30pm Tue–Sun (to 3pm Sun).

Around the Island
The rugged island is scattered with volcanic relics and long stretches of beach. The vast Bay of Mílos, the site of the volcano's central vent, is one of the finest natural harbours in the Mediterranean, and has some of Mílos's best sights.

West of Adámas, the small and sandy **Langáda** beach is popular with families. On the way to the beach are the municipal baths with their warm mineral waters.

South of Adámas, the Bay of Mílos has a succession of attractive beaches, including **Chivadolímni**, backed by a turquoise saltwater lake. On the south coast is the lovely beach of Agía Kyriakí, near the village of Provatás.

Situated on the northeast tip of the island is **Apollonía**, a popular resort with a tree-fringed beach. Water taxis leave here for the island of **Kímolos**, named after the chalk (*kimolía*) mined there.

Once an important centre of civilization, little remains now of **Ancient Phylakopi**, just southwest of Apollonía. You can make out the old Mycenaean city walls, ruined houses and grave sites, but a large part of the city has been submerged beneath the sea.

VISITORS' CHECKLIST

Practical Information
🗺 4,500. ℹ harbourfront, Adámas (22870 22445). 🏞 Nautical week: end Jun–beg Jul; Panagía at Zefyría: 15 Aug.

Transport
✈ 7 km (4 miles) SE of Adámas.
🚌 Adámas. 🚌 Adámas.

Geology of Milos

Due to its volcanic origins, Mílos is rich in minerals and has some spectacular rock formations. Boat tours from Adámas go to the eerie pumice moonscape of Sarakíniko, formed two to three million years ago, the lava formations known as the "organ-pipes" of Glaronísia (offshore near Philakopí), and the sulphurous blue water at Papáfragkas. Geothermal action has provided a wealth of hot springs; in some areas, such as off the Mávra Gkrémna cliffs, the sea can reach 100° C (212° F) only 30 cm (12 inches) below the surface.

Mineral mine at Voúdia, still in operation

The white pumice landscape at Sarakíniko

The sulphurous blue water at Papáfragkas

Santoríni
Σαντορίνη

Colonized by the Minoans in 3000 BC, this volcanic island erupted in 1450 BC, forming Santoríni's crescent shape. The island is widely believed to be a candidate for the lost kingdom of Atlantis. Named Thíra by the Dorians when they settled here in the 8th century BC, it was renamed Santoríni, after St Irene, by the Venetians who conquered the island in the 13th century. Despite tourism, Santoríni remains a stunning island with its white villages clinging to volcanic cliffs above black sand beaches.

One of the many cliffside bars in Firá, with views over the caldera

❶ Firá
Φηρά

🏘 1,550. 🚌 🚌 50 m (165 ft) S of main square. 🛈 22863 60100. 🚌 Monólithos 5 km (2.5 miles) E.

Firá, or Thíra, overlooking the caldera and the island of Néa Kaméni, is the island's capital. It was founded in the late 18th century, when islanders moved from the Venetian citadel of Skáros, near present-day Imerovígli, to the clifftop plains for easier access to the sea.

Devastated by an earthquake in 1956, Firá has been rebuilt, terraced into the volcanic cliffs with domed churches and barrel-roofed cave houses (skaftá).

The terraces are packed with hotels, bars and restaurants in good positions along the lip of the caldera to enjoy the magnificent views, especially at sunset. The tiny port of Skála Firón is 270 m (885 ft) below Firá, connected by cable car or by mule up the 580 steps. Firá is largely pedestrianized with winding cobbled alleys. The town's main square, Plateía Theotokopoúlou, is the bus terminal and hub of the road network. All the roads running north from here and the harbour eventually merge in Plateía Firostefáni. The most spectacular street, Agíou Miná, runs south

Firá's whitewashed buildings lining the clifftop

Sights at a Glance
❶ Firá
❷ Oía
❸ Ancient Thíra
❹ Akrotíri

```
0 kilometres        5
0 miles            3
```

For keys to symbols see back flap

along the edge of the caldera to the 18th-century church of **Agíos Minás**. With its distinctive blue dome and its white bell tower, it has become the symbol of Santoríni. The **Archae-ological Museum** houses finds from Akrotíri *(see p245)* and the ancient city of Mésa Vounó *(see p244)*, including early Cycladic figurines found in local pumice mines. The **Pre-historic Museum** contains the colourful Firá frescoes orginally thought to be from the mythical city of Atlantis.

Housed in a beautiful 17th-century mansion, the **Cultural Centre Gyzi Hall**, in the northern part of the town, holds manu-scripts from the 16th to 19th centuries, maps, paintings, and photographs of Firá before and after the earthquake.

Despite the 1956 earthquake you can still see vestiges of Firá's architectural glory from the 17th and 18th centuries, on Nomikoú and Erythroú Stavroú where several mansions have been restored.

The pretty ochre chapel of **Agíos Stylianós**, clinging to the edge of the cliff, is worth a visit on the way to the Frangika, or Frankish quarter, with its maze of arcaded streets. To the south, the Orthodox **cathedral** is dedicated to the Ypapantí (the Presentation of Christ in the Temple). Built in 1827, it is an imposing ochre building with two bell towers and murals by

Detail of bright orange volcanic cliff in Firá

the artist Christóforos Asimís. The bell tower of the **Dómos** dominates the north of town on Agíou Ioánnou. Though severely damaged in the earthquake, much of its Baroque interior has now been restored.

A donkey ride up the steps from Skála Firón to Firá

VISITORS' CHECKLIST

Practical Information

🗺 13,000. 🛈 Firá (22863 60100). 🎦 Classical Music, Firá: Aug & Sep.

Transport

✈ 5 km (3 miles) SE of Firá. 🚢 Skála Firón. 🚌

🏛 **Archaeological Museum**
Opposite cable car. **Tel** 22860 22217. **Open** 8:30am–5pm Tue–Sun. **Closed** main public hols. 🎟

🏛 **Cultural Centre Gyzi Hall**
Near cable car. **Tel** 22860 23077. **Open** May–Oct: 10am–4pm Mon–Sat. 🎟

🏛 **Prehistoric Museum**
Near Firá central square. **Tel** 22860 23217. **Open** 8:30am–3pm Tue–Sun. **Closed** main public hols. 🎟

Geological History of Santoríni

Santoríni is one of several ancient volcanoes lying on the southern Aegean volcanic arc. During the Minoan era, around 1450 BC, there was a huge eruption which began Santoríni's transformation to how we see it today.

1 Santoríni was a circular volcanic island before the massive eruption that blew out its middle.

The volcano was active for centuries, building up to the 1450 BC explosion.

Clouds containing molten rock spread over 30 km (19 miles).

Crater of 22 sq km (8.5 sq miles)

2 The eruption left a huge crater, or caldera. The rush of water into the void created a tidal wave, or tsunami, which devastated Minoan Crete.

A huge volume of lava was ejected, burying Akrotíri *(see p245)*.

Néa Kaméni and its active volcanic cone

Volcano walls up to 300 m (985 ft) high

3 The islands of Néa Kaméni and Palaiá Kaméni, visible today, emerged after more recent volcanic activity in 197 BC and 1707. They are still volcanically active.

Thirasía

Aspro Nisí

Palaiá Kaméni

Around Santoríni Island

Santoríni has much to offer apart from the frequently photographed attractions of Firá. There are some charming inland villages, and excellent beaches at Kamári and Eríssa with their long stretches of black sand. You can also visit some of Santoríni's wineries, or take a ferry or boat to the smaller islands. There are good bus services, but a car or bike will allow you more freedom to explore. Major sites such as Ancient Thíra and Akrotíri have frequent bus or organized tour services.

A blue and ochre painted housefront in Oía

❷ Oía

Oía

11 km (7 miles) NW of Firá.
🚶 400. 🚌

At the northern tip of the island, the beautiful town of Oía is famous for its spectacular sunsets. A popular island excursion is to have dinner in one of the many restaurants at the edge of the abyss as the sun sinks behind the caldera. According to legend, the atmospheric town is haunted and home to vampires.

Reached by one of the most tortuous roads in the Cyclades, Oía is the island's third port and

was an important and wealthy commercial centre before it was badly damaged in the 1956 earthquake.

Today Oía is designated a traditional settlement, having been carefully reconstructed after the earthquake. Its white and pastel-coloured houses with red pebble walls cling to the cliff face with the famous *skaftá* cave houses and blue-domed churches. Some of the Neo-Classical mansions built by shipowners can still be seen. A marble-paved pathway skirts the edge of the caldera to Firá. Staircases lead down to Arméni and the nearby fishing harbour at **Ammoúdi** with its floating pumice stones and red pebble beach. The tradition of boatbuilding continues at Arméni's small ferry dock at the base of the cliff, although the port is now mainly used by tourist boats departing daily for the small island of Thirasía.

Ancient Thíra, situated at the end of the Mésa Vounó peninsula

❸ Ancient Thíra

Αρχαία Θήρα

11 km (7 miles) SW of Firá. 🚌 to Kamári. **Open** 8:30am–3pm Tue–Sun. **Closed** main public hols. 🚢 Eríssa 200 m (600 ft) below.

Commanding the rocky head-land of Mésa Vounó, 370 m (1,215 ft) up on the southeast coast, the ruins of the Dorian town of Ancient Thíra are still visible. Recolonized after the great eruption (*see p243*), the ruins stand on terraces overlooking the sea.

Rock carving in Ancient Thíra

Excavated by the German archaeologist Hiller von Gortringen in the 1860s, most of the ruins date from the Ptolemies, who built temples to the Egyptian gods in the 4th and 3rd centuries BC. There are also Hellenistic and Roman remains. The 7th-century Santoríni vases that were discovered here are now housed in Firá's Archaeological Museum (*see p243*).

A path through the site passes an early Christian basilica, remains of private houses, some with mosaics, the agora (or market) and a theatre, with a sheer view down to the sea. On the far west is a 3rd-century-BC sanctuary cut into the rock, founded by Artemídoros of Perge, an admiral of the Ptolemaic fleet. It features relief carvings of an eagle, a lion, a dolphin and a phallus symbolizing the gods Zeus, Apollo, Poseidon and Priapus.

Ammoúdi fishing village overlooked by Oía on the clifftop above

The view from Ancient Thíra down to Kamári

To the east, on the Terrace of Celebrations, you can find graffiti which dates back as far as 800 BC. The messages praise the competitors and dancers of the *gymnopediés* – festivals in which boys danced naked and sang hymns to Apollo, or competed in feats of physical strength.

Environs
The headland of Mésa Vounó, which rises to the peak of Mount Profítis, juts out into the sea between the popular beaches of Kamári and Heríssa. **Kamári** is situated below Ancient Thíra to the north, and is the island's main resort. The beach is a mix of stone and black volcanic sand, and is backed by bars, tavernas and apartments. **Heríssa** has 8 km (5 miles) of black volcanic sand, a wide range of watersports and a camp site. A modern church stands on the site of the Byzantine chapel of Irene, after whom the island is named.

❹ Akrotíri
Ακρωτήρι

12 km (7 miles) SW of Firá. 🚍 350.
Open 10am–5pm daily. 🚗 Kókkini
Ammos 1 km (0.5 miles) S.

Akrotíri was once a Minoan outpost on the southwest tip of the island and is one of the most inspiring archaeological sites in the Cyclades. After an eruption in 1866, French archaeologists discovered Minoan pots at

Akrotíri, though it was Professor Spyrídon Marinátos who, digging in 1967, unearthed the complete city; it was wonderfully preserved after some 3,500 years of burial under tonnes of volcanic ash. The highlight was the discovery of frescoes, which are now displayed at the National Archaeological Museum in Athens *(see p290)*.

Marinátos was killed in a fall on the site in 1974 and his grave is beside his life's work. Covered by a modern roof, the excavations include late 16th-century BC houses on the Telchínes road, two and three storeys high, many still containing huge *pithoi*, or ceramic storage jars. The lanes were covered in ash and it was here that the well-known fresco of the two boys boxing was uncovered.

Further along there is a mill and a pottery. A flyover-style bridge enables you to see the town's layout, including a storeroom for *pithoi* which held grain, flour and oil. The three-storey House of the Ladies is named after the fresco of two voluptuous dark women. The Triangle Square has large houses that were originally decorated with frescoes of fisherboys and ships, now

removed to Firá's New Archaeological Museum.

The city's drainage system demonstrates how sophisticated and advanced the civilization was. No human or animal remains or treasure were ever found, suggesting that the inhabitants were probably warned by tremors before the catastrophe and fled in good time.

Outlying Islands
From Athiniós, 12 km (7 miles) south of Firá, excursion boats run to the neighbouring islands. The nearest are **Palaiá Kaméni** and **Néa Kaméni**, known as the Burned Islands. You can take a hot mudbath in the springs off Palaiá Kaméni and walk up the volcanic cone and crater of Néa Kaméni. **Thirasía** has a few tavernas and hotels. Its main town, the picturesque Manolás, has fine views across the caldera to Firá. Remote **Anáfi** is the most southerly of the Cyclades and shares the history of the other islands in the group. It is a peaceful retreat with good beaches. There are a few ancient ruins, but nothing remains of the sanctuaries of Apollo and Artemis that once stood here.

Storage jars found at Akrotíri

Frescoes of Akrotíri
Painted around 1500 BC, these Minoan-style murals are similar to those found at Knosós *(see pp276–9)*. The best known are *The Young Fisherman*, depicting a youth holding blue and yellow fish, and *The Young Boxers*, showing two young sparring partners with long black hair and almond-shaped eyes. Preserved by lava, the frescoes have kept their colour and are displayed on a rotating basis at the Archaeological Museum in Firá *(see p243)*.

CRETE

CHANIA • RETHYMNO • IRAKLEIO • LASITHI

The island of Crete is dominated by harsh, soaring mountains whose uncompromising impregnability is etched deep into the Cretan psyche. For centuries, cut off by these mountains and isolated by sea, the character of the island people has been proudly independent. Many conquerors have come and gone, but the Cretan passion for individuality and freedom has never been extinguished.

For nearly 3,000 years the ruins of an ancient Minoan civilization lay buried and forgotten beneath the coastal plains of Crete. It was not until the early 20th century that the remains of great Minoan palaces at Knosós, Phaestos, Mália and Zákros were unearthed. Their magnificence demonstrates the level of sophistication and artistic imagination of the Minoan civilization, now considered the well-spring of European culture.

Historically, the island and its people have endured occupation by foreign powers and the hardships of religious persecution. The Romans brought their administrative expertise to the island, and the ancient city-state of Górtys became capital of the Roman province of Crete in 65 BC. Byzantine rule was followed by the Venetians (1204–1669), whose formidable fortresses, such as Frangokástello, and elegant buildings in cities such as Réthymno and Chaniá testify to 400 years of foreign rule. Oppression and religious persecution by the Ottoman Turks (1669–1898) encouraged a strong independence movement. By 1913, led by Elefthérios Venizélos (1864–1936), Crete had become a province of Greece. The island was again occupied by German forces during World War II, despite valiant resistance.

Today, mountains, sparkling seas and ancient history combine with the Cretans' relaxed nature to make the island an idyllic holiday destination.

A local in Réthymno wearing traditional Cretan boots and headdress

◀ Balos beach, on Chaniá's Gramvoúsa Peninsula, on the northwest coast of Crete

The Flora and Fauna of Crete

Crete's wildlife is as varied as its landscape. In spring, flowers cover the coastal strip and appear inland in the patchwork of olive groves, meadows and orchards. Stony, arid *phrygana* habitat is widespread and pockets of native evergreen forests still persist in remote gorges. Freshwater marshes act as magnets for waterbirds, while Crete's position between North Africa and the Greek mainland makes it a key staging post for migrant birds in spring and autumn. Its comparative isolation has meant that several unique species of plant have evolved.

The Samariá Gorge *(see pp258–9)* has been carved out by winter torrents washing down from the Omalós Plateau. Visitors should look out for peonies, cyclamens and Cretan ebony. Watch out as well for wild goats, called *kri-kri*, whose sure-footed confidence enables them to scale the precipitous slopes and cliffs.

The Akrotíri peninsula offers sightings of chameleons.

Chaniá

York City

Omalós Plateau

0 kilometres 20
0 miles 10

Kourtaliótiko gorge is a good spot to look for clumps of Jerusalem sage.

Moní Préveli

Agía Galíni

The Omalós Plateau *(see p258)* is home to the lammergeier, one of Europe's largest birds of prey. With narrow wings and distinctive wedge-shaped tail, it can be seen soaring over mountains and ravines.

Agía Triáda

Agía Triáda's wetlands are the haunt of black-winged stilts.

The Gulf of Mesará has a rough, grassy shoreline that is home to butterflies like the swallowtail.

Moní Préveli *(see p264)* is visited by the migrant Ruppell's warbler between May and August. With his bold black and white head markings and beady red eyes, the male is a striking bird.

Agía Galíni *(see p267)* is an excellent spot for spring flowers, and in particular the striking giant orchid. It stands more than 60 cm (24 inches) tall and can bloom as early as February or early March.

The colourful yellow bee orchid

The catchfly with its sticky stems

Cretan ebony, endemic to Crete

Wild Flowers on Crete

Botanists visit Crete in their thousands each year to enjoy the spectacular display of wild flowers. They are at their best, and in greatest profusion, from February to April. By late May, with the sun higher in the sky, many have withered and turned brown. Most of those that undergo this transformation survive the summer as underground bulbs or tubers.

Mália** (see p281)** is one of the many coastal resorts on Crete that provide a temporary home for migrant waders in spring and autumn. This wood sandpiper will stay and feed for a day or so around the margins of pools and marshes.

Mount Díkti's slopes are covered in wild flowers in spring, including Cretan bee orchids.

Dolphins can be spotted from northern headlands.

Eloúnta has saltpans that are much favoured by avocets.

Siteía's precipitous cliffs *(see p284)* are the habitat for Cretan ebony, a shrub unique to the island, which produces pinkish-purple spikes of flowers in the spring.

Mál150

Siteía

Lasíthi's fields are feeding grounds for colourful hoopoes.

Ierápetra

Agios Nikólaos is a stopping-off place for migrants such as wagtails.

Geckos can be found on stone walls beside many roads in eastern Crete.

Ierápetra *(see p283)* attracts the migrant woodchat shrike in summer. Woodchats feed on insects and small lizards, which they sometimes impale on thorns to make them easier to eat.

Zákros *(see p285)*, with its high cliffs, is where you find Eleonora's falcons performing aerobatic displays in summer.

Exploring Crete

The most southerly of the Greek islands, Crete boasts clear blue seas, sandy beaches and glorious sunshine. Its north coast bustles with thriving resorts as well as historic towns such as Réthymno and Chaniá. Its rugged southern coast, in particular the southwest, is less developed. Four great mountain ranges stretch from east to west, forming the spine of the 250 km- (155 mile-) long island. A hiker's paradise, they offer magnificent scenery and some spectacular gorges. The island's capital, Irákleio, is famous for its Archaeological Museum and is also a good base for exploring the greatest of Crete's Minoan palaces, Knosós.

Card players in the vine-canopied streets of Réthymno's old town

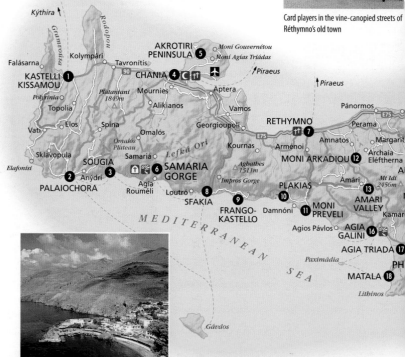

The harbour at Sfakiá

Getting Around

The provincial capitals of Chaniá, Réthymno, Irákleio and Agios Nikólaos act as the main transport hub for each region. Crete's bus service is quite well developed, with regular buses running along the north coast road. For touring the island a car is the most convenient mode of transport, though taxi fares are reasonable. Mountain roads between villages are now largely paved.

Large domed mosque inside Réthymno's Venetian Fortétsa

For hotels and restaurants in this region see pp312–13 and pp331–2

Sights at a Glance

Locator Map

The north entrance to the Palace of Knosós

A pelican in the picturesque harbour at Siteía

Key

- ━━ Motorway
- ━━ Main road
- ┄┄ Minor road
- ━ Scenic route
- – – Track
- ┈┈ High-season, direct ferry route
- △ Summit

For keys to symbols *see back flap*

The magnificent beach of Falásarna with its long stretch of sand and turquoise waters

❶ Kastélli Kissámou

Καστέλλι Κισσάμου

Chaniá. 🏔 3,000. 🚌 🚢 🚕
Kastélli Kissámou.

The small, unassuming town of Kastélli Kissámou, also known simply as Kastélli, sits at the eastern base of the virtually uninhabited Gramvoúsa Peninsula, once a stronghold of pirates. While not a tourist-oriented town, it has a scattering of hotels and restaurants along its pebbly shore. In the town square there is a fine **Archaeological Museum** housing some spectacular Roman mosaics excavated in the area. The town is also a good base from which to explore the west coast of Crete. Boat trips run to the tip of the **Gramvoúsa Peninsula**, where there are some isolated and beautiful sandy beaches.

🏛 Archaeological Museum
Platía Tzanakáki (near the bus station). **Tel** 28220 83308. **Open** 8:30am–3pm Tue–Sun.

Environs
Some 7 km (4 miles) south of Kastélli, the ruins of the ancient city of **Polyrínia** are scattered above the village of Ano Palaiókastro (also known as Polyrínia). Dating from the 6th century BC, the fortified city-state was developed by the Romans and later the Byzantines and Venetians. The present church of **Enenínta ennéa Martýron** (Ninety-Nine Martyrs), built in 1894, stands on the site of a large Hellenistic building.

On the west coast of the Gramvoúsa Peninsula, 16 km (10 miles) west of Kastélli, a winding road descends to the spectacular and isolated beach at **Falásarna**. Once the site of a Hellenistic city-state of that name, earthquakes have obliterated almost all trace of the once-thriving harbour and town. Today a few small guesthouses and tavernas are scattered along the northern end of the beach.

About 20 km (12 miles) east of Kastélli lies the picturesque fishing village of **Kolympári**. Head 1 km (0.5 miles) north of Kolympári for the impressive 17th-century **Moní Panagías Goniás**, with a magnificent seaside setting and a fine collection of 17th-century icons. Every year on 29 August (Feast of St John the Baptist), hundreds of pilgrims make the 3-hour walk up the peninsula to the church of **Agios Ioánnis** to witness the mass baptism of boys named John (Ioánnis).

❷ Palaióchora

Παλαιόχωρα

Chaniá. 🏔 1,800. 🚢 🚌 🚕
Elafónisos 14 km (9 miles) W.

First discovered in the 1960s by the hippie community, Palaióchora has become a haven for backpackers and package holiday-makers. This small port began life as a castle built by the Venetians in 1279. Today the remains of the fort, destroyed by pirate attacks in 1539, stand guard on a little headland dividing the village's two excellent beaches. To the west is a wide sandy beach with a windsurfing school, while to the east is a rocky but sheltered beach.

Environs
Winding up through the Lefká Ori (White Mountains), a network of roads passes through a stunning landscape of terraced hills and mountain villages, noted for their Byzantine churches. The closest

Moní Chrysoskalítissas near Palaióchora

The Battle of Crete (1941)

Following the occupation of Greece in World War II, German forces invaded Crete. Thousands of German troops were para-

German parachutists in Crete, 1941

chuted into the Chaniá district, where they seized Máleme Airport on 20 May 1941. The Battle of Crete raged fiercely for ten days, with high casualties on both sides. Allied troops retreated through the Lefká Ori (White Mountains) to the south where, with the help of locals, they were evacuated from the island. Four years of German occupation followed, during which time implacable local resistance kept up the pressure on the invaders, until their final surrender in 1945.

of these is **Anýdri**, 5 km (3 miles) east of Palaióchora, with the 14th-century double-naved church of **Agios Geórgios** containing frescoes by Ioánnis Pagoménos (John the Frozen) from 1323.

In summer, a daily boat service runs to **Elafonísi**, a lagoon-like beach of golden sand and brilliant blue water. From here, a 5 km (3 mile) walk north takes you to **Moní Chrysoskalítissas** (Golden Step), named for the 90 steps leading up to its church, one of which is said to appear golden, at least in the eyes of the virtuous. It can also be reached by road 28 km (17 miles) south of Kastélli Kissámou. From Palaióchora, boat trips make the rough, 64 km (40 mile) crossing (Mondays and Thursdays only) to **Gávdos** island, Europe's southernmost point.

❸ Soúgia
Σούγια

Chaniá. 🚗 270. 🚌 🚐 🚕 Soúgia; Lissós 3 km (1.5 miles) W.

Once isolated from the rest of the world at the mouth of the Agía Eiríni Gorge, the hamlet of Soúgia is now linked with Chaniá and the north coast by a good road. Still growing as a resort, the village has rooms to

rent, and a few tavernas and bars. The beach is long and pebbly. It is overlooked by the village church which is built on top of a Byzantine structure, whose mosaic floors have been largely removed.

Environs

Just over an hour's walk west of Soúgia, the ancient city-state of **Lissós** was a flourishing commercial centre in Hellenistic and Roman times. Among the remains are two fine 13th-century Christian basilicas, a 3rd-century-BC Asklepieion (temple of healing) and a sanctuary. The route to Lissós leads up through the **Agía Eiríni Gorge**. Popular with experienced hikers, the gorge's path has been improved, and plans are under way to develop the area along the lines of the Samariá Gorge.

Fresco by Ioánnis Pagoménos, Agios Geórgios

❹ Chaniá
See pp256–7.

❺ Akrotíri Peninsula
Χερσόνησος Ακρωτηρίου

6 km (3.5 miles) NW of Chaniá. 🚢 Soúda. 🚌 Chaniá & Soúda. 🚕 Stavrós 14 km (9 miles) N of Chaniá. Maráthi 10 km (6 miles) E of Chaniá.

Flat by Cretan standards, the Akrotíri Peninsula lies between Réthymno *(see pp262–3)* and Chaniá *(see pp256–7)*. At its base, on top of Profítis Ilías hill, is a shrine to Crete's national hero, Elefthérios Venizélos *(see p47)*. His tomb is a place of pilgrimage, for it was here that Cretan rebels raised the Greek flag in 1897 in defiance of the Great Powers.

There are several monasteries in the northeastern hills of the peninsula. **Moní Agías Triádas**, which has an impressive multidomed church, is 17th century, while **Moní Gouvernétou** dates back to the early Venetian occupation. Monks still inhabit both. Nearby, but accessible only on foot, the abandoned **Moní Katholikoú**, is partly carved out of the rock. Situated at the neck of the peninsula is a military base and the **Commonwealth War Cemetery**, burial ground of over 1,500 British, Australian and New Zealand soldiers killed in the Battle of Crete.

🏛 **Commonwealth War Cemetery** 4 km (2.5 miles) SE of Chaniá **Open** daily.

Goats grazing on the Akrotíri Peninsula

❹ Chaniá

Χανιά

Set against a spectacular backdrop of majestic mountains and aquamarine seas, Chaniá is one of the island's most appealing cities and a good base from which to explore western Crete. Its stately Neo-Classical mansions and massive Venetian fortifications testify to the city's turbulent and diverse past. Once the Minoan settlement of ancient Kydonia, Chaniá has been fought over and controlled by Romans, Byzantines, Venetians, Genoese, Turks and Egyptians. Following unification with Greece in 1913, the island saw yet another invasion during World War II – this time by the German army in 1941, when the Battle of Crete raged around Chaniá *(see p255)*.

The Mosque of the Janissaries

along Líthinon, a street lined with ornate Venetian doorways. Many of the finds from the site are on display in Chaniá's Archaeological Museum, including a collection of clay tablets inscribed with Minoan Linear A script.

By the inner harbour stand the now derelict 16th-century Venetian arsenals, where ships were once stored and repaired. The Venetian lighthouse, at the end of the seawall, offers superb views over Chaniá.

The Venetian Fort Firkás overlooking Chaniá's outer harbour

The Harbour

Most of the city's interesting sights are to be found in the old Venetian quarter, around the harbour and surrounding alleyways. At the northwest point of the outer harbour, the **Naval Museum**'s collection of model ships and other maritime artifacts is displayed in the well-restored Venetian Fort Firkás – also the setting for theatre and evenings of traditional dance in summer.

On the other side of the outer harbour, the **Mosque of the Janissaries** dates back to the arrival of the Turks in 1645 and is the island's oldest Ottoman building. It was damaged in World War II and rebuilt soon after. Behind the mosque rises the hilltop quarter of Kastélli, the oldest part of the city, where the Minoan settlement of **Kydonia** is being excavated. The site, closed to the public but clearly visible from the road, is approached

🏛 **Naval Museum**

Fort Firkás, Aktí Kountourióti. **Tel** 28210 91875. **Open** May–Oct: 9am–5pm Mon–Sat, 10am–6pm Sun; Nov–Apr: 9am–2pm daily. **Closed** main public hols. 🅿 🔲 **mar-mus-crete.gr**

Around the Covered Market

Connected to the harbour by Chálidon, this turn-of-the-century covered market sells local fruit and vegetables and Cretan souvenirs. Alongside the market, the bustling Skrýdlot, or Stivanádika, has shops selling leather goods, including

Chaniá's old harbour at dawn

The atmospheric backstreets of the old Splántzia quarter

traditional Cretan boots and made-to-measure sandals. The nearby **Archaeological Museum**, in the church of San Francesco, displays artifacts from western Crete including pottery, sculpture, mosaics and coins. Across a small square next to the museum is the 19th-century cathedral of **Agía Triáda**. Also nearby is the restored 15th-century **Etz Hayyim Synagogue**, which was used by Chaniá's Jewish population until the German occupation of 1941–5, when they were deported to death camps.

Dionysos and Ariadne mosaic, Chaniá Archaeological Museum

Archaeological Museum
Chálidon 21. **Tel** 28210 90334.
Open Jun–Oct: 9am–4pm Tue–Sun; Nov–May: 8am–3pm Tue–Sun.
Closed main public hols.

Etz Hayyim Synagogue
Parados Kondylaki. **Tel** 28210 86286.
Open 10am–6pm Mon–Thu (Nov–Apr: to 5pm), 10am–3pm Fri.

The Splántzia Quarter
Northeast of the market, the picturesque Splántzia quarter has houses with wooden balconies that overhang cobbled backstreets. The tree-lined square known as **Plateía 1821** commemorates a rebellion against the occupying Turks, during which an Orthodox bishop was hanged. Overlooking the square stands the Venetian church of **Agios Nikólaos**. Nearby are the 16th-century church of **Agioi Anárgyroi**, with its beautiful icons and paintings, and the church of **San Rocco**, which was built in 1630.

Outside the City Walls
South of the covered market along Tzanakáki are the **Public Gardens**. They were laid out in the 19th century by a Turkish *pasha* (governor). The gardens include a modest zoo which houses a few animals, including the *kri-kri* (the Cretan wild goat). The gardens also offer a children's play area, a café and an open-air auditorium, which is often used for cultural performances. The nearby **Historical Museum and Archives** is housed in a Neo-Classical building, and is devoted to the Cretan preoccupation with rebellions and invasions. Its exhibits include photographs and letters of the famous statesman Elefthérios Venizélos (1864–1936), as well as many other historical records.

Historical Museum and Archives
Sfakianáki 20. **Tel** 28210 52606.
Open 9am–1pm Mon–Fri.
Closed main public hols.

Environs
A series of sandy beaches stretches west from Chaniá all the way to the agricultural town of Tavronítis, 21 km (13 miles) away. A short walk west of Chaniá, the sandy beach of Agioi Apóstoloi is quieter and less developed than the city beaches.

Further west, the well-tended **German War Cemetery** stands witness to the airborne landing at Máleme of the German army in 1941 (*see p255*). Built into the side of a hill, the peaceful setting is home to over 4,000 graves whose simple stone markers look out over the Mediterranean. A small pavilion by the entrance to the cemetery houses a display commemorating the event.

German War Cemetery
19 km (12 miles) W of Chaniá.
Open daily.

The sandy beach of Agioi Apóstoloi, a short walk west of Chaniá

❻ Samariá Gorge
Φαράγγι της Σαμαριάς

The most spectacular landscape in Crete lies along the Samariá Gorge, the longest ravine in Europe. When the gorge was established as a national park in 1962, the inhabitants of pastoral Samariá village moved elsewhere, leaving behind the tiny chapels seen today. Starting from the Xylóskalo, 44 km (27 miles) south of Chaniá, a well-trodden trail leads down a tortuous 18 km (11 mile) course to the seaside village of Agía Rouméli. The walk takes from 5 to 7 hours. Water fountains can be found en route and sturdy shoes should be worn.

Facing east across the spectacular Samariá Gorge

★ Xylóskalo (wooden stairs)
The Samariá Gorge is reached via the Xylóskalo, a zigzag path with wooden handrails which drops a staggering 1,000 m (3,280 ft) in the first 2 km (1 mile) of the walk.

Omalós Plateau

KEY

① **Agios Geórgios**

② **Osía María**, a tiny church standing at the foot of a steep cliff, contains frescoes dating to the 14th century

③ **Agios Christós**

④ **Metamórphosis**

⑤ **Agía Paraskeví**

⑥ **Agios Geórgios**

⑦ **Agía Rouméli (Old Village)**

Agios Nikólaos
This tiny chapel nestles under the shade of pines and cypresses near the bottom of the Xylóskalo.

The Kri-kri (Cretan Wild Goat)

Found in only a few areas of Crete, notably the Samariá Gorge, the Cretan wild goat is thought to be a truly wild relative of the all-too-numerous feral goats that are found throughout the Mediterranean region, as well as in other parts of the world. A protected species, the Cretan wild goat is nimble and sure-footed on rugged terrain, attributes that help guard against attacks by other predators. Mature adults have attractively marked coats and horns with three rings along their length.

A kri-kri on rocky terrain

0 kilometres 2
0 miles 1

★ **Samariá Village**
Once inhabited, the village was abandoned in 1962, when the gorge was designated as a national park.

VISITORS' CHECKLIST

Practical Information
44 km (27 miles) S of Chaniá.
🌐 **sfakia-crete.com** for details.
Gorge: **Open** May–Oct:
7am–4pm daily (early May–end Oct if weather permits).

Transport
🚌 to Xylóskalo. ⛴ Agía Rouméli to Sfakiá (via Loutró); to Palaiochóra (via Soúgia); last boat back varies, check before travel.

★ **Sideróportes** (Iron Gates)
At 12 km (7 miles) along the gorge, the route squeezes between two towering rock walls, only 3 m (10 ft) apart, forming the famous Iron Gates, the narrowest part of the gorge.

Agía Rouméli (New Village)
Now equipped with tavernas and *domátia* (rooms to let), the seaside village of Agía Rouméli was once the haunt of pirates and the port used to export cypress wood to Egypt.

Key

▬▬ Asphalt road

▬▬ Park boundary

-- Path

For keys to symbols *see back flap*

❼ Réthymno
Ρέθυμνο

Once the Greco-Roman town of Rithymna, the site of today's Réthymno has been occupied since Minoan times. The city flourished under Venetian rule during the 16th century, developing into a literary and artistic centre, and becoming a haven for scholars fleeing Constantinople. Despite modern development and tourism, the city today has retained much of its charm and remains the intellectual capital of Crete. The old quarter is rich in elegant, well-preserved Venetian and Ottoman architecture. The huge Venetian Fortétsa, built in the 16th century to defend the island against the increasing attacks by pirates, overlooks the picturesque harbour with its charming 13th-century lighthouse.

The impressive 17th-century Nerantzés Mosque

Exploring Réthymno

Réthymno's bustling harbour-front serves as one great outdoor cafeteria, catering almost exclusively for tourists. It is skirted along most of its length by a good, sandy beach, but at its western end lies a small inner harbour. A restored 13th-century lighthouse stands on its breakwater.

The **Fortétsa** dominates the town, above the inner harbour. Designed by Pallavicini in the 1570s, it was built to defend the port against pirate attacks (Barbarossa had devastated the town in 1538) and the threat of expansionist Turks. The ramparts are still largely intact. Within them, a mosque, a small church and parts of the governor's quarters can still be seen, though most are in ruins. In the summer there are open-air concerts.

Traditional weaving in the Historical and Folk Art Museum

Directly opposite the main entrance to the Fortétsa, the **Archaeological Museum** occupies a converted Turkish bastion. Its collection is set out chronologically from Neolithic through Minoan to Roman times and includes artifacts from cemeteries, sanctuaries and caves in the region. Highlights include the late Minoan burial caskets (*larnakes*) and grave goods. The old town clusters behind the Fortétsa, characterized by a maze of narrow vine-canopied streets and its Venetian and Ottoman houses with wrought-iron balconies. Off Plateía Títou Peocháki is the **Nerantzés Mosque**. This is the best-preserved mosque in the city. Built as a church by the Venetians, it was converted in 1657 into a mosque by the Turks. It now serves as the city's concert hall. On Palaiológou, the 17th-century Venetian **Rimóndi Fountain** stands alongside busy cafés and shops. The elegant 16th-century Venetian **Lótzia** (Loggia) can also be seen here.

The small **Historical and Folk Art Museum** is housed in a Venetian mansion. On display are local crafts, including some weaving, pottery and lace.

🏰 Fortétsa
Katecháki. **Tel** 28310 28101.
Open 8am–8pm daily. **Closed** main public hols. 📷

🏛 Archaeological Museum
Cheimárras. **Tel** 28310 54668.
Open 9am–5pm Tue–Sun.
Closed main public hols. 📷

🏛 Lótzia
Palaiológou & Arkadíou.
Tel 28310 53270.
Open 9am–5pm Tue–Sun. 📷

🏛 Historical and Folk Art Museum
Vernárdou 30. **Tel** 28310 23398.
Open 10am–2:30pm Mon–Sat.
Closed main public hols. 📷

Tavernas and bars along Réthymno's waterfront, the focus of the town's activity

◀ The old harbour of Réthymno, with its Venetian and Ottoman buildings

Environs

East of Réthymno, towards
Pánormos, the resort developments flow one into another,
while west of the city a 20 km-
(12 mile-) stretch of relatively
uncrowded beach culminates
in the village of **Georgioúpoli**.
Despite wholesale tourist
development, this small
community still retains some
of its traditional atmosphere.
Massive eucalyptus trees line the
streets and a picturesque, turtle-
inhabited river flows placidly
down to the sea. **Lake Kournás**,
5 km (3 miles) inland from
Georgioúpoli, is set in a hollow
among the steeply rising hills.
Pedalos, sailboards and canoes
can be hired at the lake
and a few shady tavernas
offer refreshments.

In Arménoi, on the main
Réthymno–Agía Galíni road,
there is an extensive late **Minoan
cemetery** where a large number
of graves have been excavated,
some with imposingly long
entrances. Among the contents
unearthed are bronze weapons,
vases and burial caskets
(larnakes), now on view in the
archaeological museums of
Chaniá (see p257) and Réthymno.

Minoan Cemetery
9 km (6 miles) S of Réthymno.
Open Tue–Sun. **Closed** main
public hols.

❽ Sfakiá
Σφακιά

Chaniá. 400. Sweetwater 3
km (2 miles) W of Loutró.

Overlooking the Libyan Sea at
the mouth of the breathtaking
Impros Gorge, Sfakiá (also known

The magnificent shell of Frangokástello set against a dramatic backdrop

as Chóra Sfakíon) enjoys a
commanding position as the last
coastal community of any size
until Palaióchora (see pp254–5).
Cut off from the outside world
until recently, it is little wonder
that historically the local Sfakiot
clansmen enjoy their reputation
for rugged self-sufficiency and
individualism, albeit accompanied by the notorious feuding.
The village today is largely
devoted to tourism and makes a
good stepping-off point for the
southwest coast.

Environs

West of Sfakiá, almost
impregnable mountains
plummet into the Libyan Sea,
allowing space for just a couple
of tiny settlements accessible
only by boat or on foot along the
E4 coastal path. The closest of
these is **Loutró**, a charming and
remote spot whose sheltered
cove, curving beach and little
white houses with blue shutters
fulfil every traveller's fantasy of a

The quiet bay and whitewashed houses
of Loutró

"real" Greek village. In summer a
dozen tavernas and houses
provide rooms and meals for
tourists. Small boats are available
to take tourists to nearby Gávdos
island and the breathtaking bay
around Sweetwater beach.

❾ Frangokástello
Φραγκοκάστελλο

14 km (9 miles) E of Sfakiá, Chaniá.
Open daily.

Built by the Venetians as a
bulwark against pirates and
unruly Sfakiots in 1371, little
remains of the interior of
Frangokástello. However, its
curtain walls are well preserved
and, from above the south
entrance, the Venetian Lion of
St Mark looks out to sea.

Ioánnis Daskalogiánnis, the
Sfakiot leader, surrendered here
in 1770 and was flayed alive in
Irákleio by his Turkish captors.
Fifty years later Chatzimichális
Daliánis, a Greek freedom
fighter, wrested the fort from
the Turks and tried to hold it
with an army of just 385 men.
Hopelessly outnumbered, he
and all his followers were
massacred by the pitiless Turks.
Legend has it that at the end of
May at dawn, their solemn
shadows can be seen climbing
up to the castle.

Directly below the fortress
is a sandy beach whose waters
are shallow and warm, an
ideal spot for families with
young children. A scattering
of hotels and tavernas cater
for holiday-makers and
passing motorists.

Boats lining the small harbour at Plakiás

⓾ Plakiás
Πλακιάς

Réthymno. 🏔 100. 🚌 🚆 Damnóni 3 km (2 miles) E.

Once a simple fishing harbour serving the villages of Mýrthios and Selliá, Plakiás has grown into a full-scale resort with all the usual facilities. Its grey sandy beach is nearly 2 km (1 mile) long. Sited at the mouth of the Kotsyfoú Gorge, and with good road connections, Plakiás makes an excellent base for exploring the region.

Environs
A 5-minute drive, or a scenic walk around the headland, leads east to the beach of **Damnóni**. Tiny coves beyond it offer good swimming. Holiday apartments are being built on the adjoining hill. Quiet Soúda beach lies 3 km (2 miles) west of Plakiás.

⓫ Moní Préveli
Μονή Πρέβελη

14 km (9 miles) E of Plakiás, Réthymno. **Tel** 28320 31246. 🚌 **Open** 8am–8pm daily; call in winter to check. 🅿 ♿

Accessible by road through the Kourtaliótiko Gorge, the working monastery of Préveli stands in an isolated but beautiful spot overlooking the sea. It played a prominent role in the evacuation of Allied forces from nearby beaches during World War II *(see p255)*.
The buildings cluster around a large central courtyard dating from 1731. There is a 19th-century church and a small museum displaying religious artifacts, including silver

candlesticks and some highly decorative robes. Further inland, the original 16th-century **Moní Agíou Ioánnou** (now known as Káto Préveli) was founded by Abbot Préveli and abandoned in the 17th century in favour of the more strategic position of the present monastery. About 1 km (0.5 mile) east of Moní Préveli, a steep path leads to **Préveli** beach (also known as Kourtaliótiko or Palm Beach), a crystal-clear, palm-fringed oasis.

Venetian façade of the church at Moní Arkadíou

⓬ Moní Arkadíou
Μονή Αρκαδίου

24 km (15 miles) SE of Réthymno, Réthymno. 🚌 to Réthymno. **Open** daily. 🅿 ♿

The 5th-century monastery of Arkadíou stands at the top of a winding gorge, at the edge of a fertile region of fruit trees and cypresses. Largely rebuilt at the

end of the 16th century, the most impressive of its buildings is the double-naved church with an ornate Venetian façade, which dates back to 1587.
The monastery provided a safe haven for its followers in times of religious persecution by local Muslims. On 9 November 1866, when its buildings were crowded with hundreds of refugees, it came under attack by the Ottoman army. Choosing death over surrender the Cretans torched the gunpowder storeroom, killing Christian and Muslim alike. The ensuing carnage created instant martyrs for freedom whose sacrifice is not forgotten. A sculpture outside the monastery depicts the only surviving girl and the abbot who lit the gunpowder.
Today, a small museum displays sacramental vessels, icons, prayer books, vestments and tributes to the martyrs.

Environs
At **Archaía Eléftherna**, 10 km (6 miles) northeast of Moní Arkadíou, lie the ruins of the ancient city-state of Eléftherna. The remains of a necropolis, a Roman villa, an early basilica, a Hellenistic bridge and a watchtower can all be seen. Northeast of Eléftherna the village of **Margarítes** is well-known for its pottery.

The isolated buildings of Moní Préveli, nestled into the rocks

⑬ Tour of the Amári Valley

Dominated by the peaks of Mount Idi to its east, the Amári Valley offers staggering views over the region's rock-strewn peaks, broad green valleys and dramatic gorges. Twisting but well-paved roads link the many small agricultural communities of the Amári where, even today, moustachioed men in knee-high boots and baggy trousers *(vrákes)* can be seen outside the local tavernas. The area is dotted with shrines, churches and monasteries harbouring Byzantine frescoes and icons. Traditionally an area of Cretan resistance, many of the Amári villages were destroyed during World War II.

Olive groves in the Amári Valley

① Thrónos
The beautifully frescoed church of the Panagía at Thrónos dates back to the 14th century and has traces of 4th-century Christian mosaics. A key is available from the nearby taverna.

⑧ Méronas
At the centre of Méronas is the Venetian-style church of the Panagía with its early 14th-century frescoes.

② Moní Asomáton
The Venetian buildings of **Moní Asomáton**, now an agricultural college, stand in a lush oasis of palm, plane and eucalyptus trees.

③ Amári
Sweeping views of Mount Idi can be seen from the Venetian clock tower in the centre of Amári. Just outside the village, the church of Agía Anna shelters the island's oldest frescoes, dated 1225.

⑦ Gerakári
Gerakári is famous for its fresh and bottled cherries and cherry brandy.

④ Vizári
Just west of the village of Vizári are the ruins of an early Christian basilica dating from the 6th century.

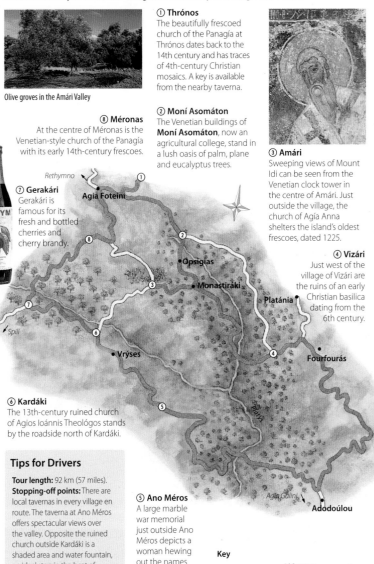

⑥ Kardáki
The 13th-century ruined church of Agios Ioánnis Theológos stands by the roadside north of Kardáki.

Map labels: Rethymno, Agía Foteiní, Opsigiás, Monastiráki, Platánia, Spili, Vrýses, Fourfourás, Ano Méros, Plariá, Agía Galíni, Adodoúlou

Tips for Drivers

Tour length: 92 km (57 miles).
Stopping-off points: There are local tavernas in every village en route. The taverna at Ano Méros offers spectacular views over the valley. Opposite the ruined church outside Kardáki is a shaded area and water fountain, an ideal stop in the heat of summer (see also p362).

⑤ Ano Méros
A large marble war memorial just outside Ano Méros depicts a woman hewing out the names of World War II Resistance heroes.

Key
— Tour route
= Other roads

0 kilometres 5
0 miles 2

⑭ Mount Idi
Ψηλορείτης

Réthymno. 🚌 to Anógeia & Kamáres.

At 2,456 m (8,080 ft) the soaring peaks of **Mount Idi** (or Psiloreítis) are the crowning glory of the massive Psiloreítis range. The highest mountain in Crete, it is home to many sanctuaries including the famous Idaian Cave.

From Anógeia, a paved road leads to the **Nída Plateau**, a journey of 23 km (14 miles) through rocky terrain, punctuated by the occasional stone shepherd's hut. Here a lone taverna caters to visitors en route to the **Idaian Cave**, a further 20-minute hike up the hill. This huge cavern, where Zeus was reared, has yielded artifacts, including some remarkable bronze shields, dating from c.700 BC. Some of the artifacts can be seen in the Irákleio Archaeological Museum *(see pp274–5)*. From the plateau, marked trails lead up to the peak of Mount Idi. The summit trek to the chapel of Timios Stavros is, approximately, an 8-hour round trip.

On the mountain's southern face, a 3-hour scramble from Kamáres village leads to the **Kamáres Cave**. Here the famous Minoan pottery known as Kamáres ware was discovered and examples are now on display in the Irákleio Archaeological Museum.

Cretan Caves and the Myth of Zeus

The island of Crete is home to 4,700 caves and potholes of which some 2,000 have been explored. Since Neolithic times, caves have been used as cult centres by successive religions and have yielded many archaeological treasures. Bound up with ancient Cretan

mythology, the Diktian *(see p281)* and Idaian caves are two of the island's most visited. According to legend, Rhea gave birth to the infant god Zeus in the Diktian Cave, where he was protected by *kourítes* (warriors) and nurtured by a goat. He was then concealed and raised in the Idaian Cave to protect him from his father, Kronos, who had swallowed his other offspring after a warning that he would be dethroned by one of his sons. The Idaian Cave was an important pilgrimage centre during Classical times.

Stalagmites in the Diktian Cave *(see p281)*, Lasíthi

⑮ Anógeia
Ανώγεια

Réthymno. 🏠 2,300. 🚌

High up in the Psiloreítis mountain range, the small village of Anógeia dates back to the 13th century. The village has suffered a turbulent past, having been destroyed by the Turks in 1821 and 1826, and then completely rebuilt after destruction by the German army in 1944.

Modern Anógeia runs along a rocky ridge, with its own square and **war memorial** – a bronze

statue of a Cretan hero in traditional dress. Inscribed on the memorial are the most significant dates in Crete's recent past: 1821, Greek Independence; 1866, slaughter of Christian refugees at Moní Arkadíou *(see p264)*; 1944, liberation from German occupation. Tavernas, shops and banks are also situated in this part of town.

The old village tumbles down the steep slopes into a warren of narrow stepped alleys, ultimately converging on a little square of stalls and tavernas. Here, a marble bust of local politician

The Nída Plateau between Anógeia village and the Idaian Cave, Mount Idi

Woman selling locally made rugs and lace in Anógeia

Vasíleios Skoulás stands next to a less formal woodcarving of his friend Venizélos *(see p47)*, by local artist Manólis Skoulás.

The stalls in the old part of the village abound in locally made embroidery, lace and brightly coloured rugs, forming one of Crete's main centres for woven and embroidered goods. Nearby tavernas serve grilled goats' meat and other Cretan specialities. Music enthusiasts can pay their respects at the shrine of Níkos Xyloúris, a 1970s folk singer who died at an early age and whose little whitewashed house overlooks the main square.

⓰ Agía Galíni
Αγία Γαλήνη

Réthymno. ⛰ 1,040. 🚌
🚊 Agía Galíni.

Formerly a fishing village situated at the southern end of the Amári Valley, Agía Galíni is today a full-blown tourist resort. The original village, now only a handful of old houses and narrow streets, is dwarfed by the mass of holiday apartments stretching up the coast. The harbourfront is alive with busy tavernas snuggled between the water and cliffs. Just beyond the harbour, the small sandy beach is popular with sunbathers.

Environs
Taxi boat trips sail daily from Agía Galíni's harbour to the neighbouring beaches of **Agios Geórgios** and **Agios Pávlos** and, further still, to **Préveli** beach at Moní Préveli *(see p264)*. There are also daily excursions to the **Paximádia islands** where there are good sandy beaches.

⓱ Agía Triáda
Αγία Τριάδα

3 km (2 miles) W of Phaestos, Irákleio. 🚌 to Phaestos. **Tel** 28920 91564. **Open** Apr–Oct: 8am–8pm daily; Nov–Mar: noon–5pm Mon, 8:30am–3pm Tue–Sun. **Closed** main public hols. 🚫 🚊 Kómo 10 km (6 miles) SW; Mátala 15 km (9 miles) SW.

The Minoan villa of Agía Triáda was excavated by the Italians from 1902 to 1914. An L-shaped structure, it was built around 1700 BC, the time of the Second Palace period *(see p279)*, over earlier houses. Its private apartments and public reception rooms are in the angle of the L, overlooking a road that may have led to the sea. Gypsum facing and magnificent frescoes used to adorn the walls of these rooms. Rich Minoan treasures, including the carved stone Harvester Vase, Boxer Rhyton (jug) and Chieftain Cup, were all found in this area and are on display at the Irákleio Archaeological Museum *(see pp274–5)*. Evidence of the villa's importance is provided by a find of clay seals and rare tablets bearing the undeciphered Minoan Linear A script.

Following the villa's destruction by fire in around 1400 BC, a Mycenaean *megaron* (hall) was built on the site. The ruined settlement to the north, with its unique porticoed row of shops, dates mostly from this period, as does the magnificent painted sarcophagus that was found in the cemetery to the north. The paintwork on the sarcophagus depicts a burial procession; it can be seen in the Irákleio Archaeological Museum.

Agía Triáda archaeological site

Environs
At the village of Vóroi, 6 km (4 miles) northeast of Agía Triáda, is the fascinating **Museum of Cretan Ethnology**. Displayed here is a collection of tools and materials used in the everyday life of rural Crete up to the early 20th century.

🏛 **Museum of Cretan Ethnology**
Tel 28920 91110. **Open** 10am–6pm daily; times may vary in winter, so call first. **Closed** main public hols. 🚫

Agía Galíni resort, nestled into the rocks at the foot of the Amári Valley

Mátala's town beach flanked by sandstone cliffs

⑱ Mátala
Μάταλα

Irákleio. 🏛 132. 🚌 🚕 Kalamáki
5 km (3 miles) N; Léntas 24 km
(15 miles) SE.

Clustered around an idyllic
sweeping bay, Mátala remained
a small fishing hamlet until the
tourist boom of the 1960s,
when it was transformed into
a pulsating resort. Hotels, bars
and restaurants abound in
the lively town centre and
development here is steadily
on the increase.

Despite present appearances,
Mátala has not passed un-
touched by history. Homer
described Menelaos, husband of
Helen of Troy *(see p58)*, being
shipwrecked here on his way
home from Troy. During
Hellenistic times, around 220 BC,
Mátala served as the port
for the ancient city-state of
Górtys. The resort's
pitted sandstone
cliffs, looming drama-
tically over the town
beach, were originally
carved out for use as
tombs in the Roman
era. Later they were
extended as cave
dwellings for early
Christians, shepherds
and even hippies.

Environs
The area around
Mátala has some
beautiful beaches,
including the bay of
Kaloí Liménes to the
southeast. This was

said to have been the landing
place of St Paul the Apostle
on his way to Egypt. To the
north, a sandy track leads
to **Kommós**, one of the
best sandy beaches on
the south coast. In this
magnificent setting lay
the Minoan settlement
of Kommós, thought
to have been a major
port serving Phaestos
(see pp270–71).
The extensive
site is currently
under excavation.

Boat excursions run
daily from Mátala to
the Paximádia islands in
the bay and to palm-
fringed Préveli beach
(see p264) further west.
There are also several bus tours
to the important archaeological
sites of Phaestos, Agía Triáda
(see p267) and Górtys.

⑲ Phaestos
See pp270–71.

⑳ Górtys
Γόρτυς

Irákleio. **Tel** 28920 31144. 🚌
Open 8am–6pm daily. **Closed**
main public hols. 📷 ♿

A settlement from Minoan
through to Christian times,
the ancient city-state of Górtys
began to flourish under Dorian
rule during the 6th century
BC. Following its defeat of
Phaestos in the 2nd century
BC, Górtys became the most
important city on Crete. Its
pre-eminence was sealed
following the Roman invasion
of 65 BC, when Górtys
was appointed capital of
the newly created Roman
province of Crete
and Cyrene (modern-
day Libya). Górtys
continued to flourish
under Byzantine
rule, strategically
sited at the point
where a tributary of
the ancient River Lethe
(today's Mitropolianós)
flowed into the fertile
Messará Plain, with
coastal ports to the west
and south. It was not until
the late 7th century AD
that the great city was
destroyed by Arab invaders.
Today, the most-visited ruins of
this extensive site lie to the
north of the main road.

Statue at the
ancient site of
Górtys

The Law Code of Gortys

Section of the Law Code of Górtys, housed in
the odeion, Górtys

The most extensive set of early written
laws in the Greek world was found at
the archaeological site of ancient Górtys
and dates from c.500 BC. Each stone
slab of the Górtys Code contains 12
columns of inscriptions in a Doric
Cretan dialect. There is a total of 600
lines which read alternately from left to
right and from right to left (a style
known as *boustrophedón*, literally "as
the ox-plough turns"). The laws were
on display to the public and related to
domestic matters including marriage,
divorce, adoption, the obligations and
rights of slaves, and the sale and division
of property.

The *bema* (area behind altar) of Agios Títos basilica, Górtys

Exploring the Ruins

A car park, ticket booth and café are located near the entrance to the site. Immediately beyond stand the remains of the 6th-century basilica of **Agios Títos**, once an impressive, three-aisled edifice whose floor plan is still clearly visible. In its heyday it was the premier Christian church of Crete, traditionally held to be the burial place of St Titus, first bishop and patron saint of Crete, who was sent by St Paul to convert the heathens. Behind the basilica is an area thought to be a Greek **agora** (marketplace). Beyond this stand the semicircular tiered benches of the Roman **odeion**, originally used for concerts and now home to the famous stone slabs inscribed with the Law Code of Górtys.

Behind the odeion, a path leads up to the **acropolis** hill above Górtys, where a post-Minoan settlement was built

around 1000 BC. Parts of the fortifications still remain. On the east slope of the hill are the foundations of the 7th-century-BC **Temple of Athena**. A statue and other votive objects found at a sacrificial altar lower down are in Irákleio Historical Museum *(see p272)*.

To the south of the main road, an extensive area of Roman Górtys remains only partially excavated. Standing in a grove of old olive trees is the 7th-century-BC **Temple of Pythian Apollo**, to which a monumental altar was added in Hellenistic times. The temple was converted into a Christian basilica in the 2nd century AD and remained important until AD 600, when it was superseded by the basilica of Agios Títos. At the far end of the site are the ruins of the 1st-century-AD **praetorium**, the grand palace of the Roman provincial governor.

Environs

East of Górtys, in the nearby village of **Agioi Déka**, is the 13th-century Byzantine church of the same name. It was built on the spot where ten Early Christian Cretans were martyred in AD 250 for their opposition to the Roman Emperor Decius. In the nave of the church is an icon portraying the ten martyrs. North of Górtys, a scenic drive

13th-century icon of the ten martyrs, Agioi Déka church

heads to the mountain village of **Zarós**, a surprisingly green oasis famous for its clear spring water. From here, a clearly marked trail leads north through the spectacular **Zarós Gorge**. About 3 km (2 miles) northwest of Zarós lies **Moní Vrontisíou**. The monastery's icons by Michaíl Damaskinós (c.1530–91), a famous painter of the Cretan School, are now on display in the Museum of Religious Art in Irákleio *(see p272)*.

The ruins of the *praetorium*, the once-grand palace complex of the governor of the province, Górtys

⑲ Phaestos

Το Ανάκτορο της Φαιστού

Spectacularly situated on a ridge overlooking the fertile Messará Plain, Phaestos was one of the most important Minoan palaces in Crete. Excavations by the Italian archaeologist Frederico Halbherr, in 1900, unearthed two palaces. Remains of the first palace, constructed around 1900 BC and destroyed by an earthquake in 1700 BC, are still visible. However, most of the present ruins are of the second palace, which was severely damaged around 1450 BC, possibly by a tidal wave. The city-state was finally destroyed by Górtys *(see pp268–9)* in the 2nd century BC. Today, the superimposed ruins of both palaces make interpretation of the site difficult.

The Messará Plain from the north court

KEY

① **Storage pits** dating from around 1900 BC. These circular walled pits were used for storing the palace's grain.

② **West courtyard and theatre area**, ruins of the west court date to c. 1900 BC, the first palace period. The seats on its north side were used for viewing rituals and ceremonies.

③ **First Palace shrine complex**

④ **North court**

⑤ **The peristyle hall**, a colonnaded courtyard, bears traces of an earlier structure dating from the Prepalatial period (3500–1900 BC).

⑥ **The archives room** consists of a series of mudbrick chests. It was here that the famous Phaestos disc was discovered.

⑦ **Northeast quarter**

⑧ **Workshops**

⑨ **The main hall** is where clay seals dating to c.1900 BC were found.

⑩ **Storerooms**

⑪ **First palace remains**, dating from c.1900 BC, are concentrated in the southeast of the site, fenced off for protection.

⑫ **A Classical temple** shows that the site was still occupied after Minoan times.

The Phaestos Disc

This round clay disc, 16 cm (6 inches) in diameter, was discovered at Phaestos in 1903. Inscribed on both sides with pictorial symbols that spiral from the circumference into the centre, no one has yet been able to decipher its meaning or identify its origins, though it is possibly a sacred hymn. The disc is one of the most important exhibits at the Irákleio Archaeological Museum *(see pp274–5)*.

★ **Grand Staircase**
This monumental staircase, which leads up to a *propylon* (porch) and colonnaded light-well, was the main entrance to the palace.

Royal Apartments
Now fenced, these rooms were the most elaborate, consisting of the Queen's Megaron or chamber (left), the King's Megaron, a lavatory and a lustral basin (covered pool).

VISITORS' CHECKLIST

Practical Information
65 km (40 miles) SW of Irákleio. **Tel** 28920 42315. **Open** Apr–Oct: 8am–8pm daily; Nov–Mar: noon–5pm Mon, 8:30am–3pm Tue–Sun. **Closed** main public hols.

Transport

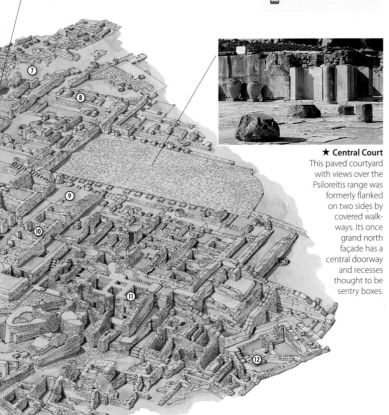

★ **Central Court**
This paved courtyard with views over the Psiloreítis range was formerly flanked on two sides by covered walk-ways. Its once grand north façade has a central doorway and recesses thought to be sentry boxes.

Reconstruction of Second Palace

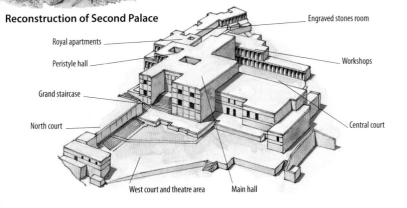

Royal apartments

Peristyle hall

Grand staircase

North court

West court and theatre area

Main hall

Engraved stones room

Workshops

Central court

㉑ Irákleio

Ηράκλειο

A settlement since the Neolithic era, Irákleio served as the port for Knosós in Roman times. Under Venetian rule in the 13th century, it became known as Candia, the capital of the Aegean territories. Today the sprawl of traffic-jammed streets and concrete apartment buildings detracts from Irákleio's appeal. Yet, despite first impressions, the island's capital harbours a wealth of Venetian architecture, including the city walls and fortress. Its Archaeological Museum houses the world's greatest collection of Minoan art, and the city provides easy access to the Palace of Knosós *(see pp276–9)*.

Façade of the Venetian church of Agios Títos

Exploring Irákleio

At the heart of Irákleio is Plateía Eleftheríou Venizélou, a pedestrian zone with cafés and shops grouped around the ornate 17th-century **Morosini Fountain**. Facing the square, the restored church of **Agios Márkos** was built by the Venetians in 1239 and is now used as a venue for concerts and exhibitions. From here, 25 Avgoústou (25 August Street) leads north to the Venetian harbour. On this street, the elegantly restored 17th-century **Loggia** was a meeting place for the island's nobility and now serves as Irákleio's city hall. Beyond the Loggia, in a small square set back from the road, is the refurbished 16th-century church of **Agios Títos**, dedicated to the island's patron saint. On the other side of 25 Avgoústou, the tiny **El Greco Park** is named after Crete's most famous painter.

At the northern end of 25 Avgoústou, the old harbour is dominated by the Venetian **fortress**, whose dauntingly massive structure successfully repulsed prolonged assaults by the invading Turks in the 17th century. Named the *Rocca al*

Lion of St Mark detail, fortress

Mare (Fort on the Sea) by the Venetians and *Koulés* by the Turks, it was erected by the Venetians between 1523 and 1540. Opposite the fortress are the arcades of the 16th-century Venetian **Arsenali** where ships were built and repaired. West along the waterfront, the **Historical Museum** traces the history of Crete since early Christian times. Its displays include Byzantine icons and friezes, sculptures, and archives of the Battle of Crete *(see p255)*. Pride of place is given to the only El Greco painting in Crete, *The Landscape of the Gods-Trodden Mount Sinai* (c.1570).

A short walk two blocks southwest of Plateía Venizélou, on Plateía Agías Aikaterínis, is the 16th-century Venetian church of Agía Aikateríni of Sinai. Once a monastic foundation famous as a centre of art and learning, it now houses the **Museum of Religious Art**, a magnificent collection of Byzantine icons, frescoes and manuscripts. The

El Greco

Domínikos Theotokópoulos (alias El Greco) was born in Crete in 1545. His art was rooted in the Cretan School of Painting, an influence that permeates his highly individualistic use of dramatic colour and elongated human forms. In Italy, El Greco became a disciple of Titian before moving to Spain. He died in 1614, and his works can be seen in major collections around the world. Ironically, only one exists in Crete, at Irákleio's Historical Museum.

El Greco's *The Landscape of the Gods-Trodden Mount Sinai* (c.1570), Historical Museum

most significant exhibits are six icons by Michaíl Damaskinós, a 16th-century Cretan artist who learned his craft here. The museum is currently closed for renovation. Next door, the 19th-century cathedral of **Agios Minás** towers over the square.

To the east, the street market in 1866 Street leads south to Plateía Kornárou. Here, coffee is served from a charming converted Turkish pumphouse, next to which a headless Roman statue graces the 16th-century **Bembo Fountain**. East, along Avérof, Plateía Eleftherías (Freedom Square) is dominated by a

Irákleio's boat-lined harbour, dominated by the vast Venetian fortress

The Bembo drinking fountain, Plateía Kornárou

VISITORS' CHECKLIST

Practical Information
Irákleio. 116,000.
Xanthoudídou 1 (28134 46100,
dtkritis@otenet.gr). Sat.
Summer Festival: Jul–Sep.
Amoudára 10 km (6 miles) W.

Transport
5 km (3 miles) E. E of
Venetian harbour. Leofóros
Papadimitríou (for Réthymno,
Chaniá, Agios Nikólaos and
Ierápetra); Plateía Kóraka
(for Mátala).

statue of Elefthérios Venizélos
(1864–1936), the politician
central to Crete's union with
Greece. Off the square, the
pedestrianized Daidálou is
good for shops and restaurants.
Just to the north is the **Irákleio
Archaeological Museum**
(see pp274–5) and main
tourist office.

South of town, beyond the
old city walls, the small **Museum
of Natural History** deals with
the natural environment of the
Aegean. Exhibits include fossils
and live animals.

Loggia
25 Avgoústou. **Tel** 28103 99399.
Open Mon–Sat. **Closed** main
public hols.

Fortress
Venetian harbour. **Tel** 28102 88484.
Open 8:30am–7pm Tue–Sun (Nov–
Mar: 3pm). **Closed** main public hols.

Historical Museum
Lysimáchou Kalokairinoú 7. **Tel** 28102
83219. **Open** 9am–5pm Mon–Sat
(Nov–Mar: 3:30pm). **Closed** main
public hols.

Museum of Religious Art
Agía Aikateríni of Sinai, Plateía Agías
Aikaterínis. **Tel** 28102 46100. **Open**
9am–5pm Mon–Sat.

Museum of Natural History
Sofokli Venizélou. **Tel** 28103 93276.
Open 9am–8pm Sun–Fri.

Environs
Travelling west by the main
Irákleio–Réthymno road, a turn-

off to Anógeia (see pp266–7)
climbs to the village of **Týlissos**,
where the remains of three
Minoan villas were found in
1902. West of Irákleio, the road
leads to the village of **Fódele**,
claimed to be the birthplace of
El Greco. His house lies above
the Byzantine church to the
north-west. The **CretAquarium**,
15 km (9 miles) east from Irákleio,
exhibits around 200 species of
fish and invertebrates.

CretAquarium
Near Gournes. **Tel** 28103 37788.
Open May–Sep: 9:30am–9pm daily;
Oct–Apr: 9:30am–5pm daily.

Irakleio City Centre

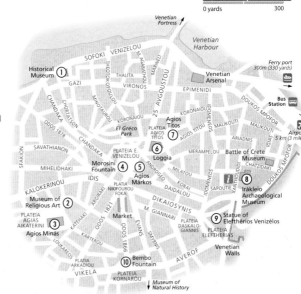

For keys to symbols see back flap

Irákleio Archaeological Museum
Αρχαιολογικό Μουσείο Ηρακλείου

The Irakleio Archaeological Museum houses the world's most important collection of Minoan artifacts, giving a unique insight into a highly sophisticated civilization that existed on Crete over 3,000 years ago. On display are exhibits from all over Crete, including the famous Minoan frescoes from Knosós *(see pp276–9)* and the Phaestos Disc *(see p270)*. Finely carved stone vessels, jewellery, Minoan double axes and other artifacts make up only part of the museum's vast collection. There are state-of-the-art interactive exhibits, as well as galleries that house archaeological treasures that are on display for the first time.

Gold Bee Pendant
Found in the Chrysólakkos cemetery at Mália *(see p281)*, this exquisite gold pendant of two bees joined together dates from the 17th century BC.

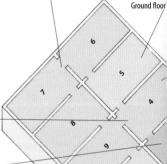

Ground floor

★ **Bull's Head** *Rhyton*
This 16th-century BC vessel *(see p67)* was used for the pouring of ritual wines. Found at Knosós, it is carved from steatite, a black stone, with inset rock crystal eyes and a mother-of-pearl snout.

★ **Phaestos Disc**
Made of clay, the disc was found at the Palace of Phaestos in 1903.

Octopus Vase
This fine late Minoan vase from Palaíkastro *(see p285)* is decorated with images from the sea.

Stairs to first floor

Minoan vase with double axe motif

The Minoan Double Axe

The Minoan double axe served both as a common tool used by carpenters, masons and shipbuilders, and as an extremely powerful sacred symbol thought to have been a cult object connected with the Mother Goddess. The famous Labyrinth at Knosós *(see pp276–9)* is believed to have been the "dwelling place of the double axe", the word *labrys* being the ancient Greek name for double axe. Evidence of the importance of the axe for the Minoans is clear from the many vases, *larnakes* (clay coffins), seals, frescoes and pillars that were inscribed or painted with the ceremonial "double axe", including the walls of the Palace of Knosós. The ceremonial axe is often depicted between sacred horns or in the hands of a priest. Votive axes (ritual offerings) were highly decorated and made of gold, silver, copper or bronze. A stylized version of the double axe also features in early Linear A and B scripts.

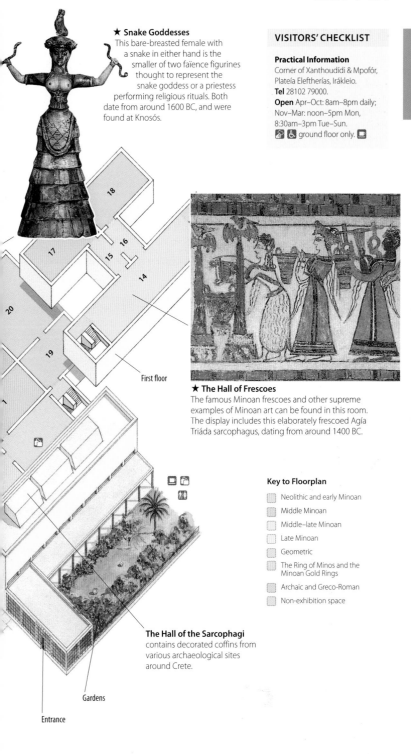

★ **Snake Goddesses**
This bare-breasted female with a snake in either hand is the smaller of two faïence figurines thought to represent the snake goddess or a priestess performing religious rituals. Both date from around 1600 BC, and were found at Knosós.

VISITORS' CHECKLIST

Practical Information
Corner of Xanthoudídi & Mpofór, Plateía Eleftherías, Irákleio.
Tel 28102 79000.
Open Apr–Oct: 8am–8pm daily; Nov–Mar: noon–5pm Mon, 8:30am–3pm Tue–Sun.
ground floor only.

First floor

★ **The Hall of Frescoes**
The famous Minoan frescoes and other supreme examples of Minoan art can be found in this room. The display includes this elaborately frescoed Agía Triáda sarcophagus, dating from around 1400 BC.

Key to Floorplan
- Neolithic and early Minoan
- Middle Minoan
- Middle–late Minoan
- Late Minoan
- Geometric
- The Ring of Minos and the Minoan Gold Rings
- Archaic and Greco-Roman
- Non-exhibition space

The Hall of the Sarcophagi contains decorated coffins from various archaeological sites around Crete.

Gardens

Entrance

㉒ The Palace of Knosós

Ανάκτορο της Κνωσού

Built around 1900 BC, the first palace of Knosós was destroyed by an earthquake in about 1700 BC and was soon completely rebuilt. The restored ruins visible today are almost entirely from this second palace. The focal point of the site is its vast north–south aligned Central Court, off which lie many of the palace's most important areas *(see pp278–9)*. The original frescoes are in the Archaeological Museum of Irákleio *(see pp274–5)*.

The Central Court facing towards the northeast

KEY

① **The South House**, partly restored, was once three storeys high. It was probably the residence of a palace official.

② Corridor of the Procession

③ Bust of Sir Arthur Evans

④ *Kouloúres* (storage pits)

⑤ West Court

⑥ West Magazines

⑦ Stairs to Piano Nobile (upper floor)

⑧ To Theatre and Royal Road

⑨ North Lustral Basin

⑩ *Charging Bull* fresco

⑪ North Pillar Hall (Customs House)

⑫ The magazines of the giant *pithoi* contain jars dating from the First Palace period (c.1800 BC).

⑬ King's Megaron (Hall of the Double Axes)

⑭ Hall of the Royal Guard

⑮ Queen's Megaron

⑯ Grand Staircase

⑰ Central Court

⑱ **The Tripartite Shrine**, formerly protected by a roof, was one of many shrines facing on to the Central Court.

Horns of Consecration
Sitting on the south façade, these restored horns are a symbol of the sacred bull, and would once have adorned the top of the palace.

Modern entrance

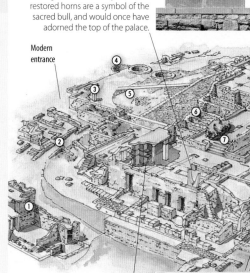

South Propylon
Entrance to the palace was through this monumental, pillared gateway, decorated with a replica of the cup-bearer figure, a detail from the *Procession* fresco.

★ **Priest-King Fresco**
This replica of the *Priest-King* fresco, also known as the *Prince of the Lilies*, is a detail from the *Procession* fresco and depicts a figure wearing a crown of lilies and feathers.

★ **Throne Room**
With its adjoining antechamber and lustral basin, the Throne Room is believed to have served as a shrine. The original stone throne, thought to be that of a priestess, is guarded by a restored fresco of griffins, sacred symbols in Minoan times.

VISITORS' CHECKLIST

Practical Information
5 km (3 miles) S of Irákleio.
Tel 28102 31940. **Open** Apr–Oct: 8am–8pm daily; Nov–Mar: noon–5pm Mon, 8:30am–3pm Tue–Sun.
Closed main public holidays.

Transport

★ **Giant *Pithoi***
Over 100 giant *pithoi* (storage jars) were unearthed at Knosós. The jars were used to store palace supplies.

North entrance

★ **Royal Apartments**
These rooms include the King's Megaron, also known as the Hall of the Double Axes; the Queen's Megaron, which is decorated with a copy of the famous dolphin fresco and has an en suite bathroom; and the Grand Staircase.

Exploring the Palace of Knosós

Unlike other Minoan sites, the Palace of Knosós was imaginatively restored by Sir Arthur Evans between 1900 and 1929. While his interpretations are the subject of academic controversy, his reconstructions of the second palace do give the visitor an impression of life in Minoan Crete that cannot so easily be gained from the other palaces on the island.

Restored clay bath tub adjacent to the Queen's Megaron

Around the South Propylon

The palace complex is entered via the **West Court**, the original ceremonial entrance now marked by a bust of Sir Arthur Evans. To the left are three circular pits known as *kouloúres*, which probably served as granaries. Ahead, along the length of the west façade, are the **West Magazines**. These contained numerous large storage jars (*pithoi*), and, along with the granaries, give an impression of how important the control of resources and storage was as a basis for the power of the palace.

At the far right-hand corner of the West Court, the west entrance leads to the **Corridor of the Procession**. Now cut short by erosion of the hillside, the corridor's frescoes, depicting a series of gift-bearers, seem to reflect the ceremony that accompanied state and religious events at the palace. This is further revealed in the frescoes of the **South Propylon**, to which one branch of the corridor led. From the South Propylon, steps lead up

Shield motif, Knosós

to the reconstructed **Piano Nobile**, the name given by Sir Arthur Evans to the probable location of the grand state apartments and reception halls. Stone vases found in this part of the palace were used for ritual purposes and indicate the centrality of religion to palace life. The close link between secular and sacred power is also reinforced by the **Throne Room**, where ritual bathing in a lustral basin (sunken bath) is thought to have taken place. Steps lead from the Throne Room to the once-paved **Central Court**. Now open to the elements, this would have once been flanked by high buildings on all four sides.

The Royal Apartments

On the east side of the Central Court lie rooms of such size and elegance that they have been identified as the Royal Apartments. The apartments are built into the side of the hill and accessed by the **Grand Staircase**, one of the most impressive surviving architectural features of the

palace. The flights of gypsum stairs descend to a colonnaded courtyard, providing a source of light to the lower storeys. These light-wells were a typical feature of Minoan architecture.

A drainage system was provided for the toilet beside the **Queen's Megaron**, which enjoyed the luxury of an en suite bathroom complete with clay bathtub. Corridors and rooms alike in this area were decorated with frescoes of floral and animal motifs. The walls of the **Hall of the Royal Guard**, a heavily guarded landing leading to the Royal Apartments, were decorated with a shield motif. The **King's Megaron**, also known as the Hall of the Double Axes, takes its name from the fine double-axe symbols incised into its stone walls. The largest of the rooms in the Royal Apartments, the King's Megaron could be divided by multiple doors, giving it great flexibility of space. Remains of what may have been a plaster throne were found here, suggesting that the room was also used for some state functions.

North and West of the Central Court

The north entrance of the Central Court was adorned with remarkable figurative decoration. Today, a replica of the *Charging Bull* fresco can be

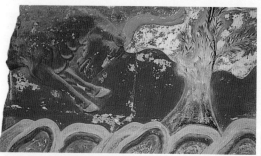

Replica of the celebrated *Charging Bull* fresco

For hotels and restaurants in this region see pp312–13 and pp331–2

seen on site. The north entrance leads to the **North Pillar Hall**, named as the Customs House by Sir Arthur Evans who believed merchandise was inspected here. The hall is an addition of the second palace period (c.1700 BC). Immediately to the west is a room with restored steps leading into a pool, known as the **North Lustral Basin**. Traces of burning and finds of oil jars suggest that those coming to the palace were purified and annointed here before entering. Further west is the **Theatre**, a stepped court whose position at the end of the Royal Road suggests that rituals connected with the

The stepped court of the theatre

reception of visitors may have occurred here. The **Royal Road**, which leads away from the Palace to the Minoan town of Knosós, was lined with houses. Just off the Royal Road lies the so-called **Little Palace**. This building has been excavated, but is not open to the public. It is architecturally very similar to the main palace and was destroyed at the same time.

The History of Knosós

The capital of Minoan Crete, Knosós was the largest and most sophisticated of the palaces on the island. It contained over 1,000 rooms and enjoyed the comforts of an elaborate drainage system, flushing toilets and paved roads. In legend, Knosós was believed to be the setting of an underground labyrinth designed to imprison the Minotaur. This half-man, half-bull was born of King Minos's wife, Pasiphaë, and slain by Theseus. This reconstruction shows the second palace as it might have looked in about 1700 BC.

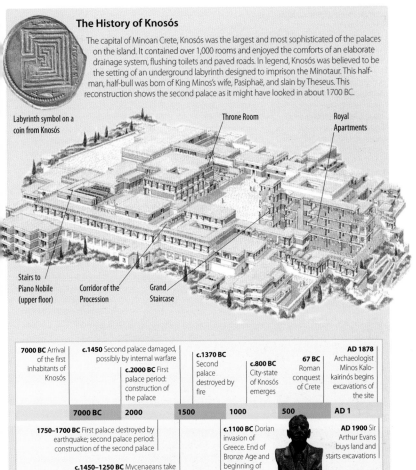

Labyrinth symbol on a coin from Knosós

Throne Room

Royal Apartments

Stairs to Piano Nobile (upper floor)

Corridor of the Procession

Grand Staircase

7000 BC								AD 1

7000 BC Arrival of the first inhabitants of Knosós

c.1450 Second palace damaged, possibly by internal warfare

c.2000 BC First palace period: construction of the palace

c.1370 BC Second palace destroyed by fire

c.800 BC City-state of Knosós emerges

67 BC Roman conquest of Crete

AD 1878 Archaeologist Mínos Kalokairinós begins excavations of the site

1750–1700 BC First palace destroyed by earthquake; second palace period: construction of the second palace

c.1450–1250 BC Mycenaeans take control of Knosós

c.1100 BC Dorian invasion of Greece. End of Bronze Age and beginning of Dark Ages

AD 1900 Sir Arthur Evans buys land and starts excavations

Sir Arthur Evans

The modern seafront of Chersónisos, the busiest of Crete's package-holiday resorts

㉓ Archánes

Αρχάνες

Irákleio. △ 4,000. 📧 🚹 28102
46299 (Irákleio office).

A way from Crete's coastal
holiday resorts, Archánes is a
down-to-earth farming centre,
where olive groves and small
vineyards chequer the rolling
landscape. Lying at the foot of
the sacred **Mount Gioúchtas**
(burial place of Zeus according
to local tradition), Archánes
was a thriving and important
settlement in Minoan times.

In 1964, the remains of a
Minoan **palace** were found in
the town of Tourkogeitoniá.
A short walk out of town, on
Fourní hill to the north, lies an
extensive **Minoan cemetery**.
Among the treasures un-
earthed here was the tomb
of a princess with mirror and
gold diadem in place, as well
as exquisitely engraved signet
rings. Some of these are
now on display at the
Archaeological Museum
of Archánes.

🏛 **Minoan cemetery**
Fourní hill. **Open** daily. **Closed** main
public hols.

🏛 **Archaeological Museum**
Kalochristianáki. **Open** 8:30am–
2:30pm Wed–Mon. **Closed** main
public hols. ♿

Environs
On the north slope of Mount
Gioúchtas is the site of a Minoan
sanctuary at **Anemospiliá**.
Excavations unearthed a shocking scene of human sacrifice
here, seemingly interrupted by
an earthquake around 1700 BC
which killed all four participants.
Though little remains to be seen
today, there are sensational views
of Mount Idi *(see p266)*.

The **Kazantzákis Museum** at
Myrtiá has memorabilia of the
author of *Zorba the Greek*. Nearby,
in Scalani, the **Boutari Winery**
offers guided tours and tastings.

🏛 **Kazantzákis Museum**
Myrtiá, 14 km (9 miles) E of Archánes.
Tel 2810 742451. **Open** Mar–Oct:
9am–5pm Wed–Sun; Nov–Feb: 10am–
3pm Sun. **Closed** main public hols. 🖼

㉔ Chersónisos

Χερσόνησος

Irákleio. △ 4,050. 📧 🚋 Chersónisos.

A flourishing and busy port
from Classical to early Byzantine
times, Chersónisos (strictly
Liménas Chersonísou) is today
the centre of the package-
holiday business. Amid the
plethora of tavernas, souvenir
shops and discos, the harbour
still retains faint intimations of
the old Chersónisos. Along the
waterfront a pyramid-shaped
Roman **fountain** with fish
mosaics dates from the 2nd–3rd
century AD. Some remains of
the **Roman harbour**, now
mostly submerged, can also
be seen here.

On the coast, at the eastern
edge of town, traditional Cretan
life is recreated at the **Cretan
Open-Air Museum** or
"Lychnostátis", where exhibits
include a windmill, a stone house
and a gallery. The **Crete Golf
Club** in Chersónisos is the only
golf course on the island. Clubs
can be hired and the clubhouse
has a bar and restaurant. To cool
off, the **Acqua Plus Water Park** is
a playground of pools, water-
slides and waterfalls.

🏛 **Cretan Open-Air Museum**
Lychnostátis. **Tel** 28970 23660.
Open 9am–2pm Sun–Fri.
Closed main public hols. 🖼 ♿

⛳ **Crete Golf Club**
7 km (4 miles) S of Chersónisos.
Tel 28970 26000. **Open** daily.
🌐 crete-golf.gr

🎢 **Acqua Plus Water Park**
5 km (3 miles) S of National Highway.
Tel 28970 24950. **Open** May–Oct:
10am–6pm daily (Jun & Sep: to
6:30pm; Jul & Aug: to 7pm). 🖼 ♿
🌐 acquaplus.gr

Nikos Kazantzákis

From the village of Myrtiá, Níkos Kazantzákis (1883–1957) was
Crete's greatest writer. Dedicated to the Cretan struggle for
freedom from Turkish rule,
he wrote poems,
philosophical essays, plays
and novels including *Zorba
the Greek* and *The Last
Temptation of Christ* (both
made into films).
Excommunicated by the
Orthodox Church, the
epitaph on his grave in
Irákleio consists of his own
words: "I hope for nothing. I
fear nothing. I am free."

Poster of the 1960s film version of
Zorba the Greek

㉕ Mália
Μάλια

36 km (22 miles) E of Irákleio.
🏘 2,700. 🚌 Stalída 3 km
(2 miles) NW.

The Mália of package-holiday
fame bustles noisily with sun-
seekers hellbent on enjoying the
crowded beaches by day and the
cacophony of competing discos
by night.

In marked contrast, the less
visited Minoan **Palace of Mália**
lies in quiet ruins along the
coastal plain to the east. The first
palace was built in 1900 BC, but,
like all the other major palaces, it
suffered destruction in 1700 BC
and again in 1450 BC *(see p279)*.
The site incorporates many
features characteristic of other

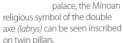

Minoan palaces –
the great central
court with its
sacrificial altar,
royal apartments,
lustral basins
(water pools) and
light-wells (court-
yards). In a small
sanctuary in the
west wing of the

**Giant *pithos*
at the Palace
of Mália**

palace, the Minoan
religious symbol of the double
axe *(labrys)* can be seen inscribed
on twin pillars.

Beyond the palace, remains
thought to be of a town are
currently under excavation,
while further north lies the
burial site of **Chrysólakkos**
(pit of gold). Important
treasures were recovered here,
including the famous gold
bee pendant displayed in
the Irákleio Archaeological
Museum *(see pp274–5)*.

A small shrine in the multichambered Mílatos Cave

The chequered landscape of the agricultural plateau of Lasíthi

🏛 Palace of Mália
3 km (2 miles) E of Mália. **Tel** 28970
31597. **Open** Apr–Oct: 8am–7:30pm
Mon–Fri (from 1pm Mon); Nov–Mar:
8:30am–3pm daily. **Closed** main
public hols. 🚫 ♿

Environs
The fast-developing village of **Sísi**
is 6.5 km (4 miles) east of Mália.
Continuing eastwards, there are
stunning views down to Mílatos.
From here a well-signposted trail
leads to the **Mílatos Cave**, where
a shrine and glass-fronted casket
of bones are a memorial to those
massacred by the Turks in 1823 in
the Greek War of Independence.

㉖ Lasíthi Plateau
Οροπέδιο Λασιθίου

Díkti mountains, Irákleio.
🚌 to Tzermiádo.

High up in the formidable Díkti
mountains, the bowl-shaped
plain of Lasíthi was for centuries
shut off from the outside world.
A row of stone windmills at the
Séli Ampélou Pass marks the

main entry to the plateau, a flat
agricultural area lying 800 m
(2,600 ft) above sea level and
encircled by mountains. Fruit,
potatoes and cereals are the
main crops here, thanks to the
fertile alluvial soil washed down
from the mountains. A few cloth-
sailed windmills are still used
today to pump irrigation water.

Along the perimeter of the
plain are several villages, the
largest of which is **Tzermiádo**
with good tourist facilities. A
path from Tzermiádo to the
Trápeza Cave (also known as
Króneion Cave) is signposted
from the village centre. At the
west end of the village, a rough
road (just over an hour's walk)
leads up to the archaeological
site of **Karfí**, the last retreat of
Minoan civilization. On the
southern edge of the plain, the
village of **Agios Geórgios** has a
small **Folk Museum** set in two
old village houses and
displaying a collection of
embroidery, paintings and
Kazantzákis memorabilia.

The highlight of a visit to Lasíthi
is the climb to the **Diktian Cave**
at Psychró, birthplace of Zeus *(see
p266)*. A wealth of artifacts have
been unearthed here, including
votive offerings, double axes and
bronze statuettes, now in the
Irákleio Archaeological Museum
(see pp274–5).

🏛 Folk Museum
Agios Geórgios. **Open** Apr–Oct:
10am–3pm daily. 🚫 ♿

🏛 Diktian Cave
Psychró. **Tel** 28440 31316. **Open**
10am–3pm daily. **Closed** main
public hols. 🚫

The fortified islet of Spinalónga off the coast of Eloúnta

❷ Eloúnta

Ελούντα

Lasíthi. 🗠 1,500. 🚌 🚊 Tue.
🚆 Eloúnta.

Once the site of the ancient city-state of Oloús, the town of Eloúnta was developed by the Venetians in 1579 as a fortified port. Today, the town is a well-established holiday resort idyllically situated on the Mirabéllou Bay. The town is blessed with attractive sandy coves and offers a good range of accommodation.

East of the village an isthmus joins the mainland to the long strip of land forming the Spinalónga peninsula. Here, remains of the Greco-Roman city-state of **Oloús**, with its temples of Zeus and Artemis, can be discerned just below the water's surface. To the north of the peninsula is the small island of **Spinalónga** where a forbidding 16th-century Venetian fortress now stands deserted. Having withstood assault from the Turks for many years, its last function was as a leper colony until the mid-1950s. Today, boats regularly ferry tourists to the island from Eloúnta and elsewhere.

Environs

The small hamlet of **Pláka**, 5 km (3 miles) north of Eloúnta, makes for a pleasant retreat from the bustle of Eloúnta. Dine on fresh fish at the waterfront, where boat trips are available to Spinalónga island.

Skull and wreath, Archaeological Museum, Agios Nikólaos

❷ Agios Nikólaos

Αγιος Νικόλαος

Lasíthi. 🗠 10,000. 🚌 🚌 *i*
Koundoúrou 21 (28410 82071). 🚊
Wed. 🚆 Almyrós 2 km (1.5 miles) E;
Chavánia 3 km (2 miles) W.

One of the most delightful holiday centres in Crete, Agios Nikólaos boasts a superb setting on the Mirabéllou Bay. In Hellenistic times, according to inscriptions dating back to 193 BC, this was one of two flourishing cities called Lató: Lató pros Kamára (towards the arch)

and Lató Etéra (Other Lató). Having declined in importance under Venetian rule, it was not until the 19th century that modern Agios Nikólaos began to develop.

Now a thriving resort, its centre is the harbour and, with a depth of 64 m (210 ft), the Almyr Lake or Voulisméni. Overlooking the lake, the **Folk Museum** houses a colourful display of traditional Cretan crafts and domestic items. Just north of town, in the grounds of the Mínos Palace Hotel, is the tiny 10th- to 11th-century church of **Agios Nikólaos**, after which the town is named.

Close to several important Minoan sites, the **Archaeological Museum** at Agios Nikólaos possesses a treasure-trove of artifacts from Lasíthi Province. Pieces housed here include carved stone vases, gold jewellery from the Minoan site of Móchlos near Gourniá and pottery, including the drinking vessel known as the Goddess of Mỳrtos. One unique exhibit is the skull of a man thought to be an athlete, complete with a wreath made of gold laurel leaves and a silver coin for his fare across the mythical River Styx.

In summer, boat trips run to Spinalónga island and Agioi Pántes, an island refuge for the Cretan wild goat, the kri-kri *(see p258)*.

▥ Folk Museum
Koúndourou 23. **Tel** 28410 25093.
Open 10am–2pm & 5–8:30pm daily.
Closed main public hols. 🚫 🚻

▥ Archaeological Museum
Palaiológou 68. **Tel** 28410 24943.
Closed for renovation; call to check times before visiting. 🚫

The inner harbour of Agios Nikólaos, with Lake Voulisméni in the foreground

Section of the *Paradise* fresco at Panagía Kerá in Kritsá

㉙ Kritsá

Κριτσά

Lasíthi. 🚗 2,500. 🚌 🏠 Mon.
🚕 Ammoudára 11 km (7 miles) E;
Istro 15 km (9 miles) SE.

Set at the foot of the Lasíthi mountains, Kritsá is a small village known throughout Crete for its famous Byzantine church. Also a popular centre for Cretan crafts, its main street is awash with lace, elaborately woven rugs and embroidered table-cloths during the summer months. From the cafés and tavernas along the main street, fine views of the valley leading down to the coast can be enjoyed. By November, Kritsá reverts back to life as a workaday Greek village.

East of Kritsá, situated just off the road among olive groves, the hallowed 13th-century church of **Panagía Kerá** contains some of the finest frescoes in Crete, dating from the 13th to mid-14th century. The building is triple-aisled, with the central aisle being the oldest. Beautiful representations of the life of Christ and the Virgin Mary cover the interior.

Environs

North of Kritsá lie the ruins of a fortified city founded by the Dorians in the 7th century BC. **Lató Etéra** flourished until Classical times, when its fortunes declined under Roman rule: it was superseded by the more easily reached port of Lató pros Kamára (today's Agios Nikólaos).

Sitting perched on a saddle between two peaks, the site offers fine views of the Mirabéllou Bay. A paved road, with workshops and houses clustered on the right, climbs up to a central agora, or marketplace, with a cistern to collect rainwater and a shrine. On the north side of the agora, a staircase flanked by two towers leads to the place where the city's archives would once have been stored. To the south of the agora a temple and a theatre can be seen.

🏛 Lató

4 km (2 miles) N of Kritsá. **Open** Tue–Sun. **Closed** main public hols.

Lace shop on Kritsá's main street

㉚ Ierápetra

Ιεράπετρα

Lasíthi. 🚗 15,000. 🚌 🏠 Sat. 🚕
Agiá Fotiá 17 km (11 miles) E; Makrýs Gialós 30 km (19 miles) E.

Situated on the southeast coast of Crete, Ierápetra boasts of its position as the most southerly city in Europe. A settlement since pre-Minoan times, trade and cultural connections with North Africa and the Middle East were an important basis of the city's existence. Sir Arthur Evans *(see p278)* declared it the "crossroads of Minoan and Achaian civilizations". Once a flourishing city with villas, temples, amphitheatres and imposing buildings, the town today has an air of decline. Gone are all signs of its ancient history, thanks partly to past pillage and, more recently, to modern "development".

The entrance to the old harbour is guarded by an early 13th-century Venetian **fortress**. West of the fortress is the attractive Turkish quarter, where a restored **mosque** and elegant Ottoman fountain can be seen. Also in this area, on Kougioumtzáki, is the 14th-century church of **Aféntis Christós** and, off Samouíl, **Napoleon's House**, where he is said to have spent a night en route to Egypt in 1798. Today it is not open to the public.

The small **Archaeological Museum** in the centre of town displays a collection of local artifacts that managed to survive marauders and various archaeological predators. The exhibits date from Minoan to Roman times and include *larnakes* (burial caskets), *pithoi* (storage jars), statues, bronze axes and stone carvings.

An almost unbroken line of sandy beaches stretches eastwards from Ierápetra, overlooked by the inevitable plethora of hotels and restaurants. From Ierápetra's harbour, a daily boat service runs to the idyllic white sands and cedar forests of the uninhabited **Chrysí** island from mid-May to late October. Here, snorkelling in the shallow waters is popular. The island boasts Minoan ruins and a naturist beach.

🏛 Fortress

Old port. **Open** daily. **Closed** main public hols. 🚫

🏛 Archaeological Museum

Adrianoú Koustoúla. **Tel** 28420 28721. **Open** 8:30am–3pm Tue–Sun. **Closed** main public hols. ♿

Mosque and Ottoman fountain in Ierápetra's old Turkish quarter

Gourniá archaeological site

❸ Gourniá
Γουρνιά

18.5 km (11 miles) E of Agios Nikólaos, Lasíthi. Open 8:30am–3pm Tue–Sun. Closed main public hols. Istro 8 km (5 miles) W.

The Minoan site of Gourniá stands on a low hill overlooking the tranquil Mirabéllou Bay. Excavated by the American archaeologist Harriet Boyd-Hawes between 1901 and 1904, Gourniá is the best-preserved Minoan town in Crete. A mini-palace (one-tenth the size of Knosós) marks its centre, surrounded by a labyrinth of narrow, stepped streets and one-room dwellings. The site was inhabited as early as the 3rd millennium BC, though what remains dates from the second palace period, around 1700 BC *(see p279)*. A fire, caused by seismic activity in around 1450 BC, destroyed the settlement at Gourniá.

Environs
Along the National Highway, 2 km (1.5 miles) west of Gourniá, an old concrete road turns left up a spectacular 6-km (4-mile) climb to **Moní Faneroménis**. Here, the 15th-century chapel of the **Panagía** has been built into a deep cave and is the repository for sacred (and some say miraculous) icons.
East along the National Highway, a left turning from Sfáka leads down to the delightful fishing village of **Móchlos**. The small island of Móchlos, once joined to the mainland by a narrow isthmus, is the site of a Minoan settlement and cemetery.

❸ Siteía
Σητεία

Lasíthi. 7,500. Tue. Siteía.

Snaking its way through the mountains between Gourniá and Siteía, the National Highway traverses some of the most magnificent scenery in Crete. Towards Siteía, the landscape gives way to barren hills and vineyards.
Although there is evidence of a large Greco-Roman city in the region, modern Siteía dates from the 4th century AD. It flourished under Byzantine and early Venetian rule but its fortunes took a downturn in the 16th century as a result of earthquakes and pirate attacks. When rebuilding took place in the 1870s Siteía began to prosper once again.
Today, the production of wine and olive oil is important to the town's economy and the mid-August Sultana Festival celebrates its success as a sultana exporter.
At the centre of Siteía's old quarter lies a picturesque

harbour, with tavernas and cafés clustering around its edges. Above the north end of the harbour the restored Venetian **fort** (now used as an open-air theatre) is all that remains of the once extensive fortifications of the town. The Kornaria Festival is a cultural event held in the fort from the beginning of July until mid-August and is a great way for visitors to learn about the customs and traditions of Siteía. Events include music and dance events, theatre, exhibitions and sports events.
On the southern outskirts of town, the **Archaeological Museum** displays artifacts from the Siteía district. Exhibits range from Neolithic to Roman times and include an exquisite Minoan ivory statuette known as the *Palaíkastro Koúros*. There are pottery finds from all over the region, including a large collection of material from Zákros Palace.

🏛 **Archaeological Museum**
Piskokefálou 3. Tel 28430 23917.
Open 8:30am–3pm Tue–Sun.
Closed main public hols.

Siteía's old quarter on the hillside overlooking the tree-lined harbour

㉝ Moní Toploú

Μονή Τοπλού

6 km (10 miles) W of Siteía, Lasíthi.
Tel 28430 61226. ▥ to Váï. Site &
Museum: **Open** daily. ▨ ▥ Itanos
7.5 km (4.5 miles) NE.

Founded in the 14th century,
Moní Toploú is now one of the
wealthiest and most influential
monasteries in Crete. The
present buildings date from
Venetian times, when the
monastery was fortified against
pirate attacks. The Turkish name
"Toplou" refers to the cannon
installed here. During World War
I, Resistance radio broadcasts
were transmitted from the
monastery, an act for which
Abbot Siligknákis was executed
by German forces near Chaniá.

Three levels of cells overlook
the inner courtyard, where a
small 14th-century church
contains frescoes and icons.
The most famous of these is
the *Lord, Thou Art Great*
icon, completed in
1770 by the
artist Ioánnis
Kornáros. On
the façade of
the church, an
inscription
records the
Arbitration
of Magnesia
in 132 BC. This was
an order that settled
a dispute between the rival city-
states of Ierapytna (today's
Ierápetra) and Itanos, over the
control of the Temple of Zeus
Diktaios at Palaíkastro. The
inscription stone was used
originally as a tombstone. The
monastery's small museum
houses etchings and 15th- to
18th-century icons.

Lord, Thou Art Great icon by
Ioánnis Kornáros, Moní Toploú

㉞ Váï Beach

Παραλία Βάι

28 km (17 miles) NE of Siteía,
Lasíthi. ▥

The exotic Váï Beach is a tropical
paradise of dense palm trees
known to have existed in
Classical times and reputedly
unique in Europe. This inviting
sandy cove is tremendously
popular with holiday-makers.

Zákros archaeological site, situated behind the hamlet of Káto Zákros

Although thoroughly commercialized, with overpriced
tavernas and the constant arrival
of tour buses, great care is taken
to protect the palm trees.

Environs
In the desolate landscape
2 km (1 mile) north of Váï, the
ruins of the ancient city-state
of **Itanos** stand on a hill
between two sandy
coves. Minoan,
Greco-Roman
and Byzantine
remains
have been
excavated
(the scant
traces of
which
can be
seen today),
including a Byzan-
tine basilica and the ruins of
some Classical temples.

The agricultural town of
Palaíkastro, 10 km (6 miles)
south of Váï, is the centre of an
expanding olive business. At the
south end of Chióna beach,
2 km (1 mile) to the east, the
Minoan site of Palaíkastro is
presently under excavation.

㉟ Zákros

Ζάκρος

Káto Zákros, Lasíthi. **Tel** 28430 26897.
▥ **Open** Tue–Sun. **Closed** main
public hols. ▨ ▥ Káto Zákros;
Xerókampos 13 km (8 miles) S.

In 1961, Cretan archaeologist
Nikólaos Pláton discovered the
unplundered Minoan palace of
Zákros. The fourth largest of the
palaces, it was built around
1700 BC and destroyed in the
island-wide disaster of 1450
BC. Its ideal location made it
a centre of trade with the
Middle East.

The two-storey palace was
arranged around a central
courtyard, the east side of
which contained the royal
apartments. Remains of a
colonnaded cistern hall can still
be seen, and a stone-lined well
in which some perfectly
preserved 3,000-year-old olives
were found in 1964. The main
hall, workshops and storerooms are in the west wing.
Finds from the palace include
an exquisite rock crystal jug
and numerous vases, now in
the Irákleio Archaeological
Museum *(see pp274–5)*.

Váï Beach with its calm waters and native palms

A SHORT STAY IN ATHENS

A vast, sprawling metropolis surrounded by rocky mountains, Athens covers 457 sq km (176 sq miles) and has a population of four million people. The city prides itself on being home to the 2,500-year-old temple of Athena – the Parthenon – as well as some superb museums. A stopover in Athens en route to the islands offers the ideal opportunity to visit the best sights in the city.

The birthplace of European civilization, Athens has been inhabited for 7,000 years, since the Neolithic era. Ancient Athens reached its high point in the 5th century BC, when Perikles commissioned many fine new buildings, including some of the temples on the Acropolis. Other relics from the Classical period can be seen in the Ancient Agora, a complex of public buildings dominated by the reconstructed stoa of Attalos, a long, covered colonnade.

There is little architectural evidence of the city's more recent history of occupation. With the exception of some fine Byzantine churches, particularly those in historic Pláka, one of the oldest areas of Athens, nothing of importance has survived from the years of Frankish, Venetian and Ottoman rule. In 1834, inspired by the Classical buildings of the Acropolis, King Otto declared Athens the new capital of Greece, and his Greek, German and Danish town planners and architects created a modern city of Neo-Classical municipal buildings, wide boulevards and elegant squares around the ancient "Sacred Rock".

The rich cultural heritage of Athens can be appreciated in some magnificent museums, including the National Archaeological Museum, where an unrivalled collection of Neolithic, Cycladic and Mycenaean artifacts, through to treasures from Roman and Hellenistic times, beautifully illustrates the glories of ancient Greece. The contemporary Acropolis Museum houses a fabulous collection of statues and reliefs from the ancient Acropolis site, while the excellent Benáki Museum hosts Greek art and crafts. The National Gallery of Art includes well-known works by both Greek and European artists.

The nightlife in Athens is excellent, with tavernas, clubs and bars open until the early hours. Open-air cinemas and theatres, such as the Theatre of Herodes Atticus at the foot of the Acropolis, are popular in summer. There is music for every taste, from traditional Greek to pop and classical. Shopping ranges from flea markets and antique and bric-a-brac shops in Monastiráki, to designer boutiques in Kolonáki. Pedestrianization of the city centre makes Athens a pleasant place to explore on foot.

The Acropolis viewed from Filopáppos Hill

◄ Lykavittós Hill rising above the Kolonáki neighbourhood of Athens

Exploring Athens

Even with only an afternoon to spend in Athens, it is possible to visit a number of the main sights. The Acropolis is the most popular attraction, along with the Ancient Agora. The National Archaeological Museum houses many finds from these sites in its fine collection of ancient Greek art. The Benáki Museum houses a glittering array of jewellery, costumes and ceramics from Greece and the Middle East, as well as many temporary exhibitions. Shopping provides an alternative to sightseeing, from the bric-a-brac in Pláka to the designer stores in Kolonáki. For information on getting around Athens, see pp296–9.

Avyssinías in Monastiráki
(see p290)

The Central Market
(Varvákios Agorá) has a fine array of foods, herbs and spices.

Panepistimíou is lined with some of the best examples of Neo-Classical architecture in Athens.

Mitrópoli is Athens' cathedral. It towers over the tiny Byzantine Panagía Gorgoepíkoös (or Little Cathedral) next to it.

Figure from the Museum of Cycladic Art *(see p295)*

| 0 metres | 250 |
| 0 yards | 250 |

The Tower of the Winds *(see p291)*

Locator Map

Athens' Top Sights

Museums and Galleries
❶ National Archaeological Museum
❼ Benáki Museum
❽ Museum of Cycladic Art
❾ National Gallery of Art

Historic Districts
❷ Monastiráki
❸ Psyrrí
❻ Pláka

Ancient Sites
❹ Ancient Agora
❺ Acropolis *pp292–4*

Kolonáki is a fashionable district with designer stores.

Plateía Syntágmatos is the home of the Tomb of the Unknown Soldier. The famous *évzones* (national guard) are on parade in front of the tomb.

The National Gardens were planted by order of Queen Amalía in the 19th century. Semi-tropical, they provide pleasant relief from the heat of the city.

For keys to symbols *see back flap*

❶ National Archaeological Museum

Εθνικό Αρχαιολογικό Μουσείο

Tositsa 1, Exárcheia. **Tel** 21321 44800. Ⓜ Omónoia, Viktória. **Open** Apr–Oct: 1:30–8pm Mon, 8am–8pm Tue–Sun; Nov–Mar: 1–8pm Mon, 9am–4pm Tue–Sun. 🅿 📷 🖥 ♿ 🅦 namuseum.gr

When it was opened in 1891, this museum brought together a collection that had previously been stored all over the city. New wings were added in 1939, but during World War II this priceless collection was dispersed and buried underground to protect it from any possible damage. The museum reopened in 1946, but it has taken another 50 years of renovation and reorganization to finally do justice to its formidable collection. With its comprehensive assembly of pottery, sculpture and jewellery, it definitely deserves ranking as one of the finest museums in the world. It is a good idea to plan ahead and be selective when visiting the museum and not attempt to cover everything in one visit.

The museum's exhibits can be divided into five main collections: Neolithic and Cycladic, Mycenaean, Geometric and Archaic, Classical sculpture, Roman and Hellenistic sculpture and the pottery collections. There are also other smaller collections that are well worth seeing. These include the stunning Eléni Stathátou jewellery collection and the Egyptian rooms.

High points of the museum include the unique finds from the grave circle at Mycenae, in particular the gold *Mask of Agamemnon*. Also not to be missed are the Archaic *koúroi* statues and the unrivalled collection of Classical and Hellenistic statues. Two of the most important and finest of the bronzes are the *Horse with the Little Jockey* and the *Poseidon*. Also housed here is one of the world's largest collections of ancient ceramics

comprising elegant figure vases from the 6th and 5th centuries BC *(see pp66–7)* and some Geometric funerary vases that date back to 1000 BC. The Library of Archaeology holds a large collection of rare books, including the diaries of Heinrich Schliemann, who uncovered the remains of Troy.

Shoppers browsing in Athens' lively Monastiráki market

The *Mask of Agamemnon* in the National Archaeological Museum

❷ Monastiráki

Μοναστηράκι

Ⓜ Monastiráki. Market: **Open** daily.

This area, named after the little monastery church in Plateía Monastirakíou, is synonymous with Athens' famous fleamarket. Located next to the Ancient Agora, it is bounded by Sari in the west and Aiólou in the east. The streets of Pandrósou, Ifaístou and Areos leading off Plateía Monastirakíou are full of shops, selling a range of goods from antiques, leather and silver to tourist trinkets.

The heart of the flea market is in Plateía Avyssinías, west of Plateía Monastirakíou, where

every morning junk dealers arrive with pieces of furniture and various odds and ends. During the week and on Sunday mornings, the shops and stalls are filled with antiques, second-hand books, rugs, leatherware, taverna chairs, army surplus gear and tools.

The market flourishes particularly along Adrianoú and in Plateía Agíou Filíppou. There are always numerous bargains to be had. Items particularly worth investing in include some of the colourful woven and embroidered cloths and an abundance of good silver jewellery.

❸ Psyrrí

Ψυρρί

Ⓜ Monastiráki.

For a taste of Athens as it was through most of its modern history, wander the warren of streets comprising the Psyrrí district. Bordered by the Central Market (Varvákios Agorá), Athinas and Ermou Streets, this neighbourhood is becoming the city's trendiest area. Many of the handsome Neo-Classical buildings have been renovated for art galleries and restaurants. There are also theatres, wine bars and boutiques. Tiny stores specialize in unique, handmade items like copper kitchenware, belt buckles, wickerwork and icons. At night the district's transformed, commercial buzz is replaced by the gentle pleasures of cafés, restaurants and wine bars. The food here is some of the most interesting in the city and prices are reasonable. This is very much an Athenian part of town.

❹ Ancient Agora
Αρχαία Αγορά

Main entrance at Adrianoú, Monastiráki. **Tel** 21032 10185. Ⓜ Thiseío, Monastiráki. Museum and site: **Open** Apr–Oct: 8am–7pm daily; Nov–Mar: 8am–3pm daily. **Closed** main public hols. 🅿 ♿ limited.

The American School of Archaeology commenced excavations of the Ancient Agora in the 1930s, and since then a complex array of public buildings and temples has been revealed. The democratically governed Agora was the political and religious heart of Ancient Athens. Also the centre of commercial and daily life, it abounded with schools and elegant stoas filled with shops. The state prison was here, as was the mint, which was used to make the city's coins inscribed with the famous owl symbol. Even the remains of an olive-oil mill have been found here.

The main building standing today is the impressive two-storey stoa of Attalos. This was rebuilt between 1953 and 1956 on the original foundations and using ancient building materials. Founded by King Attalos of Pergamon (ruled 159–138 BC), it dominated the eastern quarter of the Agora until it was destroyed in AD 267. It is used today as a museum, exhibiting the finds from the Agora. These include legal finds, such as a *klepsydra* (a water clock that was used for timing plaintiffs' speeches), bronze

The rooftop of the church of Agios Nikólaos Ragavás, Pláka

ballots and items from everyday life such as some terracotta toys and leather sandals. The best-preserved ruins on the site are the Odeion of Agrippa, a covered theatre, and the Hephaisteion, a temple to Hephaistos, which is also known as the Theseion.

❺ Acropolis
See pp292–3.

❻ Pláka
Πλάκα

Ⓜ Monastiráki. 🚌 1, 2, 4, 5, 9, 10, 11, 12, 15, 18.

The area of Pláka is the historic heart of Athens. Even though only a few buildings date back further than the Ottoman period, it remains the oldest continuously inhabited area in the city. One probable explanation of its name comes

from the word used by Albanian soldiers in the service of the Turks who settled here in the 16th century – *pliaka* (old) was how they used to describe the area. Despite the constant swarm of tourists and Athenians, who come to eat in old-fashioned tavernas or browse in the antique and icon shops, Pláka still retains the atmosphere of a traditional neighbourhood. The only choregic monument still intact in Athens is the **Lysikrates Monument** in Plateía Lysikrátous. Built to commemorate the victors at the annual choral and dramatic festival at the Theatre of Dionysos, these monuments take their name from the sponsor (*choregos*) of the winning team.

Detail from a terracotta roof, Pláka

Many churches are worth a visit: the 11th-century **Agios Nikólaos Ragavás** has ancient columns built into the walls.

The **Tower of the Winds**, in the far west of Pláka, lies in the grounds of the Roman Agora. It was built by the Syrian astronomer Andronikos Kyrrestes around 100 BC as a weather vane and waterclock. On each of its marble sides, one of the eight mythological winds is depicted.

🏛 **Tower of the Winds**
Plateía Aéridon. **Tel** 21032 45220. **Open** 8am–7pm daily (Nov–Mar: to 3pm). **Closed** main public hols. 🅿

The façade of the Hephaisteion in the Ancient Agora

❺ Acropolis
Ακρόπολη

In the mid-5th century BC, Perikles persuaded the Athenians to begin a grand programme of new building work in Athens that has come to represent the political and cultural achievements of Greece. The work transformed the Acropolis with three contrasting temples and a monumental gateway. The Theatre of Dionysos on the south slope was developed further in the 4th century BC, and the Theatre of Herodes Atticus was added in the 2nd century AD.

The Acropolis with the Temple of Olympian Zeus in the foreground

★ **Porch of the Caryatids**
These statues of women were used in place of columns on the south porch of the Erechtheion. The originals, four of which can be seen in the Acropolis Museum, have been replaced by casts.

★ **Temple of Athena Nike**
This temple to Athena of Victory is on the west side of the Propylaia. It was built in 427–424 BC.

KEY

① **Pathway to Acropolis from ticket office**

② **The Belué Gate** was the first entrance to the Acropolis.

③ **The Propylaia** was built in 437–432 BC to form a new entrance to the Acropolis.

④ **An olive tree** now grows where Athena first planted her tree in a competition against Poseidon.

⑤ **Acropolis Museum** *(p290)*

⑥ **Two Corinthian columns** are the remains of choregic monuments erected by sponsors of successful dramatic performances.

⑦ **Panagía Spiliótissa** is a chapel cut into the Acropolis rock itself.

⑧ **Shrine of Asklepios**

⑨ **Stoa of Eumenes**

⑩ **The Acropolis rock** was an easily defended site. It has been in use for nearly 5,000 years.

Theatre of Herodes Atticus
Also known as the Odeion of Herodes Atticus, this superb theatre was originally built between AD 161 and 174. It was restored in 1955 and is used today for outdoor concerts.

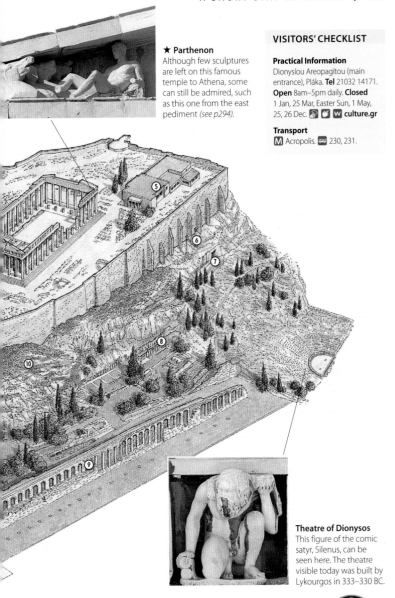

★ Parthenon
Although few sculptures are left on this famous temple to Athena, some can still be admired, such as this one from the east pediment *(see p294).*

VISITORS' CHECKLIST

Practical Information
Dionysíou Areopagítou (main entrance), Pláka. **Tel** 21032 14171.
Open 8am–5pm daily. **Closed** 1 Jan, 25 Mar, Easter Sun, 1 May, 25, 26 Dec. 🅿 💳 🅦 culture.gr

Transport
Ⓜ Acropolis. 🚌 230, 231.

Theatre of Dionysos
This figure of the comic satyr, Silenus, can be seen here. The theatre visible today was built by Lykourgos in 333–330 BC.

3000 BC First settlement on the Acropolis during Neolithic period

AD 51 St Paul delivers sermon on Areopagos hill

480 BC All buildings of Archaic period destroyed by the Persians

AD 267 Germanic Heruli tribe destroy Acropolis

St Paul

3000 BC	2000 BC	1000 BC	AD 1	AD 1000

1200 BC Cyclopean wall built to replace original ramparts

510 BC Delphic Oracle declares Acropolis a holy place of the gods, banning habitation by mortals

447–438 BC Construction of the Parthenon under Perikles

Perikles (495–429 BC)

AD 1687 Parthenon damaged by Venetians

AD 1987 Restoration of the Erechtheion completed

Exploring the Acropolis

Once through the Propylaia, the grand entrance to the site, the Parthenon exerts an overwhelming fascination. The other fine temples on "the Rock" include the Erechtheion and the Temple of Athena Nike. Since 1975, access to all the temple precincts has been banned. However, it is a miracle that anything remains at all. The ravages of war, the removal of treasures and pollution have all taken their irrevocable toll on the Acropolis.

A section from the north frieze of the Parthenon

🏛 The Parthenon

One of the world's most famous buildings, the Parthenon was commissioned by Perikles as part of his rebuilding plan. Work began in 447 BC, when the sculptor Pheidias was entrusted with supervising the building of a magnificent new Doric temple to Athena, the patron goddess of the city. It was built on the site of earlier Archaic temples, and was designed primarily to house the *Parthenos*, Pheidias's impressive 12 m- (39 ft-) high cult statue of Athena covered in ivory and gold.

Taking nine years to complete, the temple was dedicated to the goddess during the Great Panathenaia festival of 438 BC. Designed and constructed in Pentelic marble by the architects Kallikrates and Iktinos, the complex architecture of the Parthenon replaces straight lines with slight curves. This is generally thought to have been done to prevent visual distortion or perhaps to increase the impression of grandeur. All the columns swell in the middle and all lean slightly inwards, while the foundation platform rises towards the centre.

For the pediments and the friezes which ran all the way round the temple, an army of sculptors and painters was employed. Agorakritos and Alkamenes, both pupils of Pheidias, are two of the sculptors who worked on the frieze, which represented the people and horses in the Panathenaic procession.

Despite much damage and alterations made to adapt to its various uses, which include a church, a mosque, and even an arsenal, the Parthenon remains a powerful symbol of the glories of ancient Greece. It is currently being restored.

The *Moschophoros* (or *Calf-Bearer*) in the Acropolis Museum

🏛 Acropolis Museum

Dionysiou Areopagitou 15. Tel 21090 00900. **Open** Apr–Oct: 8am–4pm Mon, 8am–8pm Tue–Sun (to 10pm Fri); Nov–Mar: 9am–5pm Tue–Thu, 9am–8pm Fri–Sun (to 10pm Fri). **Closed** 1 Jan, 1 May, Easter Sun & Mon, 25 & 26 Dec. 🅿 ⛑ 🕸 theacropolismuseum.gr

Located in the historic Makrigiánni district at the foot of the Acropolis, this €130 million showpiece has been designed by Bernad Tschumi to house the treasures found on the Acropolis hill. It is constructed over excavations of an early Christian settlement and a glass walkway hovers over the ruins.

The collection has been installed in chronological order and begins with finds from the slopes of the Acropolis, including statues and reliefs from the Shrine of Asklepios.

The **Archaic Collection** is set out in a magnificent double-height gallery and contains fragments of pedimental statues such as the statue of *Moschophoros*, or the *Calf-Bearer* (c.570 BC).

The sky-lit **Parthenon Gallery** on the top floor is the highlight. Here, looking out onto the Acropolis hill itself, the remaining parts of the Parthenon frieze are displayed in their original order.

View of the Parthenon from the southwest at sunrise

For hotels and restaurants in this region see p313 and pp332–3

❼ Benáki Museum

Μουσείο Μπενάκη

Corner of Koumpári & Vasilíssis Sofías, Kolonáki. **Tel** 21036 71000. 🚍 3, 7, 8, 13. **Open** 9am–5pm Wed & Fri, 9am–midnight Thu & Sat, 9am–3pm Sun. **Closed** main public hols. 🎫 (free Thu). ♿ limited. 🌐 **benaki.gr**

This outstanding museum contains a diverse collection of Greek art and crafts, jewellery, regional costumes and political memorabilia from the 3rd century BC to the 20th century. It was founded by Antónios Benákis (1873–1954), the son of Emmanouíl Benákis, a wealthy Greek who made his fortune in Egypt. Antónios Benákis was interested in Greek, Persian, Egyptian and Ottoman art from an early age and started collecting while living in Alexandria. When he moved to Athens in 1926, he donated his collection to the Greek State, using the family house as a museum, which was opened to the public in 1931. The elegant Neo-Classical mansion was built towards the end of the 19th century by Anastásios Metaxás, who was also the architect of the Kallimármaro stadium.

The Benáki collection consists of gold jewellery, some dating as far back as 3000 BC, as well as icons, pieces of liturgical silverware, Egyptian artifacts, Greek embroideries and the work of the late artist Chatzikyriákos-Gkíkas.

❽ Museum of Cycladic Art

Μουσείο Κυκλαδικής και Αρχαίας Ελληνικής Τέχνης

Neofýtou Doúka 4 (New Wing at Irodótou 1), Kolonáki. **Tel** 21072 28321. 🚍 3, 7, 8, 13. **Open** 10am–5pm Mon, Wed, Fri & Sat; 10am–8pm Thu; 11am–5pm Sun. **Closed** main public holidays. 🎫 ♿ 📷 📱 🌐 **cycladic.gr**

Opened in 1986, this modern museum offers the visitor the world's finest collection of Cycladic art. Assembled by Nikólaos and Dolly Goulandrís and helped by the donations of other Greek collectors, it has

brought together a fine selection of ancient Greek art, spanning 5,000 years of history.

Spread over five floors, the displays start on the first floor, which is home to the Cycladic collection. Dating back to the 3rd millennium BC, the Cycladic figurines were found mostly in graves, although their exact usage remains a mystery. One of the finest examples is the *Harp Player*. Ancient Greek art is exhibited on the second floor and the Charles Polítis collection of Classical and Prehistoric art on the fourth floor, high-lights of which include some terracotta figurines of women from Tanágra, central Greece. The third floor of the museum is used for temporary, visiting exhibitions.

Seated Cycladic figure

A new wing was opened in the adjoining Stathátos Mansion in 1992, named after its original inhabitants, Otto and Athiná Stathátos. It houses the Greek Art Collection of the Athens Academy. Temporary exhibitions are also on display on the first floor of the Stathátos Mansion.

❾ National Gallery of Art

Εθνική Πινακοθήκη

Vasiléos Konstantínou 50, Ilísia. (During renovation work, enter via Panagiotis Kanellapoulos avenue.) **Tel** 21072 35857. 🚍 3, 13. **Open** 9am–4pm Wed–Mon. **Closed** main public hols. 🎫 ♿ 🌐 **nationalgallery.gr**

This gallery holds a permanent collection of European and Greek art. One gallery is devoted mainly to European art and includes works by Van Dyck, Cézanne, Dürer and Rembrandt, as well as Picasso's *Woman in a White Dress* (1939) and Caravaggio's *Singer* (1620). Most of the collection is made up of Greek art from the 18th to 20th centuries. The 1800s feature paintings of the War of Independence *(see pp46–7)*. There are also some excellent portraits, including *The Loser of the Bet* (1878) by Nikólaos Gýzis (1842–1901), *Waiting* (1900) by Nikifóros Lýtras (1832–1904) and *The Straw Hat* (1925) by Nikólaos Lýtras (1883–1927).

Icon of the *Adoration of the Magi* from the Benáki Museum

Getting Around Athens

The sights of Athens' city centre are closely packed, and almost everything of interest can be reached on foot. Walking is the best way of sightseeing in the city, especially in view of the traffic congestion, which can make both public and private transport slow and inefficient. The bus and trolleybus network provides the majority of public transport in the capital for Athenians and visitors alike; the three-line metro system offers a good alternative to the roads for some journeys. Taxis are another option and, with the lowest tariffs of any EU capital, they are worth considering even for longer journeys.

Tickets can be bought individually or in a book of ten. The same ticket can be used on any bus, trolleybus, metro or tram, but must be validated in a ticket machine upon boarding. There is a penalty fine for not stamping your ticket, and tourists who are unfamiliar with this may be caught out by inspectors who board buses to carry out random checks. Tickets are valid for one ride only, regardless of the distance travelled.

One of the fleet of yellow, blue and white buses

Bus Services

Athens is served by an extensive bus network. Buses are white, yellow and blue. The network covers over 300 routes, connecting various districts to each other and to the city centre.

All buses are ecologically friendly and run on natural gas. Bus journeys are inexpensive but can be slow and uncomfortably crowded, particularly in the city centre and during rush hours; the worst times are from 7am to 8:30am, from 2pm to 3:30pm and from 7:30pm t o 9pm.

Note that to reach Piraeus port, metro Line 1 (green line) is infinitely faster and more convenient than the bus. Timetables and route maps (only in Greek) are available from **OASA**, the Athens Urban Transport Organization.

Tickets must be purchased in advance from a metro station or a *períptero* (street kiosk).

Trolleybuses

Athens has a good network of trolleybuses, which are yellow and purple in colour and run on electricity. There are more than 20 routes that crisscross the city centre and connect many of the main sights. Routes 7 and 8 are useful to reach the National Archaeological Museum from Sýntagma Square, while route 1 links Laríssis railway station with Omónoia Square and Sýntagma Square.

Tickets can be bought at ticket machines at metro stations and at most street kiosks and must be validated upon boarding the trolleybus.

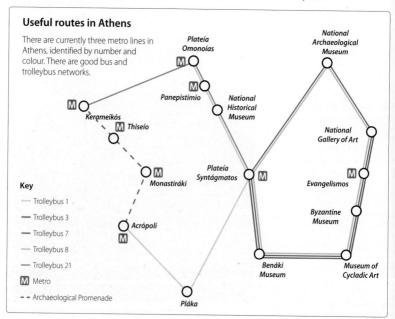

Useful routes in Athens

There are currently three metro lines in Athens, identified by number and colour. There are good bus and trolleybus networks.

Plateía Omonoías Ⓜ

National Archaeological Museum Ⓜ

Ⓜ **Panepistímio**

National Historical Museum

Ⓜ **Kerameikós**

Ⓜ **Thiseío**

National Gallery of Art Ⓜ

Ⓜ **Monastiráki**

Plateía Syntágmatos Ⓜ

Evangelismos Ⓜ

Byzantine Museum

Acrópoli Ⓜ

Benáki Museum

Museum of Cycladic Art

Pláka

Key

— Trolleybus 1
— Trolleybus 3
— Trolleybus 7
— Trolleybus 8
— Trolleybus 21
Ⓜ Metro
- - Archaeological Promenade

An Athens trolleybus

Trams

Athens' tram system, a project inaugurated for the 2004 Olympics, connects the city centre and the coast. There are just three lines, confusingly numbered 3, 4 and 5. Line 3 runs along the coast between Néo Fáliro and Voúla; Line 4 runs from Sýntagma Square to Néo Fáliro; and Line 5 runs from Sýntagma Square to Voúla. The trams operate 5:30am–1am daily (until 2:30am on Fridays and Saturdays).

Tickets can be bought at ticket machines at tram stops and must be validated at a machine at the tram stop before entering the tram.

Metro

The metro is fast and reliable. It was a key element in the restructuring of urban public transport for the 2004 Olympics, leading to the reduction in the number of private cars, as well as buses, in the city centre.

The metro has three lines: Line 1 (green line) runs from Kifissiá in the north to the port of Piraeus in the south, with central stops at Thiseío, Monastiráki, Omónoia and Victória. Most of the line is overland and only runs underground between Attikí and Monastiráki stations. The green line is used mainly by commuters who live in the northern suburbs, but it also offers visitors the fastest way of reaching Piraeus.

Line 2 (red line) and Line 3 (blue line) form part of a huge expansion of the system, most of which was completed in time for the 2004 Olympic

Games. These two lines, which intersect at Sýntagma, were built 20 m (66 ft) underground in order to avoid interfering with material of archaeological interest. Both Sýntagma and Acrópoli stations feature impressive displays of archaeological finds uncovered during construction work.

Line 2 (red line) runs from Agios Antónios in northwest Athens to Agios Dimitrios in the southeast. Line 3 (blue line) runs from Egaleo to Doukissis Plakentias in the northeast, with some trains continuing on to Eleftheríos Venizélos airport. There are plans to extend Line 2 northwest to Peristeri and Anthoupoli, and southeast to Ilioupoli, Alimos, Argyroupoli and Elliniko, and to extend Line 3 west to Agia Marina. There are no fixed dates for their completion.

One ticket allows travel on any of the three lines and is valid for 90 minutes in one direction. You cannot exit a station and then go back to continue your journey with the same ticket. Day tickets are available at €3 or a weekly ticket at €10, for use on all metro lines, together with trams and city buses. Tickets must be validated before boarding the train; use the machines at the entrances to all platforms. Trains run every five minutes, 5am–midnight on Line 1, and

Evangelismos metro sign

5:30am–midnight on Lines 2 and 3. At weekends trains run from 5am to 2am.

Walking

The centre of Athens is very compact, and most major sights and museums are within a 25-minute walk of Sýntagma Square, which is regarded as the city's centre.

Since the 2004 opening of the Archaeological Promenade, Athens has become infinitely more pleasant to navigate on foot.

A broad car-free walkway running 4 km (2.5 miles), the Promenade skirts the foot of the Acropolis to link the city's main ancient sites, as well as four metro stations. The streets of Dionissiou Areopagitou and Apostolou Pavlou run between Acrópoli metro station and Thiseío metro station (passing the Acropolis and the New Acropolis Museum); Adrianou street runs from Thiseío metro station to Monastiráki station (passing the Ancient Agora); and Ermou runs from Thiseío metro station to Kerameikós metro station (passing the Kerameikós archaeological site).

By day, Athens is still one of the safest European cities in which to walk around. However, it pays to be vigilant at night.

Archaeological remains on display at Sýntagma metro station

A yellow taxi on a street in Athens

Taxis

Swarms of yellow taxis can be seen cruising the streets of Athens at most times of the day or night. However, trying to persuade one to stop can be a difficult task, especially between 2pm and 3pm, when taxi drivers usually change shifts. Then, they will only pick you up if you happen to be going in a direction that is convenient for them.

To hail a taxi, stand on the edge of the pavement and shout out your destination to any cab that slows down. If a cab's "taxi" sign is lit up, then it is definitely for hire (though often a taxi is also for hire when the sign is not lit). It is common practice in Athens for drivers to pick up extra passengers along the way, so it is worth flagging the occupied cabs, too. If you are not the first passenger on board, make a note of the meter reading immediately;

there is no fare-sharing, so you should be charged for your portion of the journey only.

Despite a rise in prices, Athenian taxis are still very cheap by European standards – depending on traffic, you should not have to pay more than about €5 to travel to any destination within the downtown area, and between €6 and €9 from the centre to Piraeus. Higher tariffs come into effect between midnight and 5am and for journeys that exceed certain distances from the city centre. Fares to the airport, which is out of town at Spata, are now fixed at €35 in the daytime and €50 at night (midnight–5am).

There are also small surcharges for extra pieces of luggage weighing more than 10 kg (22 lb), and for journeys from the ferry or railway terminals. Taxi fares are increased during holiday periods, such as Christmas and Easter.

For an extra charge (€3.50–€6), you can make a phone call to a radio taxi company and arrange for a car to pick you up at an appointed place and time. Radio taxis are plentiful in the Athens area. Telephone numbers of a few companies are listed in the Directory box.

Driving

Driving in Athens can be a nerve-racking experience and best avoided, especially if you are not accustomed to Greek road habits. Many streets in the centre are pedestrianized, and there are also plenty of

one-way streets, so you need to plan your route carefully.

Finding a parking space can be very difficult, too. Despite appearances to the contrary, parking in front of a no-parking sign or on a single yellow line is illegal. There are pay-and-display machines for legal on-street parking, as well as underground car parks, though these usually fill up quickly.

In an attempt to reduce dangerously high air pollution levels, there is an "odd–even" driving system in force. Cars that have an odd number at the end of their licence plates can enter the central grid, also called the *daktýlios*, only on dates with an odd number, and cars with an even number at the end of their plates are allowed into it only on dates with an even number. To avoid being unable to access the *daktýlios*, some people have two cars – with odd and even plates. The "odd–even" rule does not apply to foreign cars; however, if possible, avoid taking your car into the city centre.

Sign for a pedestrianized area

No parking on odd-numbered months (Jan, Mar, etc)

No parking on even-numbered months (Feb, Apr, etc)

DIRECTORY

Public transport

Metro & Trams
Tel 21441 46400. 🅦 stasy.gr

OASA (buses/trolleybuses)
Metsovou 15. **Tel** 21082 00999; or call centre: 11 185.
🅦 oasa.gr

Taxis

Athina 1
Tel 21092 17942 (central Athens).

Asteras
Tel 21061 44000 (central Athens).

Glyfada
Tel 21096 05603 (south Athens).

Parthenon
Tel 21052 23300 (north Athens).

Athens Transport Links

The hub of Athens' public transport is the area around Sýntagma Square and Omónoia Square. From this central area, the metro and various buses can be taken to Elefthérios Venizélos International Airport, the port at Piraeus and Athens' Laríssis train station. In addition, three tram lines connect the city centre with the Attic coast.

Bus X95 runs between the airport *(see p355)* and Sýntagma Square, and bus X96 links the airport to Piraeus *(see pp358–9)*. The airport is also served by Line 3 (blue line) of the metro, from Sýntagma and Monastiráki. Metro Line 1 (green line, from Omónoia and Monastiráki) extends to Piraeus; the journey from the city centre to the port takes about half an hour. Trolleybus route 1 goes past Sýntagma metro station and Laríssis train and metro stations. Laríssis train station is also served by metro Line 2 (red line), from both Sýntagma and Omónoia.

Tram line 3 runs along the coast from Néo Fáliro to the seaside suburb of Voúla; tram line 4 runs from Sýntagma Square in the city centre to Néo Fáliro; and tram line 5 runs from Voúla to Sýntagma Square. These lines are especially useful if you are staying in a hotel along the coast, or if you wish to have a day on the beach.

Though more expensive than public transport, the most convenient way of getting to and from any of these destinations is by taxi. The journey times vary greatly, but if traffic is free-flowing, the journey from the city centre to the airport takes about 40 minutes; the journey from the city centre to Piraeus takes around 30 minutes; and the journey from Piraeus to the airport takes about 60 minutes.

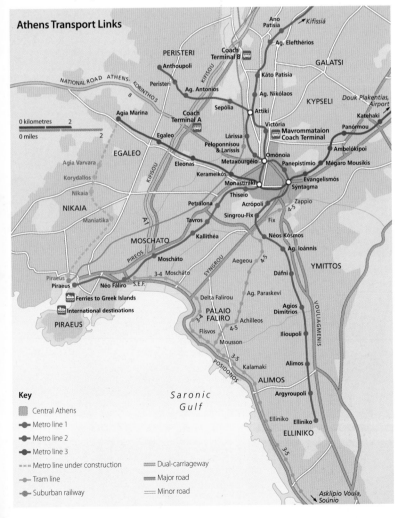

Athens Transport Links

Key

- Central Athens
- Metro line 1
- Metro line 2
- Metro line 3
- Metro line under construction
- Tram line
- Suburban railway
- Dual-carriageway
- Major road
- Minor road

TRAVELLERS' NEEDS

WHERE TO STAY

You can easily spend a large amount on some of the most luxurious hotel suites in the Greek islands, and further down the market Greece no longer enjoys a price advantage over other Mediterranean destinations. However, due to the ongoing economic crisis, many hotels are offering good-value deals, either direct or through booking websites. Despite the effects of mass tourism, hospitality off the beaten track can still be warm and heartfelt. Various types of accommodation, including campsites, are described over the next four pages. The listings section on pages 306–13 includes over 120 places to stay, ranging from pensions or rented rooms to luxury/boutique establishments, by way of accommodation in restored buildings and self-catering units.

Booking

Many visitors to the Greek islands choose to travel with a package-holiday company, reserving their flights and accommodation in advance. But increasingly travellers are assembling their own packages online, booking transport and lodging separately – without, however, the protections afforded by a bonded tour operator. By contrast, a reputable tour operator is required to look after you and make remedial arrangements.

Almost all accommodation providers, except for the smallest pension/room outfits, can be contacted direct via their own websites, and many subscribe to the various accommodation booking engines on the Internet. Always look first, however, at the establishment's own website – they can, surprisingly often, match offers made on third-party booking engines. For either booking engines or the hotels' own sites, credit card details are normally required, with at least one night's lodging charged to your card if you fail to arrive. For some heavily discounted "no cancellation" room rates, prepayment is demanded in full.

Prices

The **EOT (Greek National Tourist Office)** sets guideline prices each year for all classes of accommodation except deluxe hotels. However, market forces mean rates vary widely. In the islands, peak season prices only really apply from about 20 July to 20 August, while milder summer prices kick in when Greek school holidays begin, ending around mid-September. Rates for all types of accommodation are up to 50 per cent lower during the "shoulder" seasons: April–early June, except Easter (*Páscha* in Greek) and Pentecost weekends (*Agion Pnévma*), and mid-September to late October. Those stopping over in Athens will again find that summer rates carry a slight premium (even though the city itself empties out during August), while a Thessaloníki stopover will see room rates vary little over the year except during the trade fair (early September) and the two film festivals (mid-March and early November).

Grading

Hotels, pensions, villas and apartments are all categorized by the EOT. Hotels are graded from one star to five stars, and then deluxe. In practice, there are now very few surviving one-star hotels on the islands. Pensions/rooms and apartments are rated from one to three keys. Some of the posher restoration inns are placed in a special category.

The categorization system places more emphasis on common amenities, fixtures and

Spectacular views from the terrace at Phaedra hotel, Ydra *(see p307)*

◀ Souvenirs on offer in the village of Fiskárdo, Kefalloniá

A charming balcony at one of Patra's Apartments in Foúrnoi *(see p308)*

fittings than on service ethic, style and quality; thus, a multi-room complex with a restaurant, pool and beachside position can be awarded more stars than a really exquisite boutique inn.

As an overview, two-star hotels must have en-suite bathrooms and at least one basic bar-restaurant, probably serving breakfast and snacks only. Three-star hotels must have a full-service restaurant, pool or patch of beach, and some other facilities, perhaps a basic gym. Four- and five-star hotels offer the full range of in-room services plus more than one on-site bar-restaurant, conference/wedding facilities, multiple pools, tennis court, spa and beachside watersports (often an affiliated independent outfit rather than in-house).

Wi-Fi signal, or much more rarely cable-Internet connection, is now almost universal, even at quite basic pension/room establishments. What varies is payment policy – the more stars awarded, the more likely there will be stiff daily charges for Internet use.

Opening Seasons

On the islands, almost all hotels, pensions and apartments work only from some time in April or early May (the exact opening date being determined by when Orthodox, or Western, Easter falls) and close in late October. Where no opening months are given for an establishment in the listings, the reader should assume this season. When a venue works a longer calendar, or even all year, this has been clearly indicated.

As a rule, expect significant numbers of hotels to function during winter only in places close to Athens, like Aígina or Ýdra, or sizeable island towns such as Corfu Town, Ermoúpoli on Sýros, Irákleio or Chaniá. The listings for Athens and Thessaloníki stay open all year. But if you plan a visit to any of the smaller, more remote islands between November and early April inclusive, choices will be limited.

Dodecanese window

Hotel Chains

On the most popular resort-islands, like Corfu, Skiáthos, Crete, Kos, Rhodes and Zákynthos, large hotels purpose-built to serve the needs of package-holiday companies prevail. These are usually located on or near the best beaches and are block-booked by tour operators during high season. However, subject to the terms of their

contracts with the operators, they may have rooms available to individual visitors – often at bargain prices – during the off-peak spring and autumn weeks. Such seaside-resort hotels have a wide range of facilities including outdoor pools, well-landscaped grounds, and fully equipped rooms, sometimes family-sized.

Top-end seaside hotels offer more luxurious facilities such as wellness spas, fully equipped gyms, watersports facilities, tennis courts, indoor and outdoor pools, multiple restaurants (buffet or à la carte) and bars, plus more attentive service. Wi-Fi is still often charged for, especially in-room, but even in such cases it may be free in the lobby or at the pool-bar.

Major international hotel brands have made few inroads into Greek-island territory. However, several well-managed, contemporary Greek hotel chains, small and large, have luxury properties on the most popular islands. These include **Bluegr Mamidakis**, active on Crete; **Aldemar**, with hotels on Crete and Rhodes; **Grace**, with properties on Mýkonos, Santoríni and Kéa; **Grecotel**, with seaside resorts on Kos, Corfu, Crete, Rhodes and Mýkonos; **Santikos**, appearing on Skiáthos and Alónnisos; and **Yades**, with lodgings on Santoríni, Ýdra, Chíos, and Crete.

The luxurious 5-star Amirandes Resort in Crete *(see p312)*

Boutique Hotels

Luxury or boutique hotels, sometimes occupying restored traditional buildings, have added a much-needed dash of character to Greece's accommodation portfolio. These design-led hotels put a premium on style, technological widgets, good food (if perhaps only at breakfast), quality materials and a high degree of customer service. They are equally likely to be found in the heart of historic towns and villages as near the beach. You should also expect to find solid-wood furniture, expensive Italian tiling on floors and walls, recessed lighting, commissioned wall art, butler sinks, rain showers, complimentary bathrobes and slippers, and so on. Such lodgings are now widespread across all the island groups, especially the Cyclades, Crete, Rhodes, the Sporades and the Ionians. Few have more than 30 or so rooms, so advance reservations are essential. Two useful websites detailing many of them, with booking links, are www.smallhotelsingreece.com and www.greatsmallhotels.com.

Restoration Inns

Perhaps the most interesting places to stay in the Greek islands are restoration inns, either a single handsome old building rescued from dereliction, or even an entire abandoned village, respectfully converted into accommodation. In especially sensitive heritage areas, such as Rhodes' old city, Sými or other architecturally

The spectacular pool at Amirandes Resort in Crete *(see p312)*

protected island communities, the local archaeological service (Byzantine/post-Byzantine division) is obliged to supervise works. In the listings, there is inevitably some overlap of Restoration Inns with Luxury/Boutique – many of these inns have all mod cons – but equally often rooms are decidedly rustic, as at some of the rural Cretan inns and at Volissós on Chíos. Besides the islands mentioned above, other recommended restoration inns can be found on Ýdra, Sýros, Lésvos, Santoríni and Kálymnos.

Pensions

Veteran Greek-island-hoppers may remember sign-waving proprietors pitching their *enoikiazómena domátia*, or rented rooms. Moreover, many pension owners are listed on the Internet, with advance booking the norm.

In the early years of tourism, islanders often rented spare bedrooms in their own dwelling. Nowadays, pension *domátia* are almost always in small, purpose-built blocks with solar- or boiler-heated hot water, en-suite bathrooms, and usually air-conditioning. Some

now offer designer bedrooms and bathrooms. Most rooms also have a balcony or ground-level terrace, as well as basic self-catering facilities.

A critical difference between a hotel and a pension/rooms establishment is that the latter is not expected to provide breakfast. So rates given for pension/rooms are room only, with breakfast where available at an extra charge.

Villas

Self-catering villas and apartments can be excellent-value accommodation in the Greek islands. Most apartments are in small complexes of 20–30 units, built by local owners to meet the demands of the big package-holiday companies. You can expect a balcony, a kitchen, an en-suite bathroom and a communal area. Studios sleep two, usually in twin beds. Many apartments on offer accommodate four to six people, with one or two separate double bedrooms, as well as a twin bedroom or a sofabed.

More luxurious, villas are usually strictly controlled by specialist holiday companies overseas, or a designated agency in the nearest town. The largest and most expensive villas sleep up to eight, offering facilities such as a private pool, satellite TV, DVD player, a fully equipped kitchen and maid service (but rarely Wi-Fi). Some even come complete with a cook and household staff. Smaller, more basic villas usually have similar facilities, but without luxury extras such as a pool.

The widest range of apartment and villa accommodation is to be found on Corfu, Paxos, Lefkáda,

A traditional-style self-catering apartment on Santoríni

Kefaloniá, Alónnisos, Skópelos, Sámos, Chíos, Chálki and Crete. Travel à la Carte *(see p339)*, Cachet Travel and Sunvil all offer good-quality self-catering apartments, plus a few villas. Specialist villa companies include Simpson Travel *(see p339)* and The Villa Collection Greek Islands Club (www. gicthevillacollection.com).

Youth Hostels

Given the presence of in-expensive pension rooms and hotels, there's little demand for hostels in the Greek islands. The only youth hostels remaining are on Íos (The Purple Pig), Corfu (The Pink Palace), Santoríni (Oía Youth Hostel) and Crete (at Irákleio, Réthymno and Plakiás). None is affiliated with Hostelling International, so presenting a YHA card will net few advan-tages. Most now appear on third-party booking engines or have their own sites, easily searchable online.

Lakka Paxi Camping, on the Ionian Islands

The tranquil living areas at Adrina Beach Resort in Skópelos *(see p308)*

Camping

The **Panhellenic Camping Association** lists officially recognized camp sites on its website. Many island campsites are fairly basic, offering just showers, toilets and grounds on which to pitch a tent. Others are quite sophisticated, with swimming pools, restaurants, volleyball courts, Wi-Fi, laundry and mini-market. Most are better geared towards motor caravans than tents, but power hook-ups might not be guaranteed.

Disabled Travellers

One UK company providing useful information about access to hotels and places of interest in Greece is **Accessible Travel and Leisure**, which currently features a disabled-friendly resort on Crete. Only the largest and most modern hotels in Greece have even the most basic facilities (such as lifts, wide room/bathroom doors and wheelchair-accessible common areas) for people with disabilities.

Recommended Hotels

The hotels listed have been chosen using a variety of criteria. The Greek islands have a large number of luxury and boutique hotels, while others are consistently rated highly for service, breakfast and amenities. Look out for those marked as a DK Choice; it is our judgment that the hotel particularly excels in that designated theme or category – providing real luxury, being a remarkably sensitive restoration of an historic building, a family-friendly or adults-only beachfront resort, or having exceptional hosts and hotel-grade features, in the case of a pension or apartment. With very few exceptions, we have not singled out members of hotel chains, whether international or Greek.

Where to Stay

The Ionian Islands

CORFU: Bella Venezia €€
Boutique
N. Zambéli 4, 49100 Corfu town
Tel *26610 46500*
🇼 bellaveneziahotel.com
Located in the atmospheric Pórta Remoúnda area, this Neo-Classical former girls' school offers stylish doubles and top-floor suites.

CORFU: Delfino Blu €€
Luxury
Agios Stéfanos Avliotón, 49081
Tel *26630 51629*
🇼 delfinoblu.gr
Alongside views of the sunset beyond Mathráki Island, Delfino Blu has contemporary suites with DVD players, music systems and rain showers. Gourmet restaurant.

CORFU: Corfu Palace €€€
Luxury
Dimokratías 2, 49100 Corfu town
Tel *26610 39485*
🇼 corfupalace.com
The grand dame of Corfu town hotels since 1954 overlooks Garítsa Bay. Rooms feature wood-veneer floors and marble bathrooms.

DK Choice

CORFU: The Merchant's House €€€
Restoration Inn
Paleá Períthia, 49081
Tel *26630 98444*
🇼 themerchantshousecorfu.com
A medieval tradesmen's bazaar has been meticulously restored as cutting-edge suites, where taste and attention to detail are reflected in natural bedding and huge bathrooms. The Anglo-Dutch owners are brilliant hosts, serving breakfast to order on the lawn terrace.

ITHACA: Captain's €
Apartments
Hillside above Kióni, 28301
Tel *26740 31481*
🇼 captains-apartments.gr
The apartments at this small, family-run complex offering a warm welcome boast sweeping bay views and neutral decor.

ITHACA: Perantzada 1811 Art Hotel €€€
Boutique
Odysséos Androútsou, 28300 Vathy
Tel *26740 33496*
🇼 perantzadahotel.com

This 1811-vintage mansion is now a carefully appointed inn. Rooms, often with full port views, contrast bold colour accents against white walls.

KEFALLONIA: Linardos €
Apartments
28085 Asos
Tel *26740 51563*
🇼 linardosapartments.gr
Smart apartments have dark-wood furniture, kitchen facilities and stunning views of Asos Bay and the pebble beach below.

KEFALLONIA: White Rocks Hotel & Bungalows €€€
Seaside Resort
Platys Gialós, 28200 Argostóli
Tel *26710 28332*
🇼 whiterocks.gr
Ideally located, these bungalows and suites with sea views are surrounded by great beaches.

LEFKADA: Ostria €
Pension
Top approach road, Agios Nikítas, 31080
Tel *26450 97483*
Charming balconied rooms have terracotta floors and objets trouvés. There are unobstructed Ionian views from all units.

LEFKADA: Porto Fico €€
Seaside Resort
Pónti beach, 31100 Vasilikí
Tel *26450 31402*
🇼 portoficohotel.com
Porto Fico's well-tended front lawn leads from the pool to the beach and windsurf school.

LEFKADA: Serenity €€€
Luxury
Just south of Atháni, 31082
Tel *26450 33639*
🇼 serenity-th.com

Suites at The Merchant's House in Corfu – a former medieval tradesmen's bazaar

This retreat hotel focuses on its spa, which uses Dead Sea products. Stunning views from the infinity pool and suites; the restaurant fare is locally sourced.

PAXOS: Paxos Beach Hotel €€
Seaside Resort
1.5 km southeast of town, 49082 Gáios
Tel *26620 32211*
🇼 paxosbeachhotel.gr
Gáios' tavernas are just a 15-minute walk from this stone-clad bungalow complex with a private pebble beach.

ZAKYNTHOS: Villa Katerina €
Villas
Pórto Róma, Vasilikós peninsula, 29100
Tel *26950 35456*
🇼 villakaterina.com
The studios and apartments on the upper floors look over the olive trees. Close to many tavernas and beaches, including Gérakas.

ZAKYNTHOS: Paliokaliva Village €€
Villas
Tragáki district, 29100 Tsilívi, 1.5 km (1 mile) from beach
Tel *26950 63770*
🇼 paliokaliva.gr
Family-friendly self-catering units inside garden- and orchard-set cottages. Rustic decor.

The Argo-Saronic Islands

AIGINA: Brown €
Restoration Inn
Aktí Hatzí 3–4, 18010
Tel *22970 22271*
🇼 hotelbrown.gr
This converted 1880s sponge factory stands opposite Aígina town's southerly beach.

AIGINA: Rastoni €€
Boutique
Stratigoú Dimítri Petrití 31, 18010
Tel *22970 27039*
🇼 rastoni.gr
Enjoy breakfast in a courtyard garden at this hotel on a slope north of the port. Every unit looks over pistachio trees to the Saronic.

KYTHIRA: Spitia Vasili €€
Apartments
80100 Kapsáli
Tel *27360 31125*
W *kythirabungalowsvasili.gr*
These simply furnished two-,
three- and four-bedroom
bungalows with sea views
are scattered amid the olive
trees just outside Kapsáli.

KYTHIRA: Venardos €€
Boutique
80100 Agía Pelagía
Tel *27360 34100*
W *venardos-hotels.gr*
Plain rooms with dark-wood
touches overlook the sea or
pool courtyard. Jacuzzi and
sauna on site. Located a short
walk to village centre and beach.

POROS: Pavlou €
Seaside Resort
Megálo Neório
Tel *22980 22734*
W *pavlouhotel.gr*
A family-friendly establishment
at the edge of one of Kalávria's
better beaches offering tennis
courts, a pool and a restaurant.

POROS: Sto Roloï €€
Villas
Kostelénou 34–36, 18020
Tel *22980 25808*
W *storoloi-poros.gr*
A cluster of three houses has
been divided into impeccably
restored, antique-filled apart-
ments. Several are suitable
for families.

SPETSES: Armata €€
Boutique
Agíou Antoníou district, 18050
Tel *22980 72683*
W *armatahotel.gr*
Romantic hideout with lush
pool garden and plush interiors.
Luxury toiletries and fluffy
towels are supplied.

SPETSES: Poseidonion Grand €€€
Luxury
West of Dápia, 18050
Tel *22980 74553*
W *poseidonion.com*
Opened in 1914, this characterful
hotel has a two-level tower suite.
Atmospheric common areas
include the period-tiled Library
Brasserie and the garden spa.

YDRA: Kirki €
Pension
Miaoúli Street, 80 m from quay, 18040
Tel *22980 53181*
W *kirkihotel.com*
The simple but serviceable
rooms at this friendly inn feature
colourful wall art and balconies.

One of the spacious rooms at the Phaedra hotel in Ydra

DK Choice

YDRA: Phaedra €€
Luxury
*220 m inland; start at Tompázi,
fork left at Amalour bar; 18040*
Tel *22980 53330*
W *phaedrahotel.com*
In a secluded locale, the
Phaedra has just six antique-
filled rooms, each with its
own entrance, and striking
mountain views from a shared
terrace. Uniquely for this area,
it is open all year round.

YDRA: Orloff €€€
Boutique
*Rafália 9, crn Vótsi (main inland
square), 18040*
Tel *22980 52564*
W *orloff.gr*
Antiques and original features
abound at this 1796-built
mansion-hotel. Enjoy breakfast
in the lemon-tree garden.

The Sporades and Evvoia

**ALONNISOS: Konstantina
Studios** €
Apartments
*Eastern edge of Palaiá Alónissos,
37005*
Tel *24240 66165*
W *konstantinastudios.gr*
The studios here have separate
kitchen alcoves for self-catering
guests and views east over the
village to Skántzoura.

ALONNISOS: Yalis €€€
Luxury
Vótsi, 37005
Tel *24240 66186*
W *yalishotel.gr*
Some of the suites at this clifftop
hotel offer sweeping views and
exposed beams. All have a small
living room and free Wi-Fi.

EVVOIA: Mousiko Pandoheio €€
Boutique
West entrance to village, 34014 Stení
Tel *22280 51202*
W *mousikopandoxeio.gr*
At the "Musical Lodge", musical
motifs dominate the decor
and the ground-floor bar hosts
regular weekend performances.

**EVVOIA: Thermae Sylla
Spa Wellness Hotel** €€€
Luxury
Posidónos 2, 34300 Aidipsós
Tel *22260 60100*
W *thermaesyllaspa-hotel.com*
Honeymooners and wedding
parties pamper themselves at
this refurbished *belle époque*
spa-hotel. The hot springs have
flowed here since antiquity.

SKIATHOS: Atrium €€
Seaside Resort
*Hillside above Agía Paraskeví beach,
37002*
Tel *24270 49345*
W *atriumhotel.gr*
A four-star complex where the
six grades of rooms/family suites
are tastefully minimalist. Good
food, especially breakfast.

SKIATHOS: Aegean Suites €€€
Luxury
*Megáli Ammos, 1.5 km west of town,
37002*
Tel *24270 24069*
W *santikoshotels.com*
Huge suites offer sound systems
and sea views at this adults-only
hillside hotel within walking
distance from Skiáthos town.
There is a poolside restaurant, too.

SKOPELOS: Kyr Sotos €
Pension
*Just inland from mid-quay,
37003 Skópelos town*
Tel *24240 22549*
W *skopelos.net/sotos/*
Centrally located pension with a
cult following. The most relaxing
rooms face the interior courtyard.

For more information on types of hotels *see pp303–305*

DK Choice

**SKOPELOS: Adrina
Beach Resort & Spa** €€
Seaside Resort
Adrína cove, 37003 Pánormos
Tel *24240 23371*
🆆 adrina.gr
Two adjacent hillside complexes
with a pebbly beach. Adrina
Beach, featuring white rooms
and maisonettes, embraces a
wood-decked saltwater pool,
while the Spa Resort comprises
grey-to-beige rooms and a
basement spa. Both hotels have
beachfront bars/restaurants.

SKYROS: Nefeli €€€
Luxury
Southern ouskirts of Chóra, 34007
Tel *22220 91964*
🆆 skyros-nefeli.gr
Standard doubles, suites and
traditional houses are arranged
around a saltwater pool. A
copious breakfast is also offered.

The Northeast
Aegean Islands

CHIOS: Chios Rooms €
Pension
Aigéou 110, 82100
Tel *22710 20198*
🆆 chiosrooms.gr
Stay in high-ceilinged rooms
in a *belle époque* building. The
penthouse has a private terrace;
other rooms have small balconies.

CHIOS: Medieval Castle Suites €€
Restoration Inn
Mesta
Tel *22710 76345*
🆆 mcsuites.gr
Expect vaulted ceilings, mini-
kitchens and either a patio or a
roof terrace in the lavish suites
inside this medieval mansion.

**CHIOS: Argentikon
Luxury Suites** €€€
Luxury
82100 Kámbos
Tel *22710 33111*
🆆 argentikon.gr
Suites of up to three bedrooms
are distributed across several
buildings, including two from the
16th century. There is also a pool,
Jacuzzi, sauna, spa and gardens.

FOURNOI: Patra's Apartments €
Villas
83400
Tel *22750 51268*
🆆 fourni-patrasrooms.gr
These spacious hillside studios
and apartments have a pleasant

decor and the occasional
fireplace. Self-catering equipment
is provided, but breakfast is also
served at the family's tree-
shrouded harbour café.

IKARIA: Akti €
Pension
*Just above the east quay, 83300
Agios Kírykos*
Tel *22750 22694*
🆆 pensionakti.gr
Most rooms at this family-run inn,
perfectly located for catching pre-
dawn ferries, have a balcony or
terrace facing Foúrnoi. The owner
is a font of island knowledge.

IKARIA: Erofili Beach €€
Boutique
*Entrance to the village, 83301
Armenistís*
Tel *22750 71058*
🆆 erofili.gr
Comfortable rooms gaze towards
the sandy Livádi beach. Common
areas include the airy breakfast
room and the bay-view bar.

**LESVOS:
Lesvos Accommodation** €€
Villas
*Scattered locations around the
village, 81108 Mólyvos*
Tel *22530 71128*
🆆 lesvosaccommodation.com
Studios and family units sleeping
up to eight in stone-built houses
refurbished to a high standard.

LESVOS: Vatera Beach €€
Seaside Resort
*West end of beach development,
81300 Vaterá*
Tel *22520 61212*
🆆 vaterabeach.gr
Now in its second generation
of friendly management, this
rambling hotel-restaurant
abuts a superior patch of
Lesvian beach.

SAMOS: Ino Village €€
Boutique
*Kefalopoúlou Street, Kalámi district,
83100 Vathý*
Tel *22730 23241*
🆆 inovillagehotel.com
A well-run three-star hotel with
parking, a large pool and a good
restaurant. "Deluxe" and "Comfort"
rooms are boutique standard.

**SAMOS:
Sirena Residence & Spa** €€€
Luxury
*200 m inland, Votsalákia, 83102
Marathókampos*
Tel *22730 31035*
🆆 sirena.gr
Boasting the best spa on Sámos,
this stone-clad apart-hotel is set
in a mock-traditional village.

The striking *belle époque* exterior of Villa
Melina in Kálymnos

**SAMOTHRAKI: Samothraki
Village** €€
Boutique
*Coast road Kamariótissa–Loutrá,
68002 Palaiópoli, Samothráki*
Tel *25510 42300*
🆆 samothrakivillage.gr
Distributed around the pool,
units have views of either the
sea or the mountains. Sauna
and gym. Open all year.

THASOS: Thassos Inn €
Pension
Treis Pigés district, 64004 Panagía
Tel *25930 61612*
🆆 thassosinn.gr
A calm, inland alternative to
beach resorts, Thassos Inn has
a terrace café that provides
breakfast, drinks and snacks.

**THASOS: Alexandra Beach
Thassos Spa Resort** €€
Seaside Resort
*1 km (0.5 miles) northwest of
Potós, 64002*
Tel *25930 5800*
🆆 alexandrabeach.gr
Set amid vast gardens, this spa
resort boasts two *hammams*, a
sauna and a Jacuzzi, as well as
a full treatment menu.

DK Choice

**THASOS: Alexandra
Golden Boutique** €€€
Luxury
*Towards the north end of the
beach, 64004 Chrysí Ammoudiá*
Tel *25930 58212*
🆆 alexandragolden.com
Choose from a variety of high-
standard units (including
galleried maisonettes with
their own pools) at this adults-
only resort. The decor is
contemporary, with a slightly
unorthodox colour range. Spa
with free *hammam* and sauna.

The Dodecanese

ASTYPALAIA:
Kilindra Studios　　　　€€
Boutique
West edge of Chóra, 85900
Tel *22430 61131*
W astipalea.com.gr
Traditionally styled studios have views of Livádia Bay. The spa offers hot-stone massages and acupuncture; there is a pool, too.

ASTYPALAIA:
Maltezana Beach　　　　€€€
Seaside Resort
85900 Análipsi (Maltezána)
Tel *22430 61558*
W maltezanabeach.gr
Spacious, modern rooms and suites overlook a garden with a pool. There is also a restaurant.

CHALKI: Captain's House　€
Restoration Inn
Inland lane, 85110 Nimporió
Tel *22460 45201*
The captain is gone now, but his ever-helpful widow keeps this intimate guesthouse in shipshape. 1 double, 2 twin en suite rooms.

KALYMNOS: Acroyali　　　€
Apartments
Lower road, 85200 Myrtiés
Tel *22430 47521*
W acroyali.gr
Book ahead to stay at these cheery sea-view apartments just steps away from Myrtiés beach. Perfect for small families.

DK Choice

KALYMNOS: Villa Melina　€
Restoration Inn
Next to the archaeological museum, Evangelístria district, 85200 Póthia
Tel *22430 22682*
W villa-melina.com
Part of Póthia's best hotel is housed in the converted mansion of a belle époque sponge magnate, with modern studios and apartments behind a large, salt-water pool. Wood-floored rooms have been modernized with large bathrooms and tasteful furnishings.

KARPATHOS: Glaros　　　€
Villas
South hillside above the port, 85700 Diafáni
Tel *22450 51501*
W hotel-glaros.gr
These spacious lodgings have unbeatable views over the bay to the mountains. Self-catering accommodation, but the hosts also offer an optional breakfast.

KASTELLORIZO: Karnayo　€€
Restoration Inn
50 m (55 yards) inland from the southwest corner of the quay, 85111
Tel *22460 49266*
W karnayo.gr
This is the best restoration inn on Kastellórizo. Architect-designed, it features four standard doubles and two apartments.

KASTELLORIZO: Kastellorizo　€€€
Boutique
West quay, 85111
Tel *22460 49044*
W kastellorizohotel.gr
Studios and maisonettes provide top comforts, with solid-wood furniture, stall showers, DVD players and free Wi-Fi. Massage pool and access to the sea.

KOS: Afendoulis　　　　€
Pension
Evrypýlou 1, 85300 Kos town
Tel *22420 25321*
W afendoulishotel.com
Guests enjoy the warm welcome and the inn-like ambience at Afendoulis. Balconied rooms have fridges and dark-wood furniture.

KOS: Kos Aktis Art　　　€€
Boutique
Vassiléos Georgíou 7, 85300 Kos town
Tel *22420 47200*
W wwwkosaktis.gr
Rooms and suites face the water, and bathrooms have light-wells and butler sinks at Kos town's first and best boutique hotel. Popular waterfront restaurant.

KOS: Oceanis Beach & Spa　€€
Seaside Resort
85300 Psalídi
Tel *22420 24641*
W oceanis-hotel.gr
The most remote of the beachfront complexes on Kos, Oceanis is a four-star adults-only resort with a pool, a spa, pleasant gardens and sports facilities.

KOS: Diamond Deluxe　　€€€
Luxury
Shore road, Lámbi beach area, 85300
Tel *22420 48835*
W diamondhotel.gr
A stylish countryside hotel arranged around two pools. Mid-range units have veneer flooring, computers and private Jacuzzis.

LEROS: Archontiko Angelou　€€
Restoration Inn
400 m (440 yards) inland from Alinda beach, 85400
Tel *22470 22749*
W hotel-angelou-leros.com
Dating from the 1900s, this atmospheric mansion set in orchards has the feel of a French country hotel, with beamed ceilings and beautiful antiques.

LIPSI: Nefeli　　　　　€€
Apartments
Hillside above Kámpos cove, 85001
Tel *22470 41120*
W nefelihotels-lipsi.com
Breakfast is included at this complex of studios and one- and two-bedroom apartments.

NISYROS: Porfyris　　　€
Boutique
Just inland from the communal orchard, 85303 Mandráki
Tel *22420 31376*
W porfyrishotel.gr
Extensive renovations and the biggest pool on the island justify a boutique label at this 1970s hotel.

PATMOS: Porto Scoutari
Romantic Hotel & Suites　€€
Boutique
Hilltop above Melóï beach, 85500
Tel *22470 33124*
W portoscoutari.com
Enormous rooms and suites have antique fittings, original art and sea views. Guests can enjoy poolside breakfasts, dinners, and a small spa. Wedding packages are available.

The majestic views from the veranda at Adrina Beach Resort in Skópelos

For more information on types of hotels *see pp303–305*

DK Choice

RHODES: Andreas €€
Restoration Inn
Omírou 28/D, Old Town, 85100
Tel *22410 34156*
🌐 hotelandreas.com
A long campaign has resulted
in this inn returning to its architectural origins as a Turkish
pasha's mansion. Two rooms
have stone fireplaces. Thanks to
its location in the highest point
in the Old Town, Andreas offers
unbeatable views and a peaceful
rest away from noisy bars.

RHODES: Rodos Park Suites €€€
Luxury
Riga Feraioú 12, 85100
Tel *22410 24612*
🌐 rodospark.gr
Stay in slick contemporary rooms
and suites and enjoy a cutting-
edge basement spa, large pool
and summer roof bar.

RHODES: Spirit of
the Knights €€€
Boutique
Old Town, 85100
Tel *22410 39765*
🌐 rhodesluxuryhotel.com
Medieval chic is a good
description for this all-suite inn
in a 15th-century manor house
with Latin and Oriental touches.

SYMI: Lapetos Village €€
Apartments
Gialós, back of main square, 85600
Tel *22460 72777*
🌐 iapetos-village.gr
Spacious balconied maisonettes
and studios – with kitchens. Unique
on Sými for its covered swimming
pool and lush exotic gardens.

SYMI: Symi Visitor
Accommodation €€
Villas
Gialós & Chorió, 85600
Tel *22460 71785*
🌐 symivisitor-accommodation.com
This agency has many properties
on its books, from simple studios
to family-sized mansions. These
are second homes let in their
owners' absence. Expect libraries,
full kitchens and music systems.

TILOS: Llidi Rock €€
Seaside Resort
West slope overlooking Livádia Bay,
85002
Tel *22460 44293*
🌐 ilidirock.gr
Open all year, this hillside hotel
has no pool, but it does offer
access to a private beach. Gym,
sauna, disabled-access wing
and traditional platform beds.

The Cyclades

AMORGOS: Aegialis €€€
Boutique
North hillside, 84008 Ormos Aigiális
Tel *22850 73393*
🌐 amorgos-aegialis.com
Rooms and suites are decorated
in neutral colours. The Aegean
cuisine restaurant, saltwater pool
and indoor spa are all excellent.

ANDROS: Eleni Mansion €€
Luxury
Andréa Empiríkou 9, 84500 Chóra
Tel *22820 22270*
🌐 archontikoeleni.gr
This Neo-Classical mansion is full
of period charm. Some rooms
have original ceiling paintings;
the attic suite has sea views.

ANDROS: Paradise €€€
Boutique
Southwest end of Chóra, on the road
out of town, 84500
Tel *22820 22187*
🌐 paradiseandros.gr
This is the top hotel in Andros. It
offers a folklore museum, a
large pool, tennis courts and
sea views from most rooms.

FOLEGANDROS: Polikandia €€
Luxury
60 m from the square, 84011 Chóra
Tel *22860 41322*
🌐 polikandia-folegandros.gr
Five grades of rooms and suites,
all with bright decor. Enjoy the
pool, with its fountain-island
feature, and the Jacuzzi annexe.

FOLEGANDROS:
Anemomilos €€€
Villas
Cliff edge, 84011 Chóra
Tel *22860 41309*
🌐 anemomilosapartments.com
"Blue" cottages (with sea views)
and larger "Green" units, which

The cool, relaxing interior of the Hotel
Andreas in Rhodes

sleep four, command a higher
price here. There are stall showers
and butler sinks throughout.

IOS: Liostasi €€€
Boutique
84001 Chóra
Tel *22860 92140*
🌐 liostasi.gr
Seven grades of stylish rooms
and suites, looking west over
Gialós Bay to Síkinos.

KEA: Porto Kea Suites €€€
Boutique
Livádi district, 84002 Korissía
Tel *22880 22870*
🌐 portokea-suites.com
Five categories of units including
family grade. Traditional furniture
and soothing tones throughout.

KOUFONISSI:
Geitonia tis Irinis €€
Villas
East edge of Chóra, near Pórta cove,
84300
Tel *22850 71674*
🌐 koufonisia-diakopes.gr
The "Neighbourhood of Peace"
offers cave-style accommodation
in cheery Cycladic colours.

KYTHNOS: Porto Klaras €€
Apartments
Yacht marina, 84006 Loutrá
Tel *22810 31276*
🌐 porto-klaras.gr
Studios, apartments and double
rooms with marina views and
mock-Cycladic decor. Beaches
and thermal springs nearby.

MILOS: Aeolis €€
Pension
Quiet street west of the ferry jetty,
84800 Adámas
Tel *22870 23985*
🌐 hotel-aeolis.com
Basic Cycladic cottages located
a short walk from the beach.
Breakfast option only in season.

MILOS: Kapetan Tasos Suites €€€
Villas
84800 Pollónia
Tel *22870 41287*
🌐 kapetantasos.gr
One-, two- and three-room suites
with natural-fibre bedding and
Korres toiletries. Room service is
available until midday.

MYKONOS: Cavo Tagoo €€€
Boutique
Tagoó hillside, 1.5 km (1 mile)
north of Chóra, 84600
Tel *22890 23692*
🌐 cavotagoo.gr
The first of Mýkonos' boutique
hotels offers knowledgeable staff,
good food and superior units
with sea views and plunge pools.

MYKONOS: Mykonos Grand Resort €€€
Seaside Resort
Agios Ioánnis, 84600
Tel *22890 2555*
W mykonosgrand.gr
A luxury outfit with sunset views over Delos Island. Superior sea-view rooms have egg-shaped soaking tubs. There's also a spa.

MYKONOS: Semeli €€€
Luxury
Róhari district, 84600 Chóra
Tel *22890 27466*
W semelihotel.gr
This discreet retreat is popular with honeymooners. Rooms and suites are arranged around a large pool. Subdued decor and Philippe Starck bathrooms.

NAXOS: Chateau Zevgoli €€
Restoration Inn
Boúrgos district, Kástro, 84300 Náxos town
Tel *22850 26123*
W naxostownhotels.com
Stay in an antique-fitted suite or a double with sea views at this inn housed in Venetian-era buildings.

NAXOS: Grotta €€
Boutique
Grótta district, 84300 Náxos town
Tel *22850 22215*
W hotelgrotta.gr
Rooms have wall art and quality furnishings more typical of multi-starred hotels. Vast breakfasts are served in a spacious lounge.

NAXOS: Kavos Hotel Naxos €€€
Villas
Hillside above Agios Prokópios beach, 84300
Tel *22850 23355*
W kavos-naxos.com
Stone-clad suites and apartments are set around an infinity pool. Top units have private roof terraces and big gardens.

PAROS: Sofia €
Pension
Livádia district, 84400 Paroikiá
Tel *22840 22085*
W pension-sofia.gr
Breakfasts are charged extra at this inn in an orchard setting with water features and gazebos.

PÁROS: Paros Bay €€
Seaside Resort
Delfínia cove, 2 km (1.3 miles) south of Paroikiá, 84400
Tel *22840 21140*
W parosbay.com
Rooms have modern bathrooms and LCD TVs. Guests can join the yoga and massage programmes. There is also an amphitheatre, popular for weddings.

A stunning view from one of the balconies at Esperas in Santoríni

DK Choice

PAROS: Petres €€
Boutique
Agios Andréas district, 2 km (1.3 miles) south of Náousa, 84401
Tel *22840 52467*
W petres.gr
Rooms here have sea views, tiled floors and beamed ceilings. Amenities such as a gym, a huge pool and a tennis court, plus the five-star service, including transfers to and from Náoussa, give this hotel the edge. Excellent breakfasts.

SANTORINI: Afroessa €€
Restoration Inn
North end of Imerovígli, 84700
Tel *22860 25362*
W afroessa.com
A sympathetic *skaftá* (dug-out house) conversion, Afroessa has a large pool, interesting bathrooms and a wine press.

SANTORINI: Aigialos €€€
Restoration Inn
Ypapantís Street, 84700 Firá
Tel *22860 25191*
W aigialos.gr
This former 18th-century convent has exceptionally tasteful decor.

DK Choice

SANTORINI: Esperas €€€
Restoration Inn
West end of Oía, 84702
Tel *22860 71088*
W esperas.gr
Spectacular sunset views, a fair-sized pool with grotto features, and great staff give this inn the edge over its competitors. The decor in the studios, cottages and suites emphasizes wood, rattan and plaster surfaces. Studio 121 and cottage 114 have the most private terraces.

SANTORINI: Ikies Traditional Houses €€€
Restoration Inn
Perivóli district, 84702 Oía
Tel *22860 71311*
W ikies.com
The cave-house units here are cleverly tiered to ensure privacy. Breakfast and bar snacks only.

SERIFOS: Indigo Studios €€
Apartments
Inland 100 m (110 yards) from beach south end, 84005 Livádi
Tel *22810 52548*
W indigostudios.gr
Stay in high-standard self-catering units or ground-floor doubles. Breakfast is available at an affiliated café.

SIFNOS: Petali Village €€
Boutique
Hilltop, Ano Petáli district, 84003 Appollonía
Tel *22840 33024*
W sifnoshotelpetali.com
Cycladic blue woodwork and white plaster outside and light, airy interiors. Minimum stay of four nights in summer.

SIFNOS: Elies Resorts €€€
Luxury
Beachfront, 84003 Vathý Bay
Tel *22840 34000*
W eliesresorts.com
This luxurious suite-and-villa village attracts A-list celebrities. All rates include champagne, buffet breakfasts and transfers to and from Sífnos.

TINOS: Voreades €
Pension
Foskólou 7, 84200 Tínos town
Tel *22830 23845*
W voreades.gr
Bay-view rooms at this quint-essentially Tinian inn have flagstone floors, traditional textiles and mock-antique furniture, including canopy beds.

For more information on types of hotels *see pp303–305*

Crete

AGIOS NIKOLAOS:
Minos Beach Art Hotel   €€€
Boutique
Aktí Illía Sotírhou, Ammoúdi district, 72100
Tel *28410 22345*
w minosbeach.com
Edgy art is the main theme here, with works by contemporary Greek and international sculptors dotting the grounds. The best accommodation is in the waterside superior bungalows.

AGIOS NIKOLAOS:
St Nicolas Bay   €€€
Seaside Resort
Nissí peninsula, 72100
Tel *28410 25041*
w stnicolasbay.gr
Large double rooms, suites with infinity pools and villas offer low-key luxury and admirable levels of privacy. There is just one sandy cove, but plenty of activities on the watersports programme, several restaurants, a kids' club and a spa.

CHANIA: Theresa   €
Pension
Angélou 8, Tophanás district, Old Port, 73131
Tel *28210 92798*
w pensiontheresa.gr
The wood-rich rooms at this bohemian favourite can be reached by Chaniá's steepest serpentine stairway. There is a roof terrace and a small self-catering kitchen for guests.

DK Choice

CHANIA: Casa Delfino   €€€
Luxury
Theofánous 7, Tophanás district, Old Port, 73131
Tel *28210 93098*
w casadelfino.com
This hotel occupies a court-yarded 17th-century palazzo and a newer annexe. In terms of style, handmade furniture meets recycled timber beams, and there are marble-clad bathrooms and wooden floors. The annexe spa uses premium Apivita products. Impeccable service and scrumptious breakfasts. No young children.

ELOUNTA: Elounda Mare   €€€
Seaside Resort
2 km north of the centre, 72053
Tel *28410 68200*
w eloundamare.gr
Elounda Mare strikes warm, intimate notes in its bungalows

with glowing fireplaces and wooden floors. Garden terraces descend to several lidos, but note that there is no beach.

IERAPETRA: Cretan Villa   €
Restoration Inn
Oplarhigoú Lakerdá 16, 72200
Tel *28420 28522*
w cretan-villa.com
This converted 18th-century mansion is the best town-centre option in Ierápetra. Rooms boast designer furniture and feature exposed stone walls.

IRAKLEIO: Galaxy   €€
Boutique
Dimokratías 75, 1 km (0.6 miles) south of Platía Eleftherías, 71306
Tel *2810 238812*
w galaxy-hotel.com
The staff are welcoming and well-trained at this top city-centre hotel with sleekly modern rooms and suites. A generous breakfast buffet is offered.

DK Choice

IRAKLEIO: Amirandes
Grecotel Exclusive Resort €€€
Seaside Resort
Goúves, about 11 km (7 miles) east of the centre, 71110
Tel *28970 41103*
w amirandes.com
The five-star Amirandes impresses with water and stone features in communal areas, plus a bewildering range of stylishly appointed rooms, suites, bungalows and villas, some with private pools. There are eight restaurants and a large pool for when the private beaches get too windy. The spa has an Ayurvedic division.

KASTELLI KISSAMOU:
Mirtilos   €
Apartments
Plateía Tzanakáki, 73400
Tel *28220 23079*
w mirtilos.com
These tile-floored, pastel-tinted self-catering studios and apartments sleep up to four people. Breakfast offered.

LOUTRO: Blue House   €
Pension
73011, Loutró, Crete
Tel *28250 91035*
w bluehouse.loutro.gr
All rooms at the Blue House have balconies with sea views, patterned floor tiles and quirky bathrooms. There is also a popular in-house restaurant.

A typically narrow street in the old part of Chania in Crete

MAKRYGIALOS:
White River Cottages   €
Villas
1 km inland up a stream gorge, 7205.
Tel *28430 51120*
An abandoned shepherds' hamlet has been turned into rural lodgings. Food is not supplied; bring provisions from Makrýgialos

PALAIOCHORA: Zafiri 2   €
Apartments
100 m (110 yards) behind Pahiá Ammos beach, 73001
Tel *28230 41811*
w zafiri-studios.com
Studios and apartments, some with sea views. Breakfast is served on the vast front lawn.

PLATANIAS:
Minoa Palace Resort   €€€
Seaside Resort
West edge of the resort, 9 km (6 miles) west of Chaniá, 73100
Tel *28210 36500*
w minoapalace.gr
Four-star rooms and suites, but only the beachside wing has sea views. Pools, spa and several restaurants. Family-friendly.

RETHYMNO: Palazzo Vecchio   €€
Restoration Inn
Melissinoú, crn Iroön Polytechníon, 74100
Tel *28310 34351*
w palazzovecchio.gr
The best units in this 15th-century townhouse feature terracotta floors, beamed ceilings and enormous bathrooms.

RETHYMNO: Mythos Suites   €€€
Boutique
Plateía Karaolí ke Dimitríou 12, 74100
Tel *28310 53917*
w mythos-crete.gr
An intimate hotel built around a small courtyard pool. Breakfast is available picnic-style for those departing on early excursions.

SOUGIA: Syia €€
Apartments
Approach road on the left, 73009
Tel *28230 51174*
W syiahotel.com
The best lodging in Soúgia,
Syia offers spacious studios
and apartments located about
500 m (0.3 mile) inland. Some
have distant sea views.

SPILI: Heracles €
Pension
*Side street downhill from the through
road, 74053*
Tel *28320 28111*
This is the perfect base from
which to tour the Amári Valley.
Heracles offers breakfast, rental
bicycles and balconied rooms.
The helpful proprietor is very
knowledgeable about the region.

VAMOS: Vamos €€
Villas
*Midway between Réthymno and
Chaniá, 73008*
Tel *28250 22190*
W vamosvillage.gr
Stay in meticulously restored
houses with such features as
ceiling beams and heritage tile
floors. Take advantage of the
cooking lessons on offer or go
on a self-guided walk.

VLATOS:
Milia Mountain Retreat €€
Restoration Inn
*Off a secondary road between
Kastélli and Palaióchora, 73012*
Tel *29210 46774*
W milia.gr
Carefully restored stone cottages
have wood-burning stoves at
this eco-lodge. An on-site taverna
serves meals prepared using
local produce.

Athens

EXARCHEIA: Dryades/Orion €
Pension
*Crn Emmanouíl Benáki & Dryádon
streets, Lófos Stréfis, 11473*
Tel *210 33 02 387*
W orion-dryades.com
Two self-catering inns for a young
clientele. Rooms at the Dryades
have LCD TVs and balconies.

KOLONAKI: St George
Lycabettus €€
Boutique
Kleoménous 2, 10675
Tel *210 72 90 711*
W sglycabettus.gr
As well as superb views over the
Acropolis, this modern hotel has
a rooftop pool and restaurant, a
spa and an art gallery.

MAKRYGIANNI: Marble House €
Pension
*Cul-de-sac off Anastasíou Zínni 35,
11741*
Tel *210 92 28 294*
W marblehouse.gr
This welcoming, family-run
pension is situated in one of
Athens' quietest spots. It is
justifiably popular, so book early.

MAKRYGIANNI: Herodion €€€
Luxury
Rovértou Gálli 4, 11742
Tel *210 92 36 832*
W herodion.gr
The common areas – the atrium
coffee shop in a tree-shaded
conservatory and the roof bar
with two Jacuzzis and views of
the Acropolis – are impressive.
Rooms have Coco-Mat bedding
and earth-tone furnishings.

MONASTIRAKI: Attalos €€
Boutique
Athínás 29, 10554
Tel *210 32 12 801*
W attaloshotel.com
This friendly hotel retains many
interwar period features. Rooms
have parquet floors, eclectic
colours and decent bathrooms.

PLAKA: Ava €€€
Luxury
Lysikrátous 9–11, 10558
Tel *210 32 59 000*
W avahotel.gr
Upper units at this all-suite hotel
have verandas with a courtyard;
some offer views of the Acropolis.
Breakfast is included, and the
service is impeccable.

SYNTAGMA:
Grand Bretagne €€€
Luxury
Plateía Syntágmatos, 10563
Tel *210 33 30 000*
W grandebretagne.gr
Athens' premier hotel has deluxe
doubles and sumptuous common
areas, including a landscaped
pool garden and a huge spa.

Thessaloníki

CENTRE: Orestias Kastorias €
Pension
Agnóstou Stratiótou 14, 54631
Tel *2310 276517*
W okhotel.gr
The best of Thessaloníki's budget
lodgings. Rooms at the front
have small balconies overlooking
a Roman forum.

DK Choice

CENTRE: The Bristol €€
Boutique
Oplopioú 2, crn Katoúni, 54625
Tel *2310 506500*
W bristol.gr
The city's premier boutique
hotel occupies an 1870s
former post office in the
trendy Ladádika district.
Wood-floored rooms and
suites sport vibrant orange,
green and maroon textiles. An
Italian-Argentinian restaurant,
prompt service and retro
bicycles are other attractions.
Breakfast is served in the lobby.

CENTRE: Electra Palace €€
Boutique
Plateía Aristotélous 9, 54624
Tel *2310 294011*
W electrahotels.gr
This landmark 1920s hotel on the
main square remains pleasantly
old-fashioned throughout. It has
unbeatable views, particularly
from the roof-garden restaurant,
where breakfast is served.

CENTRE: Le Palace €€
Boutique
Tsimiskí 12, 54624
Tel *2310 257400*
W lepalace.gr
Another updated 1920s hotel
with colourful, spacious rooms
and suites offering premium
bedding and double glazing.
Great breakfasts and helpful staff.

The impressive 1920s façade of the Electra Palace in Thessaloníki

For more information on types of hotels *see pp303–305*

WHERE TO EAT AND DRINK

Greeks consider the best places for eating out to be where the food is fresh, good value and properly cooked, not necessarily where the setting or cuisine is the fanciest. Visitors, too, have come to appreciate the simplicity and appeal of traditional Greek cuisine – cheeses, vegetables, a little meat or seafood and some wine or *tsípouro*, always shared with friends. Entire families dine together and take ample time over their meal,

especially at weekends. The traditional three-hour lunch and siesta is still observed in the islands; only in resort areas will you find the Northern European routine of a substantial breakfast, an earlier, briefer lunch (1–2:30pm) and an earlier dinner (6:30–9pm). Greeks prefer a quick morning coffee before a baked pastry, a substantial lunch, and a sundowner drink with a *mezés* platter, before a full, late (9pm–midnight) dinner.

Types of Restaurant

Several types of restaurant appear across the Greek islands, varying in the fare they provide, the methods of preparation and the cost. Since the 1990s there has been a resurgence of pride in regional cuisines and local ingredients, and almost every island now eagerly promotes its typical dishes, dairy products, honey, fruit or vegetable varieties, and wines (increasingly, beers too). Islands with pronounced regional cooking include Corfu, Lésvos, Sífnos, Santoríni, Crete and Kos. Where there are Asia Minor refugee communities, or ethnic Turks, as on Kos and Rhodes, recipes are apt to be more adventurous and spicier than the Greek norm.

The stylish interior of Mani-Mani in Athens

The Oinomageireíon

After nearly disappearing, nostalgia gave a new lease of life to the traditional *oino-mageireíon* (literally, "wine-cookhouse"), which specializes in an array of home-style *mageireftá* (pre-cooked casserole dishes), set out before customers in a row. Visitors new to Greek fare will find it easy to point to their preferences and will occasionally still be invited into the kitchen to inspect pots and their contents. Some *oinomageireía* (in the plural) work only at lunch or close quite early in the evening after being open continuously all afternoon.

An *oinomageireíon* usually prides itself on wine from the barrel – referred to as *chýma* (in bulk) or *me to kiló* (by weight, ie by the litre); it should be at least drinkable if not very good indeed, often from the owner's own vineyards. Many *oinomageireía* have been in the same family for generations and strive to uphold their reputation.

Tavernas

There's some overlap between *oinomageireía* and the traditional taverna, which in addition to *mageireftá* of the day and bulk wine will offer a range of meat and often seafood dishes, both grilled *tis óras* (to order), and probably a bottled wine list of Greek vintners. Traditional tavernas open for lunch,

Traditional restaurant on Pátmos

and then dinner hours, or work continuously from noon until late

A *psarotavérna* (seafood taverna) relies on scaly fish, cephalopods like octopus, and shellfish starters. Ideally, the owner or other close family member has their own boat and brings in fresh catch each morning, served to tables on the sand close to the lapping waves. Presentation is usually basic: a whole grilled fish or squid delivered on an oval metal plate, doused in *ladolémono* (olive oil and lemon), with no garnish other than chopped parsley and a lemon wedge.

More large-scale, commercial-resort seafood restaurants will serve frozen, farmed or imported fish. Frozen seafood must be clearly designated on menus, but this is often done merely with inconspicuous asterisks or the abbreviation "KAT" (for *katapsygméno*, "frozen").

Prestige bream species are often farmed and are little different to what's sold at a Northern European supermarket counter. The best seafood in Greece is always caught wild, seasonally: *gónos kalamaráki* (baby squid) in early summer,

Outside diners at a taverna in Plakiás, Crete

flash-fried *atherína* (sand smelt) or *gávros* (anchovy) and later on, whole grilled *thrápsalo* (giant deep-water squid) or octopus (one tentacle per portion).

For delicious meat grills, head to a *psistariá*, specializing in spit-roasts and charcoal-grilled (*sta kárvouna*) chops or *biftéki* (Greek hamburger). *Mprizóles* are pork chops; *païdákia* are lamb chops. The standard roast meats are entire spitted chickens, *gourounópoulo* (suckling pig), *kondosoúvli* (pork chunks), whole lamb and *kokorétsi* (offal roulade).

With outstanding exceptions, tavernas attached to beach-resort hotels peddle predictable "international" menus and rather bland versions of Greek specialities. Smaller island hotels, however, may have excellent in-house restaurants, open to non-residents. As part of the general culinary renaissance, creative Greek tavernas have appeared. Chef-owners, having absorbed lessons living and working abroad, are updating traditional recipes and being more daring: less oil, more herbs and spices, unusual flavour juxtapositions, while proudly sourcing local (often organic) ingredients. Such tavernas are also more likely to stock foreign wines along with premium Greek labels, and to have an interesting dessert list.

Worth a mention, in the Cyclades, is the Aegean Cuisine chain; member tavernas commit themselves to using traditional recipes and supporting local producers, so presence of their distinctive badge at the door guarantees, in theory, that the establishment will probably prove very good if not excellent.

Exotic cooking has made few inroads except on the largest or busiest resort islands. Nouvelle French, generic Southeast Asian, rather bogus Mexican, better Argentinian and variably successful Middle Eastern are the main international cuisines encountered. The most popular, though, is Italian, either simple pizza or full-on restaurants.

The Mezedopoleío, Ouzerí and Tsipourádiko

Other categories of taverna that have come to prominence in recent decades are the *ouzerí* and the *mezedopoleío* (*mezédes* shop), the *tsipourádiko* and the *rakádiko*. The terms have to some extent become interchangeable, though the proportions of tipple to food – and which type of drink predominates – vary subtly between them. They are most fun in a small group, since the idea is to share a dozen or so small platters of savoury meats, seafood delicacies and vegetables – the *mezédes* or appetizers. More substantial mains follow, but it is not obligatory to order these; starters alone can be very rich and filling.

All serve one or more of the following spirits distilled from the grape residue of wine-making: anise-flavoured *oúzo*, *tsípouro* (available without anise and less cloying), or Cretan/Cycladic *rakí* (always without anise), along with a bucket of ice and water. Pop the ice in a glass, pour in the spirit, and adjust the strength with water. All spirits arrive in sealed bottles (typically 200 ml) or, if decanted in bulk, small jugs (150–250 ml).

Musical *mezedopoleía/rakádika*, with a price premium for quality acoustic live music several nights weekly, are found in larger centres like Athens, Thessaloníki, Rhodes and Crete.

Artemon restaurant (formerly Lembesis) on the island of Sífnos (see p330)

dion player in a
rna on Sými

A waterside restaurant at Skála Sykaminiás, Lésvos

Kafeneía and Sweet Shops

Kafeneía (coffeehouses), open from dawn until late, were traditionally the hub of Greek village life, though they have been eclipsed by trendier cafés and *frappádika*, the latter serving frappé (whipped iced instant coffee) to a younger clientele. Besides frappé, you can find Greek coffee, soft drinks, beer and brandy at a *kafeneío*, and all of these plus espresso and cappuccino at a modern café.

Baklavás, a sweet cake of wheat, honey and nuts

Those with a sweet tooth should repair to a *zacharoplasteío*, or confectioner, doing syrupy Oriental sweetmeats (*baklavás*, *kataïfí*). A *galaktopoleío*, or "milk shop", sells dairy-based desserts like *galaktoboúreko* (custard pie), *ryzógalo* (rice pudding) and *krémes* (custards). Ice cream (*pagotó*) is the quintessential summer treat, and even surprisingly small islands have excellent ice-cream parlours, often run by Italians as authentic *gelaterie*.

Wine and Beer

Along with renewed pride in regional cuisines has come an interest in fine bottled wine and obscure grape varieties. Almost every island, even smaller ones, now has at least one winery. In particular, Chíos, Kefalloniá, Páros, Náxos, Límnos, Crete, Santoríni, Sámos, Ikaría, Rhodes, Kárpathos, Kos and Zákynthos, are known for their bulk or bottled wine.

Because of poor economies of scale – few Greek microwineries produce more than 15,000 bottles annually – premium wine is expensive, well over €30 a bottle in a fancy taverna. More mid-range, affordable labels include Lazaridi from mainland Dráma, Tselepos or Skouras from the Peloponnese, Ktima Argyrou from Santoríni and anything from Límnos.

Quality white bulk wine is somewhat easier to find across the islands than quality red. If in doubt, ask for a glass to sample before committing yourself – you'll cause no offence by declining it in favour of alternatives. When those are not available, a drop of soda water added to the wine makes even the harshest ones drinkable. So-called rosé, bottled or *chýma*, is often very dark by European standards.

Beer was long the province of centralized mainland breweries, with a few bland labels and a stranglehold on national distribution; the best of these are Fix and Alpha, typically in 500ml bottles. But recently, dozens of islands have seen microbreweries set up. Their beers are generally in 330ml

bottles, relatively expensive, and vary from wonderful to insipid. The top micro-beers hail from Santoríni, Corfu, Chíos and Crete.

Fast Food, Snacks and Breakfast

Although American-style fast-food chain outlets dominate tourist centres and airports – the big names are Everest, Gregory's and Goody's – they can be avoided in favour of more traditional options. A *souvlatzídiko* provides mostly *sto chéri* (in the hand, to take away) service for *souvláki* – chunks of grilled pork, wrapped in a Middle Eastern *píta* bread stuffed with *tzatzíki*, onion and tomatoes too. A *gyrádiko* does the same for *gýros*, thin slabs of compressed pork cut from a rotisserie cone, and *kebáb*, minced spicy beef treated like *souvláki*. Most stalls offer rudimentary seating and a range of beer and soft drinks.

Wine from Límnos

Most bakeries sell a selection of savoury small stuffed turnovers (*pitákia* or *píttes*); the most popular fillings are cheese, ham and cheese, spinach, and frankfurter. The best cheese pies are the so-called *kouroú* (dry) ones without messy filo pastry. These make ideal on-the-hoof breakfasts, and most *píttes* sell out by noon.

For Greeks, breakfast is the least important meal of the day. Most make do with a coffee accompanied by *paximádia* (sliced barley rusks) or *kouloúria* (firm, sesame-seed-sprinkled rolls in rings or S-shapes). Any hotel of two stars or more is

A selection of traditional greek side dishes, salads and sauces

The breakfast terrace at Hotel Andreas in Rhodes *(see p310)*

obliged to offer breakfast – sometimes worth taking, sometimes not. If not, outside cafés will offer continental or "full English" breakfasts.

Vegetarian Food

Greek cuisine notionally has many vegetarian dishes, because of the number of meatless recipes for the numerous fasting days of the Orthodox religious calendar. However some, especially rice-stuffed vegetables, may still be prepared using meat stock. Vegans will be in for a hard time, since items topped or stuffed with cheese and other dairy products are popular.

Basket of local bread from Rhodes

Reservations

Island restaurants generally have a casual atmosphere. Bookings are best made in person; that way, you can see what the *piáta iméras* (dishes of the day) will be and also reserve some of them. Good tavernas only cook as much as they can sell on the day; when it's gone, it's gone.

Payment and Tipping

Greece is still very much a cash society; only the fanciest or most expensive restaurants (especially seafood ones where bills mount up) take credit cards. Always confirm card acceptability first. And if the phone lines suddenly go down and the machine doesn't work, you will have to have the necessary cash on hand.

Menu prices notionally include a service charge, but it is customary to leave at least

10 per cent of the bill as a tip – and a little more if service has been exceptional.

Dress Code

Greeks usually dress to at least smart-casual standard when dining out. For visitors, comfortable cotton clothing is fine, but skimpy tops, shorts or active sportswear are only acceptable at beachside tavernas. Restaurants in hotels of three stars and above usually have a policy of no shorts, no sandals, no sleeveless tops. Even in summer, a cardigan or long-sleeved top is useful for outside dining after dark.

Children

Children become Greek taverna habitués at a very early age – it is an essential part of their socialization – and are welcome everywhere except the most drink-and-music-oriented *mezedopoleía* or the more rarefied gourmet eateries. Children are expected to behave themselves, but in summer, nobody cares if kids play around outside tables, and indeed some tavernas provide small playgrounds. High chairs for infants are slowly becoming more common but cannot be assumed.

Smoking

Since mid-2010, smoking has been illegal in enclosed Greek restaurants or bars, and only permitted at terrace or garden seating. In such a tobacco-mad country, compliance has proved notably spotty, and law enforcement by roving inspectors

varies greatly by island. Few establishments set out ashtrays indoors, but asking for – and receiving – one tells you that the management is defiant. You can politely ask neighbouring diners to stop smoking, but expect a brusque reaction and to move outside yourself.

Wheelchair Access

In small island villages with level, kerbless expanses of outdoor seating, there are few problems for wheelchair users, but access to the restaurant interior (where the toilets often are) cannot be guaranteed. Doors are typically narrow, ramps absent, stairs abundant.

Recommended Restaurants

The choices reflect the diverse nature of Greek cuisine and how it varies from island to island, ranging from simple dining to more elaborate contemporary restaurants. Many are in remote spots and do not have a specific address. Those highlighted as a DK Choice succeed notably within their designated category. For example, a taverna owned by the same family for three generations will likely be genuinely traditional, take some pride in its work or be a stylish eatery along the coast serving the freshest seafood in a spectacular location. Updated Greek or foreign/fusion menus should be creative while still offering value in these straitened times. Grill and fish tavernas that always source local meat and seafood also feature.

Ntolmádes (stuffed vine leaves)

The Flavours of Greece

On some of the Greek islands, you may still find recipes and culinary methods little changed since medieval times. Elsewhere, Greek cookery has been much influenced by the Ottoman Empire, with its spices, filled pastries and vegetable casseroles. An important factor was the many fasting days in the Greek Orthodox calendar and the need to devise meat- and dairy-free recipes – the so-called *nistísima* cusine. Until the 1990s, authentic Greek cuisine was scorned as peasant food. But there has since been an upsurge of pride in country cooking and quality, seasonal local ingredients.

Oregano and thyme

Island fisherman returning to harbour with the day's catch

Athens, Thessaloníki and the North

Athens is a city of immigrants from all over Greece as well as Asia Minor and Egypt, a diversity reflected in its markets and eateries. There are tavernas and *mezedopoleía* devoted to Anatolian or island cooking, notable among the latter Santoríni and Crete *(see our "Where to Eat and Drink" listings for Athens)*. Chefs from Crete,

with its long growing season and myriad wild greens, have been one of the main drivers of recent creative Greek cuisine.

Many recipes from Thessaloníki and the mainland appear on Aegean taverna menus. The cuisine of northern Greece especially, with strong Jewish and refugee contributions, is most adventurous, being the only part of Greece with a taste for spicy food.

The Islands

Each group of islands has a distinct culinary identity reflecting its geographical location and history. Many Ionian dishes are pasta-based, a legacy of the era of Venetian occupation, though noodles are pan-Hellenic. Cycladic recipes use sharp cheeses and pungent greens. The northeast Aegean and the Dodecanese have the richest seafood

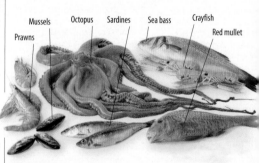

Mussels   Octopus   Sardines   Sea bass   Crayfish

Prawns   Red mullet

Selection of seafood from the clear waters of Greece

Staples and Specialities

Greece boasts the world's largest variety of olives, and larger islands – particularly Crete, Corfu, Evvoia, Lésvos and Sámos – grow them at least for oil if not for eating. Few islands support cattle, so many cheeses are made from varying proportions of blended goat and sheep milk, though tasty single-milk varieties exist. Pickled caper-greens (*kápari*) are prepared on smaller islands, adorning salads; *tzitzírafa* (young terebinth shoots), rock samphire (*krítamo*) and glasswort (*kardamídi*) garnish seafood. Every sizeable island town has at least one premium delicatessen purveying the best cheese, charcuterie, cured fish, marinated vegetables, olive oil, noodles, honey, yogurt and herbs from every corner of Greece. Butchers and tavernas proudly proclaim their locally sourced meat.

Olives

Fakés is a sour Peloponnese soup of green lentils, lemon juice or wine vinegar, tomatoes, herbs and olive oil.

Produce on sale in a typical Greek market

harvest, thanks to nutrient-rich waters around them and migrations through the Dardanelles. Cretan cooking has various unique recipes and emphasizes pork, particularly rich cured *apáki*.

Fish and Seafood

The warm and sheltered waters of the Aegean are home to assorted fish, cephalopods and shellfish, whose abundance varies seasonally. Fish are served with heads on to indicate the variety; Greeks consider the cheek flesh especially tasty.

Scaly fish featuring in taverna chiller-cases often belong to the enormous bream family, most prestigiously *fagkrí* (red porgy) or *skathári* (black bream). Equally prized are the dentexes *synagrída* and *ballás*, or the groupers *rofós* and *sfirída*. Late

spring sees squid spawn (*gónos kalamaráki*), while summer means the appearance of *sardélla* (sardine), *gávros* (anchovy), *atherína* (sand smelt) and *skáros* (parrotfish), as well as pelagic tuna and swordfish. Mussels are always farmed, and prawns are likely to be imported except in winter or spring.

Bread being baked in an outdoor communal oven

For Dessert

There may be a separate menu section for desserts (*epidórpia*), but more commonly a taverna will provide a *kérasma*, or sweet, on the house. Depending on locale and season, this might be *simigdalísio chalvás* (halva made from semolina, butter and sugar), *ekmek kataïfi* (bread pudding and custard on a *kataïfi* base), a dollop of panna cotta, *glykó koutalioú* ("spoon sweet", commonly candied green grapes, green figs or cherries) or a platter of Persian melon or watermelon.

WHAT TO DRINK

Wine-making in Greece is an ancient – and ongoing – tradition, with older wineries found on Crete, Santoríni, Rhodes, Sámos and Límnos. That said, most tavernas still stock labels, especially reds, from premier mainland wine-producing regions: Pelo-ponnesian Neméa, Macedonia and Epirus. Retsína (pine-resin-flavoured wine) is made on Límnos, Sámos, Rhodes and Crete, but purists aver that only Attica retsína is genuine.

Distilled spirits include oúzo, *tsípouro* and *rakí*. Brandy and fortified dessert wines may also be offered. "Greek" (Oriental) coffee is made from fine-ground beans boiled up in a long-handled *mpríki* pot, served in tiny cups.

Spetzofáï, from Thessalian Mount Pílio, is sautéed slices of spicy country sausage with herbs and vegetables.

Barboúnia (red mullet) has been Greece's most esteemed fish since antiquity. *Koutsomoúres* are a smaller, cheaper offering.

Loukoumádes are a street-stall snack of small deep-fried dough-nuts soaked in honey-syrup and sprinkled with cinnamon.

The Classic Greek Island Menu

The menu begins with a selection of *mezédes* (appetizers), listed separately as cold, then hot. Many are light vegetarian; others, like *loukánika* (sausages) or *apáki* (cured pork), can be combined to substitute for main dishes. Salads, simple or involved, then make their appearance. Next up are meat and seafood mains, the former usually with a side serving of rice or fried potatoes, the latter unadorned. Bread has quasi-religious symbolism for Greeks and arrives in a basket as part of the *servítzio* (cutlery and napkins). Formerly, taverna bread was perfunctory, but now many establishments serve healthier whole-grain, olive or corn breads.

Round Greek bread loaves

Souvlákia are small chunks of meat, often pork, flavoured with lemon, herbs and olive oil and grilled on skewers.

Choriátiki saláta, Greek salad, combines tomatoes, cucumber, onions, herbs, capers and feta cheese.

Psária plakí is a whole fish baked in an open dish with vegetables in a tomato, onion, garlic and olive oil sauce.

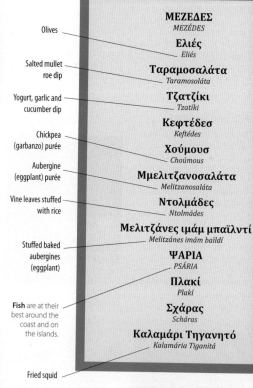

ΜΕΖΕΔΕΣ
MEZÉDES

Ελιές
Eliés
Olives

Ταραμοσαλάτα
Taramosaláta
Salted mullet roe dip

Τζατζίκι
Tzatíki
Yogurt, garlic and cucumber dip

Κεφτέδεσ
Keftédes

Χούμουσ
Choúmous
Chickpea (garbanzo) purée

Μμελιτζανοσαλάτα
Melitzanosaláta
Aubergine (eggplant) purée

Ντολμάδες
Ntolmádes
Vine leaves stuffed with rice

Μελιτζάνες ιμάμ μπαϊλντί
Melitzánes imám baïldí
Stuffed baked aubergines (eggplant)

ΨΑΡΙΑ
PSÁRIA

Πλακί
Plakí
Fish are at their best around the coast and on the islands.

Σχάρας
Scháras

Καλαμάρι Τηγανητό
Kalamária Tiganitá
Fried squid

Scháras means "from the grill". It can be applied to meat or fish, or even vegetables. Here, a swordfish steak has been marinated in lemon juice, olive oil and herbs before being swiftly chargrilled.

Mezédes

Mezédes are eaten as a first course or as a snack with wine or other drinks. *Taramosaláta* is a purée of salted mullet roe and breadcrumbs or potato. Traditionally a dish for Lent, it is now on every taverna menu. *Melitzanosaláta* and *revithosaláta* are both purées. *Melitzanosaláta* is grilled aubergines (eggplant) and herbs; *choúmous* is chickpeas (*garbanzos*), tahini, cumin and garlic. *Melitzánes imám baïldí* are aubergines filled with a purée of onions, tomatoes and herbs. *Ntolmádes* are vine leaves stuffed with pine nuts, rice and herbs.

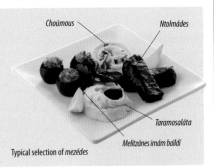

Choúmous

Ntolmádes

Taramosaláta

Melitzánes imám baïldí

Typical selection of *mezédes*

Meat is more readily available on the mainland than on the islands.

ΚΡΕΑΣ
KREÁS

Μουσακάς
Mousakás

Moussaka (minced lamb and aubergine baked in layers)

Σουβλάκια
Souvlákia

Xoirino σουβλάκι
Choirinó souvláki

Κλέφτικο
Kléftiko

Vegetables and salads often use wild produce.

Λαχανικά και Σαλάτα
Lachaniká kai salatiká

Χωριάτικη σαλάτα
Choriátiki saláta

Fried aubergines (eggplant) and courgettes (zucchini)

ζάνες και κολοκυθάκια τηγανιτά
Melitzánes kai kolokythákia tiganitá

ΕΠΙΔΌΡΠΙΑ
EPIDÓRPIA

Τιραμισού
Tiramisu

Desserts are simple affairs of pastry, fruit or yogurt.

Παγωτό
Ice-cream

Χαλβάς
Chalvás

Sweetmeat of cooked semolina, butter or oil, cinnamon, pine nuts and brown sugar

Γλυκά του κουταλιού
Glyká tou koutalioú

Candied, preserved fruit

Keftédes are fried meatballs made of minced beef, or beef and lamb, egg and breadcrumbs, flavoured with mint and parsley.

Kléftiko is lamb in chunks, less commonly goat meat, wrapped in parchment paper and cooked to seal in the juices and flavours.

Sweet pastries filled with nuts and honey are made at *zacharoplasteía* but sometimes figure as taverna desserts. The most famous are *baklavás*, with layers of filo pastry and nuts, and *kataïfi*, similar but with the filo shredded into filaments.

Giaoúrti kai méli (yogurt with honey) is a common component of hotel breakfasts, or offered as a taverna dessert.

Where to Eat and Drink

The Ionian Islands

DK Choice

CORFU: Tsipouradiko €
Mezedopoleío/Ouzerí
Prosalénou crn Gída, Spiliá district, Kérkyra town
Tel *26610 82240*
Quality fare, efficient service, a buzzing, young crowd and quaffable barrel wine (the white is better than the semi-sweet rosé), plus delicious *tsípouro*, all combine to make this place a winner. Local sausages, grilled mushrooms and fresh beets are typical offerings. For those with a smaller appetite, there are half-priced half-portions.

CORFU: Bacchus/Vachos €€
Traditional Taverna
South end of the beach strip, Mesongí
Tel *26610 75301*   **Closed** *Nov–Apr*
Established in 1978, this water-side taverna sources local, seasonal ingredients, especially seafood: baby squid, Amvrakikos gulf prawns and *koutsomoúra* (small red mullet). In the spring-time artichokes too.

CORFU: Chrysomalis €€
Traditional Taverna
Nikifórou Theotókou 6, Kérkyra town
Tel *26610 30342*
This century-old *mageireftá* near the Listón dishes out top-notch *tzatzíki*, stuffed cabbage leaves, hearty stews and lentils, accompanied by the potent but palatable local wine.

CORFU: The Old School €€
Seafood
Kassiópi, roundabout at the main port
Tel *26630 81211*   **Closed** *Nov–Apr*
Set on Kassiópi's picturesque port, this is the place for octopus and garlic-flavoured mussels, plus superior versions of *taramo-saláta* and *melitsanosaláta*. The best tables are those under the giant plane tree. Micro-beers available.

CORFU: Sinión Fagopoteion €€
Traditional Taverna
Waterfront, Agios Stéfanos Sinión
Tel *26630 82020*   **Closed** *Oct–Easter: Sun pm–Fri noon*
Fresh, non-farmed fish from around the Diapóndia islets, *saganáki* (fried starters), chard *tsigarélli* (sti-fry) and baby squid

in springtime are some of the menu highlights here.

CORFU: Etrusco €€€
Ettore Bottrini
Fine Dining
Outside Káto Korakiána, on the Dassiá road
Tel *26610 93342*   **Closed** *Nov–Mar*
Nouvelle cuisine is produced with organic ingredients at this restored country manor. Expect delights like octopus carpaccio, pappardelle with duck and truffles, or lamb in kumquat sauce.

ITHACA: Rementzo €
Traditional Taverna
Quay, Fríkes
Tel *26740 31719*   **Closed** *Nov–Mar*
Vegetarian and seafood starters – including hummus, chickpeas with caramelized onions and dill, and marinated octopus – lead the way to main courses like *savóro* (fish in sweet herb sauce).

ITHACA: Kohili €€
Traditional Taverna
East quay, near the town hall, Vathý waterfront
Tel *26740 33565*   **Closed** *Oct–Mar*
At this versatile all-rounder, the menu encompasses grills, local specialities such as onion pie, and a few spicy dishes like *tzoutzou-kákia* (baked mince rissoles in red sauce).

KEFALLONIA: Xouras €
Traditional Taverna
Petaní beach, northwest of Lixoúri
Tel *26710 97128*   **Closed** *Nov–Apr*
This Greek-American-run beach outfit offers all the standard *mageireftá*, encompassing meat, vegetables and some seafood. Good bulk wine.

An artistic menu from Tsipouradiko in Kérkyra Town

KEFALLONIA: Platanos €€
Grills
Mid-village, Asos
Tel *26740 51381*   **Closed** *Nov–Mar*
What you lose in location you gain in excellent dishes such as roast suckling pig (order it in advance), lamb and fish. Plenty for vegetarians, too, either cooked or as salads, plus cheese dishes.

KEFALLONIA: Kyani Akti €€€
Seafood
Antóni Trítsi 1 (far end of the quay), Metélas district, Argostóli
Tel *26710 26680*   **Closed** *Nov–Apr*
Try signature dishes such as Mesolóngi smoked eel, razor clams in mustard sauce, and *striftí* (twisted) cheese pie while admiring views of the bay from the enormous wooden pier.

LEFKADA: Platania €
Grill
Main plaza, Karyá village
Tel *26450 41247*   **Closed** *Nov–Apr*
Grab a table under the giant plane trees of the name, and enjoy fresh meat platters, salads and beers. This is the best of three adjacent tavernas here.

LEFKADA: Regantos €
Tradtional Taverna
Dimárhou Venióti 17, Lefkáda town
Tel *26450 22855*   **Closed** *Dec–Feb*
The menu offers a good balance between *mageireftá*, grills and seafood. Eat alfresco in the summer or in the extremely characterful interior. Regantos occasionally hosts live music, from *kantádes* to Ionian ballads.

LEFKADA: Sapfo €€
Seafood
Cove shoreline, end of the main street, Agios Nikítas
Tel *26450 97497*   **Closed** *Oct–Apr*
Savour creative and deftly executed dishes like seafood lasagne and cheese-and-broccoli pie, as well as standard fish grills. The seaside terrace offers the best view in the resort.

PAXOS: Alexandros €€
Traditional Taverna
Platía Edward Kennedy, Lákka
Tel *26620 30045*   **Closed** *Nov–Apr*
Book ahead to enjoy authentic

island cooking – own-raised grilled meats (suckling pig or chicken), rabbit stew, mushroom-based dishes and fish. Atmospheric seating.

ZAKYNTHOS: Kalas €
Traditional Taverna
Centre of Kampí village
Tel *26950 48576* **Closed** *Nov–Mar*
Forgo the sunset-view tavernas in favour of this garden eatery serving local favourites (sausages, fried potatoes, *choriátiki saláta*) exceptionally well.

ZAKYNTHOS: Malanos €
Oinomageireio
Agíou Athanasíou 38, Kípi district, Zákynthos town
Tel *26950 45936*
A deservedly popular mecca for *mageireftá*. Try the mince-rich *giouvarlákia* (rissoles) and the *fasolákia yahní* (runner beans in a red sauce), or opt for more daring dishes like salmon in an orange sauce. Live music at weekends.

ZAKYNTHOS: Alitzerinoi €€
Traditional Taverna
Entrance to Kilioméno village
Tel *26950 48552* **Closed** *Sep–May: Mon–Thu*
In an exquisite 17th-century arcaded townhouse (one of the few buildings to survive the 1953 earthquake) this place offers rich meat- and cheese-based country fare, plus its own wine.

The Argo-Saronic Islands

AGKISTRI: Parnassos €
Traditional Taverna
Metóchi hamlet
Tel *22970 91339*
Besides grills and *mezédes*, this welcoming, rustically decorated spot serves cheese from its own flocks, a few *mageireftá* per day, and palatable bulk wine. Roof terrace with views.

AIGINA: Agora (Geladakis) €
Seaood
Behind the fish market, Aígina Town
Tel *22970 27308*
The oldest seafood taverna in town offers no-nonsense fare. Their grilled octopus is famously good value. You might have to wait for a table, but the service is quick once you are seated.

AIGINA: To Deka €
Traditional Taverna
Waterfront, Pérdika
Tel *22970 61231*
After opening at dawn as the

A classic Greek cafe in the Argo-Saronic Islands

local fishermen's café, this place morphs at noon into a taverna serving exceptional *tara-mosaláta*, crunchy little shrimp, and succulent runner beans, plus fresh fish.

KYTHIRA: Pierros €
Traditional Taverna
Main street, Livádi
Tel *27360 31014*
A well-established taverna with an effusively friendly owner and all the standards – moussaka, *pastísio* (macaroni pie), lamb and pork dishes. Customers pick their choices from the kitchen steam trays.

KYTHIRA: Platanos €
Traditional Taverna
Mylopótamos, central square
Tel *27360 33397* **Closed** *Nov–Apr*
A combination *kafeneio*-taverna with a genuine vibe and plenty of old Greek gents installed. Dine outside under the giant plane tree of the name, or inside, where you can admire a gallery of old photos.

POROS: Karavolos €
Traditional Taverna
Behind Cinema Diana, Póros town
Tel *22980 26158*
Karavolos means "snail", and these are on the menu, along with casseroles and grilled dishes such as beets with *skordaliá* (mashed potatoes and garlic), swordfish, roast pork and fresh *kalamári*.

POROS: Mezedokamomata €
Mezedopoleío/Ouzerí
Galatás quay, opposite Póros
Tel *22980 43085*
Take the shuttle across the strait, then walk 150 m (500 ft) north for a *mezédes* feast: spiny chicory, beets, *tyrokafterí* (chilli-spiced cheese mash) and grills. Full view of Póros Town and jolly hosts.

SPETSES: Patralis €€
Seafood
Kounopítsa waterfront, Spétses town
Tel *22980 75380* **Closed** *Nov–Dec*
Running since 1935, this upmarket outfit is famous for its *psári a la spetsióta* (fish in a ratatouille sauce), as well as simply grilled fish.

SPETSES: Tarsanas €€€
Seafood
Old Harbour, Spétses town
Tel *22980 74490*
Romantic but expensive venue, especially if you opt for pasta with lobster. More modest but equally tasty platters include fish soup and seafood terrine.

**YDRA: Geitoniko
(Manolis & Christina)** €€
Traditional Taverna
Inland beyond Bratsera Hotel, Ydra town
Tel *22980 53615*
Popular with both locals and visitors, Geitoniko produces veggie casseroles and meat and fish grills with great aplomb. Lunch dishes may run out by 2:30pm, so arrive early.

DK Choice

YDRA: Paradosiako €€
Creative Greek
Tompázi lane, 100 m inland from Alpha Bank, Ydra town
Tel *22980 54155* **Closed** *late Nov–Orthodox Lent*
Arguably the best all-round eatery in town, Paradosiako serves dishes like seafood risotto, *kritharáki* (orzo) pasta with oyster mushrooms and, in the spring, crispy fresh shrimp. Limited and highly prized terrace seating, so be sure to book in advance. The idiosyncratically appointed interior is welcoming on cooler evenings.

For more information on types of restaurants *see pp314–15*

YDRA: Kodylenia €€€
Seafood
Kamíni port, 15-minute walk from Ydra west quay
Tel 22980 53520   **Closed** *late Oct–Orthodox Lent*
Ydra's premier waterside eatery, with great sunset views. Choose freshly caught local squid in season, or scaly fish in spring or autumn. Dishes are garnished with *krítamo* (rock samphire).

The Sporades and Evvoia

ALONNISOS: Eleonas €€
Traditional Taverna
Leftós Gialós beach
Tel 24240 66066   **Closed** *Oct–Apr*
Expect unusual platters like *ftéri* (fern shoots), *xynógalos* (a thick dairy dip) and marinated tuna, plus various *píttes* prepared by the grannies in the kitchen. Relax in the beach bar out front.

ALONNISOS: Peri Orexeos €€
Creative Greek
By the bus stop, Palaiá Alónnisos
Tel 24240 66421
The menu at this quality outfit embraces burgers, fusion snacks and tweaked Greek fare like beets with *manoúri* cheese, chickpeas with caramelized onions, and spare ribs in a honey and thyme sauce.

EVVOIA: Cavo d'Oro €
Oinomageireio
Párodos Sachtoúri (the alley west of the main square), Kárystos
Tel 22240 22326   **Closed** *Tue*
An old-fashioned place with trays of *gígantes*, *katsíki stifádo* and stuffed pork to choose from, plus typical *mezédes* like stuffed red peppers and *tzatzíki*.

EVVOIA: To Pikandiko €
Grill
Seafront, Límni
Tel 22270 31300
A curious meat haven in this historic port. On the menu, not just the usual *gýros* and *souvláki*, but also offal and mutton chunks to eat here or take away.

EVVOIA: Apanemo €€
Seafood
Ethnikís Symfíllosis 78, Fanári district, Chalkída
Tel 22210 22614
Chalkida is renowned for its shellfish, and this restaurant is a great spot to sample local delicacies like *petrosolínes* (razor clams). The tables are on the sand, close to the lighthouse.

SKIATHOS: Amfilíki €€€
Creative Greek
Opposite the public clinic, southwest edge of Skiáthos town
Tel 24270 22839   **Closed** *Nov–mid-Feb*
Fish recipes are the focus here, in dishes like *bráska kipouroú* (monkfish in a spicy red sauce) or swordfish with courgettes, peppers and capers. The food is accompanied by views of the bay from the window tables.

SKIATHOS: Ergon €€€
Creative Greek
Papadiamánti pedestrianized lane, Skiáthos town
Tel 24270 21441
This delicatessen pitching premium Greek-only products doubles as a creative *ouzerí*. Sample the yellow split peas with bacon and caramelized onions, or the Kozáni cheese turnovers in a honey sauce, and wash them down with a microbrew. They also offer cooking courses.

SKOPELOS: Stella €
Traditional Taverna
150 m before Glyfonéri beach, on the left
Tel 24240 23143   **Closed** *Oct–Apr*
The proprietress and her well-tended vegetable patch are the motors of this appealing 1980s time capsule. Her light touch is evident in the heaping platters of stuffed *biftéki*, squid, courgettes and chips.

DK Choice

SKOPELOS: Pavlos €€
Seafood
Agnóntas port
Tel 24240 22409 **Closed** *Nov–Feb*
With its waterside tables under giant trees, this popular outfit has long been considered the top fish place on Skópelos. Enjoy regional specialities like *goúnes* (sun-dried mackerel), *tzitzírafa* (wild terebinth shoots) and *krítamo* (rock samphire); the bulk wine is supplied by Apostolakis, a respected Thessalian vintner.

SKOPELOS: Ta Kymata (Aggelos) €€
Traditional Taverna
Far north end of the quay, Skópelos town
Tel 24240 22381
This old shoreline taverna is famous for its vegetables and lamb in filo pastry and *katsíki lemonáto*. For starters, try the aubergine salad and fresh beets. Go early – dishes tend to run out.

Marinated sardines (*sardélla*) with chillies and herbs

SKYROS: Maïstros €
Traditional Taverna
Behind the plane tree, Linariá quay
Tel 22220 93431
This *ouzerí*-ish taverna acquits itself well at *mezédes*, seafood titbits and local meat platters, which are best paired with their refreshing rosé wine. Good enough to justify a detour.

SKYROS: Perasma €
Grills
Start of the airport access road
Tel 22220 92911
Despite being in the middle of nowhere, Perasma draws scores of islanders with its locally produced meat, vegetables and cheeses, not to mention the quality island wine on offer.

SKYROS: Istories tou Barba €€
Creative Greek
Boundary between Magaziá and Mólos beaches
Tel 22220 91453
The seafood-strong menu here guarantees repeated visits from a loyal clientele. The excellent fare is accompanied by some of the best sea views on Skýros. The interior is decorated with old photos, and Cretan music creates an authentic atmosphere.

The Northeast Aegean Islands

CHIOS: Fabrika €
Traditional Taverna
Behind the main parking lot, Volissós
Tel 22740 22045
As the name suggests, this place is housed in an industrial setting, a former olive/flour mill, with machinery left as decor. They use own-raised meat and own-grown vegetables. The wine is from the nearby Kefala winery.

CHIOS: Hotzas €
Traditional Taverna
Georgíou Kondýli 3, Chíos town
Tel *22710 42787* **Closed** *Sun*
The oldest taverna in the city, Hotzas features irresistible courtyard seating in the summer. The menu changes regularly. Expect superior *fáva* (yellow split peas) mash, skinny local sausages, and *pastírma* and cheese quiche, plus good Kefala bulk wine and house *soúma*.

CHIOS: Passas €
Seafood
Waterfront, where it bends, Langáda
Tel *22710 74218* **Closed** *winter: Mon–Thu*
This is the go-to place on the east coast for a delicious seafood spread. There are some meat dishes, too, and a wide range of *tsípouro* and *oúzo*. Enjoy them all sitting under the eucalyptus tree.

CHIOS: Poseidonas (Neptune) €
Traditional Taverna
Emporeiós waterfront, on the left as you arrive
Touring Chíos' *mastichohória*, you're likely to pass through the tiny port of Emporeiós, where the local summer speciality is *atherína* – sand smelt, served fried with flour and onions. Try it at this typical taverna.

CHIOS: Ta Mylarakia €
Mezedopoleío/Ouzerí
Coast road, Tambákika district, just before the hospital
Tel *22710 41412*
This place boasts Chíos's most romantic setting, on a terrace overlooking the three little mills of the name. All local brands of oúzo are available to accompany tasty starters, seafood mains and some meat recipes.

FOURNOI: Kali Kardia €
Grills
Just beyond the main inland square, Foúrnoi Town
Tel *22750 51217*
If you've had your fill of seafood, come here to enjoy *kondosoúvli* (boneless pork) or, more daringly, *kokorétsi* (offal roulade) straight from the spits turning outside.

IKARIA: Arodou €
Traditional Taverna
Xylosýrti, 5 km west of Agios Kírykos
Tel *22750 22700* **Closed** *Oct–May: Mon*
Next to the village church, this taverna with sea views attracts crowds from the capital for its *mezédes* and grills. Big portions, so be sure to bring your appetite.

IKARIA: Thea €
Oinomageireio
Main street, Nas clifftop
Tel *22750 71491* **Closed** *Oct–May: weekdays*
An American-Ikarian lady produces the best *mageireftá* in the area. Crops from the family farm turn up in vegetarian dishes like *soufikó* (the island ratatouille) and squash-stuffed *píttes*.

IKARIA: Paschalia €€
Traditional Taverna
Waterfront promenade, Armenistís
Tel *22750 71302* **Closed** *early Nov–early May*
The first taverna to open in spring and the last to close in autumn, Paschalia serves grilled meat and fish (try the local *balládes*), plus a few *mageireftá*.

LESVOS: Balouhanas €
Mezedopoleío/Ouzerí
North end of the waterfront, Pérama
Tel *22510 51948*
Besides seafood, Balouhanas (from the Turkish for "fish hall") is known for unusual platters like *gioúzleme* (Anatolian crêpes) and an ample dessert list. Seating is on a wooden deck over the water.

DK Choice

LESVOS: Ermis €
Mezedopoleío/Ouzerí
Kornárou 2, crn of Ermoú, Epáno Skála district, Mytilíni town
Tel *22510 26232*
A Muslim, alcohol-free coffeehouse from 1805 to 1922, Ermis is now the oldest *ouzerí* on Lésvos. It has been known to host fashion shoots among its antique furniture, wall art and *belle époque* floor tiles. Generously portioned *mezédes* and a full spectrum of island oúzo are served indoors or on the patio, under a vine pergola.

LÉSVOS: Thalassa €
Traditional Taverna
Seafront, Pétra resort
Tel *22530 41336* **Closed** *mid-Oct–Apr*
The same family who for years kept the excellent Petrí taverna in that inland hamlet now runs this shoreline spot. The menu features the same salubrious *mageireftá* and a few grills, but there is also more seafood.

LÉSVOS: Anemoessa €€
Traditional Taverna
Closest to the harbour chapel, Skála Sykaminiás
Tel *22530 55360* **Closed** *winter: weekdays*
Given its picture-postcard setting, Anemoessa could easily rest on its laurels in terms of food offerings, but it doesn't. Tuck into fresh sardines, *chtapódi krasáto*, aubergine dishes and that perennial island favourite: stuffed courgette blossoms.

LÉSVOS: Ouzadiko tou Baboukou €€
Mezedopoleío/Ouzerí
South quay, Mólyvos port
Tel *22530 71776* **Closed** *winter: weekdays*
This long-running bohemian *ouzerí* has been going since the 1960s, and it proves a dab hand at all seafood (try the prawn pasta), salads and the usual *mezédes*. All Lésvos ouzos are available.

LÍMNOS: Man-Tella €
Traditional Taverna
Centre, Sardés village
Tel *22540 61349*
This cult taverna doubles as the central *kafeneío* of the highest Limnian village. Big portions of country cooking (rabbit *stifádo*, rooster *krasáto* with noodles) are washed down with excellent barrel wine.

Oúzo, an anise-flavoured apéritif which turns cloudy when mixed with water

For more information on types of restaurants *see pp314–15*

LIMNOS: Sozos €
Mezedopoleío/Ouzerí
Central plaza, Platý village
Tel 22540 25085 **Closed** *Nov–Apr*
Mezédes, salads, a few *mageireftá*,
steamed mussels and all manner
of roasts are present and correct
here. Booking advised.

LIMNOS: Mouragio €€
Seafood
Seafront, Kótsinas harbour
Tel 22540 41065 **Closed** *late Oct–
late Apr*
The best of three tavernas in this
little fishing port, Mouragio is
worth the slight price premium
for the service and the handling
of the abundant local fish.

SAMOS: Kallisti €
Traditional Taverna
Central Manolátes village
Tel 22730 94661 **Closed** *Nov–Apr:
weekdays*
A favourite of locals and visitors
thanks to its lovely proprietress,
and its hearty stews. Wood stove
and live music in winter.

SAMOS: Tarsanas €
Pizzeria
West waterfront, Kokkári
Tel 22730 92337 **Closed** *late Oct–
Easter*
The specialities at this 1970s
throwback are the thick-crust
pizzas (large enough for two) and
the *mageireftá*. Seafront seating.

SAMOS: Cohyli €€
Creative Greek
*West end of Iréon resort,
one block inland*
Tel 22730 95282 **Closed** *winter:
weekdays*
Count on squid, octopus and sea
urchin starters at this taverna,
but on vegetarian platters, too:
aubergine roll, yellow split peas
and fried potatoes. Occasional
weekend acoustic music.

DK Choice

**SAMOS: To Koutouki
tou Barba Dimitri** €€
Creative Greek
Village centre, Pýrgos
Tel 22730 41060
This remote taverna takes
great pride in high-quality
ingredients, including Samos
black truffle, which is grated
onto pasta dishes, goat grilled
as succulent chops and three-
cheese-stuffed peppers. Some
outdoor terrace tables have
mountain views, while indoors
a fireplace crackles during the
winter. Try the family's *liastó*
wine (made from raisins).

SAMOS: Triandafyllos €€
Seafood
*Upper (Small) Plateía, Palaiókastro,
2 km southeast of Vathý*
Tel 22730 27860 **Closed** *Sun*
This is the undisputed place for
fish and seafood near Vathý.
Abundant mixed salads and
shellfish starters precede
expertly grilled swordfish or
tuna in the summer, with less
familiar species available at other
times of year. Booking advised.

SAMOTHRAKI: 1900 €
Traditional Taverna
Near Plateía, Chóra
Tel 22510 41222 **Closed** *mid-Sep–
mid-May*
This restaurant is renowned for
its goat- and aubergine-based
dishes. The brave might want to
sample *tzigerosarmádes* (goat's
gutskin stuffed with its own
liver and offal). There are views
of the Kástro from the terrace.

SAMOTHRAKI: Karydies €
Traditional Taverna
*1 km aboven Ano Meriá village,
far east of the island*
Tel 22510 98266 **Closed** *Oct–May*
Karydies is a temple for goat
meat, which is prepared a
dozen different ways here.
Vegetarians, however, need
not fear, since there is also
plenty for them, especially
aubergine and bean dishes.

THASOS: Archodissa €
Traditional Taverna
Hillside beyond eastern cove, Alykí
Tel 25930 31552
Despite its popularity with
tourists, Archodissa remains
an enjoyable all-round taverna,
with reasonably priced fish by
weight and a few *mageireftá* and
moussakas. Attractively priced
bottled wine and great views.

Lamb *kofta* kebabs with *tzatziki* is a classic
Grecian dish

THASOS: Pigi €
Oinomageireio
*Southwest corner, main
plateía, Liménas*
Tel 25930 22941 **Closed** *Nov–Mar*
Located near a spring, this
decades-old eatery serves
generous meat-based *mageireftá*
as well as courgette-and-potato
croquettes and *gígantes* (haricot
beans). Atmospheric by night.

The Dodecanese

ASTYPALAIA: To Gerani €
Traditional Taverna
*In the stream bed, just behind
Livádi beach*
Tel 22430 61484 **Closed** *Nov–April*
The most consistently good
(and consistently open) taverna
at this beach resort is renowned
for *mageireftá* and assorted
seafood platters. Big portions.

ASTYPALAIA: Australia €€
Traditional Taverna
Just inland from the head of Skála Bay
Tel 22430 61275 **Closed** *Dec–Mar*
With vegetables from the family
farm and great seafood, this is
the oldest and most reliable
taverna in Skála. They also have
island-produced wine, which
you're unlikely to find elsewhere.

**CHALKI: To Paradosiako
Piato tis Lefkosias** €€
Creative Greek
Quayside, Nimporió
Tel 69406 67845
The owner-chef has featured
on BBC cookery shows, and
the food is, unsurprisingly, top-
notch. Elaborate baked dishes,
perfectly grilled lamb chops and
meticulous *mezédes* are all made
with own-farmed ingredients.

KALYMNOS: Kafenes €
Mezedopoleío/Ouzerí
Platía Christoú, Póthia
Tel 22430 28727
From this hole in the wall issue
such delights as crispy *maridáki*
fish, savoury local cheeses and
heaping salads with ingredients
such as purslane, rocket, sun-
dried tomatoes and cheese.

KALYMNOS: Pandelis €€
Seafood
*Lane behind the Olympic Hotel,
Agios Nikólaos district, Póthia*
Tel 22430 51508
Ignore their printed menu and,
instead, ask for scaly fish and
shellfish like *kalognómes* (ark
shell), *strídia* (round oysters) and
foúskes (a marine invertebrate
gathered by sponge divers).

KALYMNOS: Tsopanakos €€
Traditional Taverna
Hillside, Armeós
Tel *22430 47929*
The chef here relies on meat and cheese from the local flocks. Starters include *tzatzíki* and stuffed vine leaves, while the signature dish is *mourí* – lamb or goat baked in a clay pot. Fine views from the terrace.

KARPATHOS: Under the Trees (Kostas) €
Traditional Greek
West coast, about 1 km north of Finíki
Tel *6977 984791* **Closed** *Nov–Mar*
Dine under the tamarisks at this rural taverna. On the menu are expertly grilled swordfish, racks of chops and courgette chips.

DK Choice

KARPATHOS: La Gorgona (Gabriella's) €€
Italian-Greek
South end of the quay, Diafáni
Tel *22450 51509* **Closed** *Nov–Mar*
The friendly proprietress makes a varied clientele feel welcome, from breakfast (Diafáni's best, with proper coffees) until late. Platters include *makaroúnes* noodles with a gorgonzola cheese sauce, aubergine dishes and seafood. End the meal with a home-made dessert and Limoncello. A soundtrack of jazz and world music completes the experience.

KARPATHOS: Orea Karpathos €€
Traditional Taverna
Southeast end of the quay, Pigádia
Tel *22450 22501* **Closed** *Dec–Feb*
This is the best all-rounder in the port, with dishes like *trahanádes* soup and spinach pie to pair with local Othos wine. Locals use it as an *ouzerí*, ordering a few small platters.

KASTELLORIZO: Paragadi €€
Seafood
Near the ferry dock on the east quay
Tel *22460 49396* **Closed** *Dec–Mar*
Despite the offputting picture menus, this is a solid choice with a full range of seafood, plus some meat dishes.

KASTELLORIZO: Alexandra's €€€
Creative Greek
West end of the quay, main port
Tel *22460 49019* **Closed** *Dec–Mar*
The best cooking in town. Diners tuck into octopus and courgette croquettes, grilled halloumi and *soupiórizo* (cuttlefish rice) jazzed up with ouzo, chilli and parsley.

The stunning views from Italian-Greek fusion restaurant La Gorgona in Kárpathos

KOS: Palaia Pigi €
Traditional Taverna
Pigí district, Pylí village
Tel *22420 41510* **Closed** *Dec–Mar*
Enjoy flawless, simple fare under a giant ficus, overlooking a bird-filled oasis watered by a spring: sausages, marinated sardines and fresh *bakaliáros*.

KOS: Pote tin Kyriaki €
Mezedopoleío/Ouzerí
Pisándrou 9, Kos old town
Tel *22420 48460* **Closed** *Sun; winter: also Mon–Thu*
Kos' only genuine *mezedopoleío* serves up *marathópita* (fennel patty), *kavourdistí* (pork fry-up) and assorted fish dishes, plus a strong house *rakí*.

KOS: Ambavris €€
Traditional Taverna
Ampávris hamlet, Vouriná district, 700 m S of Kos town
Tel *22420 25696* **Closed** *mid-Oct–Apr*
The *mezédes* at this farmhouse-courtyard taverna include lamb meatballs in oúzo sauce, husked beans with garlic, snails, stuffed courgette blossoms and *fáva*. Bottled or bulk oúzo and *tsípouro*.

KOS: Hasan €€
Turkish
Central junction, Platáni village
Tel *22420 20230*
Hasan has fewer *mezédes* than the other Turkish restaurants in Platáni, but it does offer a decent Adana kebab and a tasty okra stew, plus a full drinks list.

KOS: Makis €€
Seafood
Just inland from the sea, Mastichári
Tel *22420 59061*
Makis is thought to offer the freshest and best-priced fish on Kos, along with salads and dips. A basic appearance and oblique sea views mean few tourists.

LEROS: Dimittris O Karaflas €€
Mezedopoleío/Ouzerí
Spiliá district, on the road between Vromólithos and Pandélli
Tel *22470 25626*
Sizeable portions of hearty food – chunky Lerian sausages, potato salad and goat's cheese – and great views over Vromólithos Bay.

LEROS: Mylos €€€
Creative Greek
Out by the sea-marooned windmill, Agía Marína
Tel *22470 24894* **Closed** *late Oct–late Mar*
Dishes like *garidopílafo* (shrimp rice), octopus carpaccio, *kalamári* with pesto and courgette patties are worth the splurge at this restaurant in a romantic setting.

LEROS: Sotos €€€
Seafood
Drymónas anchorage, west coast
Tel *22470 24546* **Closed** *Nov–Mar*
Only seasonal and local seafood is featured here: oysters, mussels, sea urchins, scallops and scaly fish. Pleasant port-view seating.

LIPSI: Dilaila €€
Fine Dining
Katsadiá, easterly bay
Tel *22470 41041 or 69728 85476*
Closed *mid-Oct–Apr*
The seasonally changing menu at Lipsí's most innovative kitchen might include octopus carpaccio with red mulberries, tuna tartare or squash and carrot salad.

NISYROS: Balkoni tou Emboriou €
Traditional Taverna
Central plateía, Emporeiós
Tel *22420 31607* **Closed** *Nov–Apr*
Enjoy *keftédes*, fresh beets, pickled caper greens and cheese-based recipes while taking in views of the volcano. When you leave, ask how the antique mirror inside shattered.

For more information on types of restaurants *see pp314–15*

NISYROS: Limenari €
Traditional Taverna
Limenári cove, west of Páli
Tel *22420 31023*
This secluded taverna attracts
islanders thanks to well-priced
portions of caper-topped salads,
delicious grilled fish and a few
daily *mageireftá*, such as
papoutsáki (aubergines stuffed
with mince). In a lovely position
overlooking the bay.

PATMOS: Panagos €
Oinomageireio
Main junction, Kámpos village
Tel *22470 31570*
This decades-old mecca of
mageireftá offers such dishes
as baked fish, lentil soup and
soupiórizo (chickpea stew), with
a small complimentary dessert.
Pleasant barrel wine.

PATMOS: Ktima Petra €€
Traditional Taverna
Pétra cove, south of Gríkos
Tel *22470 33207* **Closed** *mid-Oct–
Mar*
Most of the ingredients for the
dishes served at this beachside
outfit are sourced from their
adjacent farm *(ktíma)*. Menu
highlights include home-made
dolmádes, stewed goat or rabbit,
aubergine-based dishes and
grills. After hours, Ktima Petra
becomes a courtyard bar.

PATMOS: Tzivaeri €€
Mezedopoleío/Ouzerí
*Behind westerly Theológos town
beach, Skála*
Tel *22470 31170*
This *ouzerí* offers tasty *dolmádes*,
chilli-spiced cheese mash and
grilled mushrooms. There is also
live Greek music at weekends.

**RHODES: Locanda
Demenagas** €
Oinomageireio
*Avstralías 16, Akandiá Port
shore avenue*
Tel *22410 30060*
This harbourside venue is the
ideal spot for a little lunch
before boarding a ferry. Daily
menus might feature *pastítsio*
(macaroni pie with a mince
filling), bean soup, liver or
rooster with noodles.

RHODES: Mezes €
Mezedopoleío/Ouzerí
*Aktí Kanári 9, Psaropoúla district,
new town*
Tel *22410 27962*
Starters like chickpea stew,
breaded *mastéllo* cheese from
Híos or *lahmatzoún* (Armenian
pizza) make a filling meal.
Especially popular at weekends.

RHODES: Platanos €
Traditional Greek
*Lower plateía, Plátanos old
village centre*
Tel *22440 46027 or 69441 99991*
Closed *winter: Mon–Fri*
Enjoying one of the most
exquisite setting on Rhodes,
Platanos has tables under the
plane trees between two
Ottoman fountains. The *mezédes*
(vine leaves, hummus, *kopanistí*)
are delicious and cheap, as
are the meat mains.

RHODES: Perigiali €€
Traditional Taverna
*South end of the waterfront by the
fishing port, Stegná*
Tel *22440 23444*
Rhodians crowd Perigiali's tree-
shaded terrace for great-quality
seafood, hand-cut chips and
superior *yaprákia* (stuffed vine
leaves), plus good bulk wine.
The travertine-clad loos are
themselves an attraction.

RHODES: Indigo €€€
Middle Eastern Fusion
Stalls 105–106, Néa Agorá
Tel *6972 663100* **Closed** *Sun;
Nov–15 Mar*
Among the menu offerings at this
charming bistro are falafel with
tabouleh and a thick hummus
garnish, *hunkar beyendi* (lamb
with aubergines), lamb burger,
proper pale-grey *taramás* and
dollops of purslane in yogurt.

The enchanting courtyard of the Marco Polo
Café in Rhodes old town

SYMI: Giorgos & Maria €€
Traditional Taverna
*Top of Kalí Stráta stair-street,
Chório*
Tel *22460 71984*
At this local institution with a
pebble-mosaic courtyard, there
is a wide choice of casseroles,
plus *mezédes* like lamb liver and
spinach and aubergine rissoles.
Informal music sessions on
Fridays keep it lively.

SYMI: Marathounda €€
Traditional Taverna
Marathoúnta cove, southern Sými
Tel *22460 71425* **Closed** *mid-Sep–
mid-May*
Although the beach here isn't
the best on the island, people
keep on coming, drawn by this
taverna's well-priced fish, its wide
range of vegetarian *mezédes* and
its convivial service.

SYMI: Mythos €€€
Fine Dining
South quay, Gialós
Tel *22460 71488* **Closed** *Oct–May*
Signature dishes here include
psaronéfri (pork medallions)
served with a mushroom and
wine sauce, and *kalamári* with
basil and pine nuts. The restaurant
occupies a former summer
cinema and offers great views.

> ### DK Choice
>
> **RHODES: Marco Polo Café** €€€
> Fine Dining
> *Agíou Fanouríou 42, old town*
> **Tel** *22410 25562* **Closed** *mid-
> Oct–early Apr*
> Come here for consistently
> good, subtle dishes with a bias
> towards seafood and decadent
> puddings courtesy of the best
> Italian dessert chefs on Rhodes.
> The setting, in an ochre-and-
> blue-splashed orchard-
> courtyard with an illuminated
> fountain, is equally magical.
> There is also a good wine list,
> though most diners choose
> the refreshing bulk white.

RHODES: Mavrikos €€€
Fine Dining
Main taxi/fountain square, Líndos
Tel *22440 31232* **Closed** *Nov–Mar*
Founded in 1933, this cutting-
edge diner creatively fuses various
Mediterranean flavours. The
menu might include cured tuna
with grilled fennel, cuttlefish-ink
risotto, haricot beans in a carob
syrup and lamb's liver with chillis.

The Cyclades

**AMORGOS: Limani tis
Kyras Katinas** €
Traditional Taverna
Ormos Aigiális, one lane inland
Tel *22850 73269* **Closed** *mid-Oct–
Easter*
A candidate for best all-round
taverna on Amorgós, this place
excels at vegetable casseroles,
seafood, pulses and local cheeses.

AMORGOS: To Hyma €
Mezedopoleío/Ouzerí
Main commercial lane, Chóra
Tel *6974 786376*
Gastronomic miracles are performed on a small camp stove at this genuine *ouzerí* in an eclectically decorated former grocery store. Try the salads and the simple meat or fish platters.

ANDROS: Stamatis €€
Traditional Taverna
Batsí harbour
Tel *22820 41283*
The fish soup changes according to the daily catch; other dishes include pasta with shrimps, lamb chops, liver and pork. Booking advisable in summer.

ANDROS: Tou Josef €€€
Creative Greek
Pitrofós hamlet, 3 km (2 miles) southwest of Chóra
Tel *22820 51050*
Favourites here include baked stuffed pumpkin, onion pie, Andriot *fourtália* omelette and, for dessert, mastic meringue with bergamot. No printed menu.

ANTIPAROS: Anargyros €€
Traditional Taverna
Harbourfront
Tel *22840 61204*
This venerable taverna is famed for its *pastítsio* and moussaka. Both are thick, succulent and laden with béchamel sauce.

FOLEGANDROS: Piatsa €€
Traditional Taverna
Plateía Piátsa (the third plaza), Chóra
Tel *22860 41274* **Closed** *Nov–Apr*
Matsáta, the hand-pulled flat noodles typical of Folégandros, are served with chicken, goat or pork. This is the best spot to sample this traditional dish.

IOS: Katogi €€
Fine Dining
Main through road, Chóra
Tel *69834 40900* **Closed** *Nov–Apr*
The outlandish, colourful decor matches the eclectic menu: Jack Daniel's pork; walnut, brie and beet salad; and chocolate cake to finish. Drinks include a tiramisu Martini. Reservations advisable.

KEA: Rolando's €€
Traditional Greek
Top of main square, Ioulída
Tel *22880 22224* **Closed** *Dec–Easter*
The menu at Rolando's reflects the Kean/Corfiot origins of the jolly managing couple. Tuck into *katsíki lemonáto, pastitsáda,* rabbit stew, grilled sardines and baby squid.

KEA: Ton Kalofagadon (Yannis) €€
Traditional Taverna
Opposite the town hall, main plateía, Ioulída
Tel *22880 22118* **Closed** *winter: weekdays*
Saturday is spit-roasted meat night here – the *gourounópoulo* (whole suckling pig) is especially good. *Píttes, mageireftá* and the house wine round off the menu.

KOUFONISSI: Karnagio €€
Mezedopoleío/Ouzerí
Loutró cove, west of Chóra
Tel *22850 71694* **Closed** *Nov–Apr*
This is an exceedingly popular quayside spot, so reservations are advisable in summer. Count on a full line of *píttes,* flash-fried baby fish, grills and *rakí* from Náxos.

KYTHNOS: Ostria €€€
Seafood
Mérichas waterfront, near the yacht anchorage
Tel *22810 33017* **Closed** *Nov–Apr*
Enjoy reliably fresh and expertly grilled fish at this basic taverna: butterflied prawn, lobster noodles and whole *kalamári,* too.

MILOS: Enalion €€
Traditional Taverna
South end waterfront, Pollónia
Tel *22870 41415*
Three-generations prepare chickpea salads with sundried tomato and feta, cheese *pitarákia,* octopus and largely fishy mains. Reserve a table by the waterside.

MILOS: O Hamos €€
Traditional Taverna
Opposite Papakínou beach, southwest of Adámas
Tel *22870 21672* **Closed** *Nov–Easter*
O Hamos prides itself on serving only dishes made from their farm-raised ingredients. Try the goat roasted in embers, or the piglet baked in grape-molasses paper.

MYKONOS: Joanna & Nikos €
Traditional Taverna
Megáli Ammos beach, 1 km southwest of Chóra
Tel *22890 24251* **Closed** *mid-Oct–Apr*
This unpretentious taverna serves moreish mushroom-based dishes, onion pie, country sausages and the ubiquitous *chórta* (steamed green leafy vegetables), plus a nice range of puddings.

MYKONOS: Madoupas €€
Oinomageireio
South waterfront, Chóra
Tel *22890 22224*
This atmospheric café-restaurant ladles up hearty, generously portioned Greek fare; excellent *revytháda* made with chickpeas, crisp-fried *gávros* and grilled oyster mushrooms.

MYKONOS: Ma'ereio €€€
Traditional Taverna
Kalogéra 16, Chóra
Tel *22890 28825*
Typical Mykonian flavours inform the menu at this cosy bistro, known for its *loúza* (cured pork-loin), garlic mushrooms and *kopanistí* (tangy cheese dip).

MYKONOS: Tasos €€€
Seafood
Parága beach
Tel *22890 23002*
The bread basket at this cult spot comes with an assortment of dips. Try the roast aubergine with tomato, parsley and garlic, or the crunchy fried shrimps.

NAXOS: Giannis €
Traditional Taverna
Central plateía, Chalkí, Trageá
Tel *22850 31214*
Giannis doles out Greek comfort food like *gígantes* (large haricot beans), aubergine roulade, *píttes* (the spinach and cheese one is scrumptious) and meat grills.

A traditional Greek starter of tomato, feta, oregano, paprika and olive oil on a bread base

For more information on types of restaurants *see pp314–15*

NAXOS: Platsa €
Traditional Greek
Village centre, Kóronos
Tel *22850 51243* **Closed** *Jan–Feb*
Sit on a plant-festooned patio, and enjoy the *mageireftá* of the day – roast potato medallions, lamb *lemonáto* – with the strong but very quaffable house rosé.

NAXOS: Sto Ladoharto €
Grills
Central quay, Náxos town
Tel *22850 22178*
The decor is mock 1950s rural grocery shop at this eatery with port views. On the menu are grilled meat platters, plus salads and starters like spicy turkey meatballs and courgette rissoles.

NAXOS: Metaxy Mas €€
Traditional Taverna
Palaiá Agorá, Náxos Town
Tel *22850 26425*
Popular with locals and visitors, this place serves heaped portions of local cheese, salads, *chórta* and flash-fried *marídes*. Affordable oúzo and local barrel wine.

DK Choice

NAXOS: Axiotissa €€€
Fine dining
Km 18 of Chóra–Alykó road, near Glyfáda
Tel *22850 75107* **Closed** *Nov–Mar*
Organic ingredients dominate unusual dishes like rocket and *xinomyzíthra* cheese salad, chilli-and-onion-sautéed Portobello mushrooms, aubergine and almond bake, grilled Armenian sausage, and goat or rabbit stews. The rosé bulk wine comes from nearby Moní; there are also select Greek labels, or try the microbrewery beer. Save room for the home-made desserts.

PAROS: Palia Agora €
Mezedopoleío/Ouzerí
Centre of old Náousa marketplace, in a warren of unmarked lanes
Tel *22840 51847*
A local hangout where *soúma* accompanies tasty, copious snacks – five platters are enough to feed two. Greek music on the speakers, occasional live sessions.

PAROS: Tsitsanis €
Traditional Greek
Main car-park plateía, Pródromos
Tel *22840 42258*
Signature dishes at this venerable eatery include rabbit stew and artichokes with broad beans. The barrel wine is white, or try the rosé made from sun-dried grapes.

PAROS: Albatros €€
Seafood
On the seafront, about 100 m southwest of the windmill, Paroikiá
Tel *22840 21848*
Typical platters here include chilli-spiced cheese mash and, in season, grilled *soupiá* (cuttlefish) served with baked potato and herb mustard.

PAROS: Apoplous €€
Traditional Taverna
South quay, Alykí port
Tel *22840 91935* **Closed** *16 Dec–Jan*
Overlooking the fishing port, Apoplous offers sun-dried mackerel and seafood, but also *paxima-dokoúloura* (salad piled on barley rusk) and lamb hotpot.

SANTORINI: Ladokolla €
Grills
Main commercial street, opposite the post office, Firá
Tel *22860 21244*
Exceptionally juicy *kondosoúvli* (spit-roasted meat) are served with a mammoth garnish of green leafy vegetables, carrots, potatoes and saffron rice. Ideally located for Athiniós' ferry port.

DK Choice

SANTORINI: Metaxy Mas €€
Cretan/Cycladic Fusion
Exo Goniá, 100 m below the top church
Tel *22860 31323*
West Crete staples like *píttes*, *apáki* and *stamnagáthi* (spiny chicory) meet local *fáva* (yellow split peas) and white aubergine dishes. The yogurt-sauce lamb on bulgur wheat takes some beating. Choose to dine alfresco, on outdoor terraces gazing out to Anáfi islet, or inside the old vaulted premises.

The ambient terrace at the Metaxy Mas in Santorini

SANTORINI: Roka €€€
Creative Greek
Follow the little signs through the north flank of Oía village.
Tel *22860 71896* **Closed** *Nov–Mar*
Santorinean vegetables make guest appearances in starters such as crustless courgette pie and tomato fritters. The mains can be pricy, but do leave room for the chocolate soufflé à la mode for dessert. Located in a renovated old house.

SANTORINI: Selene €€€
Fine Dining
Centre, Pýrgos village
Tel *22860 22249*
Two restaurants in one: head upstairs for special occasions or choose the cheaper downstairs "meze and wine" deli, where the menu features squid with lemon split beans and white aubergine salad with octopus carpaccio.

SANTORINI: To Psaraki €€€
Fine Dining
Vlycháda seafront, south of the island
Tel *22860 82783* **Closed** *Nov–Mar*
Take in the lovely sea views while dining on pesto-grilled goat's cheese, smoked white aubergine salad, grilled marinated rooster and pan-seared *lakérda* (bonito).

SERIFOS: Fagopoti (Nikoulias) €
Traditional Taverna
Mid-beach, on the sand, Livádi
Tel *22810 52595*
Generous portions of *biftéki*, butterflied sardines and chunky chilli-spiced cheese mash, plus hard-to-find island bulk wine, are cheerfully served on tables on the sand. The best-value choice in the area.

SERIFOS: Petros €
Oinomageireio
Outskirts, Chóra
Tel *22810 51302* **Closed** *Nov–Apr*
Mageireftá like *melitzána sto foúrno* (baked aubergine), stews and bean dishes are accompanied by quaffable island bulk wine. A good bet before hiking across Sérifos.

SIFNOS: Artemon (ex-Lembesis) €€
Traditional Taverna
Inside Artemon Hotel, through road, northern Arteмónas
Tel *22840 31303* **Closed** *Oct–Mar*
An excellent one-stop shop for Sifniot recipes. Enjoy caper salad, *revytháda* (chickpea stew), beer-stewed beef and lamb or goat hotpot at tables under a vine-shaded patio.

SIFNOS: Leonidas €€
Traditional Taverna
Ouskirts of the inner citadel,
Kástro
Tel *22840 31153* **Closed** *Oct–Easter*
Signature dishes at this friendly
restaurant include moussaka,
warm *fáva* (yellow split beans)
with onions, baby tomatoes
and capers, lamb, pork with
mustard sauce, rabbit stew
and shrimp spaghetti.

SYROS: Apano Hora €€
Mezedopoleío/Ouzerí
Village centre, Ano Sýros
Tel *22810 80565*
Popular in the daytime as a
kafeneío, after dark Apano Hora
becomes an *ouzerí*, with terrace
tables that go quickly. The menu
might feature carrot and cheese
pittákia, local sausages and a
daily special or two.

SYROS: Lliovassilema €€
Creative Greek
Beach, Galissás
Tel *22810 43325* **Closed** *mid-*
Oct–Apr
The best taverna in this popular
resort offers imaginative versions
of island staples like *marathópitta*
(fennel pie), prawns served with
barley, and suitable crunchy
atherína (sand smelt).

TINOS: Katoi €€
Grills
Falatádos village
Tel *22830 41000*
Signature dishes at this carnivore
haven in a pretty hill-village are
superior lamb chops, pork stuffed
with cheese, peppers and baby
carrots, and *kokorétsi* (spit-roasted
roulade of lamb offal).

TINOS: Marathia €€€
Seafood
Agios Fokás beach
Tel *22830 23249*
Seafood specialists with a
twist – think scorpion fish with
mint and basil, and lobster pasta.
Also good salads and vegetable
starters, and irresistible desserts.

Crete

AGIOS NIKOLAOS:
Chrysofillis €€
Mezedopoleío/Ouzerí
Aktí Papá Pangálou, Kitroplatía cove
Tel *28410 22705*
Creative dishes such as *sfoungáta*
(onion and sausage soufflé),
aubergine with melted cheese
and beets in a yogurt and walnut
sauce. The minimalist interior
doubles as a photo gallery.

A traditional restaurant on the shore of Chania in Crete

AGKATHIAS: Agkistri €€
N O Psaras
Seafood
Village centre, 3 km (2 miles)
southeast of Palaíkastro
Tel *28430 61598*
The freshest fish in eastern Crete
is delivered daily from the family
boat, while the brother-and-sister
team grill and serve, respectively.
Great views.

CHANIA: Kalamoti €
Traditional Taverna
Venizélou 142, Halépa district, just
east of the old town
Tel *28210 59198*
This popular seaside hangout
serves generously portioned
mageireftá, vegetarian starters
and seafood. Other highlights
include *dolmádes*, *myzíthra*
cheese and grilled cuttlefish, all
accompanied by Cretan *retsína*.

DK Choice

CHANIA: Ta Halkina €
Musical Rakadiko
Aktí Tompázi 29–30, Old Port
Tel *28210 41570*
One of only a few good-value,
old-harbourside spots, the
popular Ta Halkina excels at
local dishes like aubergine
roulade, *marathópita* (fennel
pie) and rosemary snails.
Solicitous waiters explain
unfamiliar dishes. Frequent
live music sessions.

CHANIA: Tamam €€
Cretan/Middle Eastern
Zambelíou 49, Evraïki, old town
Tel *28210 96080*
A 1645-built *hammam* houses
Chania's most atmospheric
restaurant. It can get hot in the
summer, so be sure to book an
outside table. Cretan menu, plus
exotic plates like Iranian pilaf,
and plenty for vegetarians, too.

IRAKLEIO: Avli
tou Defkaliona €
Traditional Cretan
Preveláki 10, behind Historical
Museum
Tel *28102 44215* **Closed** *Sun*
The best spot in town for Cretan
home-style cooking. After a
warm welcome from the owners,
lucky diners grab a sought-after
terrace table by a medieval
fountain. When the mood strikes,
accordions and guitars come out.

IRAKLEIO: Ippokambos €€
Mezedopoleío/Ouzerí
Sofoklí Venizélou 3 (seafront)
Tel *28102 80240* **Closed** *Sun;*
Dec–Mar
This well-regarded *mezedopoleío*
overlooking the old Venetian
harbour does not accept
reservations, so be sure to go
early to beat the crowds.
Excellent *mezédes* (including
snails), fair-priced seafood and
winning bulk wine. Good bulk wine.

LERAPETRA: Levante €€
Traditional Cretan
Stratigoú Samouíl 38 (waterfront)
Tel *28420 80585*
Founded in 1936, this is the most
venerable of Ierápetra's seafront,
castle-view tavernas. On the
menu are dishes like milk-based
xýgalo, stuffed cabbage leaves,
aubergines stuffed with mince
meat and rice-and-offal sausages.
There is also a full seafood list.

MOCHLOS: Kochylia €€
Traditional Taverna
East quay, Móchlos port
Tel *28430 94432* **Closed** *Dec–Jan*
The oldest (established 1902)
and most reliable among several
eateries here has a pleasant
interior and operates year-round.
Try the springtime artichokes,
the small fish or any of the
mageireftá, which are prepared
daily. Discreet, non-pushy service.

PALAIOCHORA:
The Third Eye €€
Vegetarian Cretan/Asian
West beach, south end,
50 m inland
Tel *28230 42223*   **Closed** *Nov–Lent*
This vegetarian diner offering
Cretan, Asian and Middle Eastern
specialities holds popular
Sunday-night concerts for which
tables must be reserved. Good
range of desserts.

PLAKIAS: Ifigeneia €
Traditional Cretan
Western outskirts, Aggouselianá
village, 10 km inland from Plakieas
Tel *28320 51362*
Come to Ifigeneia for country
delicacies like *volví skordaláta*
(garlic-pickled wild hyacinth
bulbs), *stamnagáthi* (wild greens),
apáki (cured pork) and boiled
goat with sticky white rice.

RETHYMNO: Rakodikeio €
Mezedopoleío/Rakádiko
Vernádou 7, opposite
Nerantzés Mosque
Tel *69457 74407* or *28310 55491*
Popular among Réthymno's
student population, this place
serves creatively tweaked dishes
like saffron and oúzo pork, and
yogurt, walnut and beet salad.
Dine alfresco on the pedestrian
alley or inside.

RETHYMNO: Mesostrati €€
Traditional Cretan
Gerakári 1, off Plateía Martýron
Tel *28310 29375*
In addition to live Cretan
music three times a week,
this neighbourhood taverna
offers dishes like *apáki*, pies
and *keftédes*, all accompanied
by palatable bulk wine.

RETHYMNO: Veneto €€€
Fine Dining
Epimenídou 4, Old Town
Tel *28310 56634*   **Closed** *Nov–Apr*
Veneto is the olde-worlde
restaurant of the eponymous inn.
Creative dishes might include
stamnagáthi (spiny chicory), *volví*
(pickled wild hyacinth bulbs),
artichoke salad or lamb wrapped
in vine leaves. There is also a well-
stocked wine cellar.

SITIA: The Balcony €€€
Creative Cretan
Fountalídou 19, old hillside quarter
Tel *28430 25084*   **Closed** *Nov–Mar*
The widely travelled owner-
chef's menu incorporates
nouveau Cretan fare (walnut and
rosemary rabbit, snails in goat's
cheese sauce), creative salads
and Mexican and Asian dishes.
Quirky old-house interior.

SOUGIA: Anchorage €€
Traditional Taverna
Main approach road, inland
Tel *28230 51487*   **Closed** *mid-Nov–*
Easter
Despite the nautical name, there
are no sea views here. Instead, a
vine-shaded courtyard provides
the setting for decent renditions
of *angináres me koukiá* (artichoke
hearts with broad beans), stewed
rabbit and *tsigaristó*.

SOUGIA: To Kyma (Paterakis) €€
Traditional Taverna
Seafront promenade, west side
Tel *28230 51688*   **Closed** *late Oct–*
late Apr
To Kyma's terrace tables are
usually packed after dark, as
locals and visitors come here for
generous portions of own-made
taramosaláta and *marathópita*,
and the quaffable bulk wine.
Friendly service.

DK Choice

VORI: Alekos €€
Traditional Cretan
Village centre, behind Agía
Pelagía church
Tel *28920 91094*
The setting at this secluded
village taverna near Agía Triáda
and Phaestos features a
courtyard with clothed tables
and cushioned bench seating,
and an interior with a fireplace
in winter. On the menu are
hearty portions of deftly
executed dishes, such as baked
goat or kid (order in advance).

Athens

ANO PETRALONA: Santorinios €
Traditional Santorinean
Doríeon 8, Metro Petrálona
Tel *21034 51629*   **Closed** *summer:*
Mon–Fri
Housed in an old artisanal
workshop, this taverna has small
dining rooms and a lovely
courtyard. There is a limited
menu of Santoríni specialities,
mostly pork-based, and excellent
island bulk wine.

CENTRAL BAZAAR: To Diporto €
Traditional Taverna
Sokrátous 9, crn Theátrou, in the
basement (no sign)
Tel *21032 11463*   **Closed** *Sun*
Businessmen and students come
to this classic eatery (est. 1887)
near the central market for grilled
and fried fish, *mageireftá*, salads
and *retsína*. It is small and busy, so
be prepared to share your table.

Meat cooking in a traditional-style
Greek oven

EXARCHEIA: Ama Lahei €–€€
Creative Greek/French Bistro
Kallidromíou 69
Tel *21038 45978*
A two-in-one eatery with an
enchanting terrace garden.
On the menu are Dráma goat
sausages, smoked caramelized
aubergine and *píttes*. To drink,
tsípouro, *rakí* and micro-beers.

EXARCHEIA/KOLONAKI:
Boundary Il Postino €€€
Italian
Grivalon 3, pedestrian lane off Skoufá
Tel *21036 41414*
At this genuine *osteria*, an Italian
chef prepares a simple menu of
Greek starters, pasta dishes and
meaty mains. Eat alfresco in the
quiet cul-de-sac, or inside, in a
dining room featuring old photos
and a retro soundtrack.

KOLONÁKI: To Kioupi €
Oinomageireio
Plateía Kolonakioú 4, basement
Tel *21036 14033*   **Closed** *Sun; Aug*
This historic eatery with exposed
stone walls is ideal for lunch after
visiting the nearby museums.
Specialities include *keftédes*,
goat stew and suckling pig.

DK Choice

MAKRYGIANNI:
Mani-Mani €€€
Creative Greek
Falírou 10, Metro Akrópoli
Tel *21092 18180*
Interesting ingredients from
the Peloponnese, like *sýglino*
(streaky pork) and Messinian
talagáni cheese, are served
stylishly on the top floor of a
converted 1930s house. Due
to Mani-Mani's popularity
and limited seating, booking
is required all year round.

METAXOURGEIO: Funky Gourmet €€€
Fine Dining
Paramythiás 13, crnr Salamínas
Tel *21052 42727* **Closed** *Sun & Mon; Aug*
A Michelin-starred restaurant with three theatrically presented menus of improbable flavour juxtapositions, such as Mesolóngi *bottarga* (fish roe) and white chocolate. Good wine selection. Booking required.

MONASTIRAKI: Thanassis €
Grills
Mitropóleos 69, just off Plateía Monastirakíou
Tel *21032 44705*
Join the long queues here for the house speciality, the Oriental kebab (minced meat with onion and spices), which is said to be Athens' best. Beware incendiary side orders of chilli peppers. Eat in or take away.

PANGRATI: Colibri €€
Pizza/Burgers/Pasta
Empedokléous 9–13
Tel *21070 11011*
The outdoor tables on this pedestrian lane are always packed with happy customers dining on stellar salads and thin-crust pizzas and quaffing keenly priced bulk wine or imported beers.

PLAKA/MONASTIRAKI: Kapnikarea €€
Musical Ouzerí
Christopoúlou 2, corner Ermoú
Tel *21032 27394* **Closed** *summer*
Fare like sausages, *saganáki*, aubergine dishes and salads has a slight price premium to cover the *rempétika* musicians playing here. Sit at outdoor tables on this pedestrian lane (there are heaters and awnings in winter).

DK Choice

PLATEIA KANIGGOS: Kritis (O Takis) €
Traditional Cretan
Inside the stoá at Veranzérou 5, Plateía Kániggos
Tel *21038 26998* **Closed** *Sun*
Athens' premier Cretan eatery has expanded into adjacent premises, reflecting its well-earned popularity. A Sitía family purveys *tyropitákia*, *almýra hórta* (seaweed), stuffed squash flowers, *stáka* (Chaniá sheep-cream dip), *volví* (pickled wild hyacinth bulbs) and Sfakian sausages. The drinks list features Sitía wine, Brinks dark Cretan beer and strong *rakí*.

ROUF: En Gazi €
Mezedopoleío/Ouzerí
Megálou Vasileíou 32, crnr of Dyaléon 5
Tel *6977 890741*
Situated on a quiet street right behind the Benaki Pireós museum, this spot offers platters of sausages, octopus in wine, *politikí saláta* (Constantinople-style salad) and beets, with bulk wine from Neméa.

THISSEIO: To Steki tou Ilia €€
Grills
Eptachálkou 5, also 100 m away at Thessaloníkis 7
Tel *21034 58052 or 21034 22407* **Closed** *Mon*
Two outlets (open alternately) are the Athenian go-to places for juicy lamb chops. All regular starters, nicely done, also appear on the menu, and there is good barrel wine. Tables indoors or out, depending on the season.

Thessaloníki

CENTRAL BAZAAR: Agora €€
Mezedopoleío/Ouzerí
Kapodistríou 5, alley off Ionos Dragoúmi
Tel *2310 532428*
Stuffed cuttlefish and vegetarian *keftédes* (including fried courgette and potato croquettes) are all delicately spicy, like most Asia Minor-influenced food. A separate, elegant bar serves oúzo and *mezé* platters.

HARBOUR: Toumbourlika €€
Musical Ouzerí
Navmachías Límnou 14
Tel *23105 48193* **Closed** *Sun*
People come here for the excellent unamplified *rempétika* music and the atmosphere rather than the standard *ouzerí* platters.

DK Choice

KAMARA: Ta Koumbarakia €
Traditional Greek
Egnatía 140, corner of Palaión Patrón Germanoú
Tel *23102 71905* **Closed** *Mon; mid-Jul–mid-Aug*
Hidden behind the Byzantine church of Metamórfosi, this place is famed for its seafood – try the *mpakaliáros me skordaliá* (hake served with a dip of mashed garlic and potatoes). Vegetarians can tuck into pomegranate and goat's cheese salad or grilled mushrooms. Excellent bulk wine, outdoor tables and friendly service.

KAMARA: Lola €€
Seafood
Agapinoú 10, crn Michaïl Ioánnou
Tel *23102 76201*
Shrimp noodles are a highlight among the array of seafood mains and standard starters at this excellent fish eatery. Leave room for the complimentary house dessert. Smokers welcome.

PLATEIA ATHONOS: Vrotos €€€
Creative Greek
Skra 3
Tel *23102 22392*
The menu at this top restaurant includes *hunkár beyéndi* (braised lamb with an aubergine purée), smoked eel and Anatolian sausages. To drink, Santoríni beer or saffron-flavoured *tsípouro*.

PLATEIA NAVARÍNOU: Loxias €
Creative Greek
Isávron 5
Tel *23102 33925*
This bohemian bookshop-café and sometime music venue is a fun place to enjoy a good microbrewery beer, but the food is also excellent: lamb sausage, large salads and desserts.

The family-run restaurant Kritis in Athens

For more information on types of restaurants see pp314–15

SHOPPING IN GREECE

Shopping in the Greek islands can be an entertaining pastime, especially when you buy directly from the producer. This is often the case in the smaller villages, where crafts are a major source of income. Embroiderers and lacemakers can often be seen sitting outside their houses, and potters can be found in their workshops. Apart from these industries, and the food and drink produced locally, most other goods are imported to the islands and therefore carry a heavy mark-up.

Olive-wood bowls and other souvenirs from Corfu old town

VAT and Tax-Free Shopping

Usually included in the price, FPA (*Fóros Prostitheménis Axías*) – the equivalent of VAT or sales tax – is about 23 per cent in Greece.

Visitors from outside the EU staying less than three months may claim this money back on purchases over €117. A "Tax-Free Cheque" form must be completed in the store, a copy of which is then given to the customs authorities on departure. You may be asked to show your receipt or goods as proof of purchase.

Opening Hours

Allowing for plenty of exceptions, shops and boutiques are generally open on Monday, Wednesday and Saturday from 9am to 2:30pm, and on Tuesday, Thursday and Friday from 9am to 2pm and 5pm to 8pm. Supermarkets, found in all but the smallest communities, are often family-run and tend to stay open longer hours, typically Monday to Saturday from 8am or 9am to 8pm or 9pm. Sunday shopping is possible in most tourist resorts. The corner *períptero* (street kiosk), found in nearly every town, is open from around 7am to 11pm or midnight, selling everything from aspirins to ice cream.

Markets

Most towns in the Greek islands have their weekly street market (*laïkí agorá*), a colourful jumble of the freshest and best-value fruit and vegetables, herbs, fish,

Basket of herbs and spices from a market stall in Irákleio, Crete

meat and poultry – often juxtaposed with a miscellany of shoes and underwear, fabrics, household items and sundry electronic equipment.

In larger towns, the street markets are in a different neighbourhood each day, usually opening early and packing up by about 1:30pm, in time for the afternoon siesta. Prices are generally cheaper than in the supermarkets, and a certain amount of bargaining is also acceptable, at least for non-perishable items.

Food and Drink

Culinary delights to look out for in the shops and markets of the Greek islands include honey, pistachios, olives, herbs and spices. Good cheeses include the salty feta, and the sweet *anthótyro* from Crete; for something sugary, try the numerous pastries and biscuits (cookies) of the *zacharoplasteío*.

Greece is also well-known for several of its wines and spirits. These include brandy, *oúzo* (an aniseed-flavoured spirit), *retsína* (a resinated wine) and, from Crete, the firewater known as *rakí*.

Size Chart

Women's dresses, coats and skirts

Greek (size)	44	46	48	50	52	54
GB/Australian (size)	10	12	14	16	18	20
US (size)	8	10	12	14	16	18

Men's suits, shirts and jumpers

Greek (size)	44	46	48	50	52	54	56
GB/US (inches)	34	36	38	40	42	44	46
Australian (cm)	87	92	97	102	107	112	117

Women's shoes

Greek (size)	36	37	38	39	40	41
GB (size)	3	4	5	6	7	8
US/Australian (size)	5	6	7	8	9	10

Men's shoes

Greek (size)	40	41	42	43	44	45
GB/Australian (size)	7	7½	8	9	10	11
US (size)	7½	8	8½	9½	10½	11½

What to Buy in Greece

Traditional handicrafts, though not particularly cheap, do offer the most genuinely Greek souvenirs. These cover a range of items from finely wrought gold reproductions of ancient Minoan pendants to rustic pots, wooden spoons and handmade sandals. Leatherwork is particularly noted on the island of Crete, where the town of Chaniá (see p256–7) hosts a huge leather market. Among the islands renowned for their ceramics are Crete, Lésvos and Sífnos. Many villages throughout the Greek islands produce brightly coloured embroidery (kéntima) and wall hangings, which are often hung out for sale. You may also see thick flokáti rugs. They are hand-woven from sheep or goat's wool, but are more often produced in the mountainous regions of mainland Greece than on the islands themselves. In the smaller island communities, crafts are often cottage industries, which earn the entire family a large chunk of its annual income during the summer. There is usually room for some bartering when buying from the villagers.

Gold jewellery is sold mainly in larger towns. Modern designs are found in jewellers such as Lalaounis, and reproductions of ancient designs in museum gift shops.

Icons are generally sold in shops and monasteries. They range from very small portraits to substantial pictures. Some of the most beautiful, and expensive, use only age-old traditional techniques and materials.

Ornate utensils, such as these wooden spoons, are found in traditional craft shops. As here, they are often hand-carved into the shapes of figures and produced from the rich-textured wood of the native olive tree.

Kitchenware is found in most markets and in specialist shops. This copper coffee pot (mpríki) is used for making Greek coffee.

Komboloï, or worry beads, are a traditional sight in Greece; the beads are counted as a way to relax. They are sold in souvenir shops and jewellers.

Leather goods are sold throughout Greece. The bags, backpacks and sandals make useful and good-value souvenirs.

Ornamental ceramics come in many shapes and finishes. Traditional earthenware, often simple, functional and unglazed, is frequently for sale on the outskirts of Athens and the larger towns of the islands.

SPECIALIST HOLIDAYS AND OUTDOOR ACTIVITIES

If you feel you want more of a focus to your holiday in the Greek islands, there are many organized tours and courses available that cater to special interests. You can visit ancient archaeological sites with a learned academic as your guide, you can improve your writing skills, paint the Greek landscape or learn the Greek language with expert tutors, learn to cook Greek food and appreciate Greek wines, or develop your spirituality. All kinds of

walking tours, as well as botanical and bird-watching expeditions, are available in the islands. So, too, are golf, tennis, cycling, sailing and horse-riding holidays. If you prefer to be pampered or rejuvenated, Greek spas now rival the best in Europe, and there is even a naturist hotel on Crete for help with the all-over tan. Information on sailing and watersports, and advice on choosing the perfect beach, are covered on pages 340–41.

Visitors at the ancient theatre at Delos *(see pp222–3)*

Archaeological Tours

For those interested in Greece's glorious ancient past, a tour to some of the famous archaeological sites, accompanied by qualified archaeologists, can make for a fascinating and memorable holiday. In addition to visiting ruins, many tours take in Venetian fortresses, Byzantine churches, caves, archaeological museums and monasteries along the way.

Martin Randall Travel organise tours of Minoan Crete, a popular destination for archaeology enthusiasts. Its tours include sites at Knosós

(see pp276–9) and Chaniá *(see pp256–7)* among others. History and archaeology specialist **Andante Travels** also operates tours of Minoan Crete to the island's main sites, with two specialist guest lecturers accompanying the group.

Creative Holidays

With their vivid landscapes and renowned quality of light, the Greek islands are an inspirational destination for artistic endeavour. Courses in creative writing, and drawing and painting, are available at all levels.

The Skýros Centre *(see p120)*, on the island of the same name, offers two locations – one at the main town and another at the remote village of Atsítsa – for self-development and therapeutic holidays, including themes directed towards writing and painting as well as yoga.

AegeanScapes runs painting and Raku pottery holidays on Pátmos and Páros. **Simpson Travel** organises drawing and painting holidays focusing on the landscape of Crete. Their courses cater for beginners through to advanced level. For the flexible, yoga holidays on Paxós and Corfu are available through **Travel à la Carte**.

Greek Language Courses

Immersing yourself in a language is the best and most enjoyable way to learn. Greek language courses at all levels are available in Límni on the island of Evvoia and on Sýros. The courses can be booked through the Greek company of **Omilo** in Athens, who offer a variety of courses from two weeks up to eight weeks.

Nature Holidays

The Greek islands are rich in natural beauty, and you need not be a fanatical botanist or ornithologist to enjoy the stunning wild flowers and variety of birdlife. Spring is the best time to explore the countryside, when the colourful flowers are in bloom, especially on lush islands such as Corfu and on mountainous Crete. It is a good time to see the influx of migrating birds, which rest and feed in Greece on their journeys between Africa and Europe.

The **Hellenic Ornithological Society** details further

Tourists visiting caves near Psychró, in Crete

information on wild birds and their habitats, as well as related activities and events. **Limosa Holidays** is a specialist tour operator offering trips centred around bird-watching and botany. They have established tours to several islands, as has **The Travelling Naturalist. Honeyguide Wildlife Holidays** offers similarly themed tours on Crete. **Simpson Travel** also explores the wildlife of Crete on their specialist walking tours. The tours are adapted to the needs and abilities of the group but generally operate at a relaxed pace.

More information on the wildlife of Crete and other specialist tour operators is given on pages 250–51. Note that these types of holidays also incorporate into the tours visits to nearby historical and archaeological sites.

A chameleon, found mainly on Crete

Walking and Trekking

The hills of the Greek islands are a walker's paradise, particularly between March and June, when the countryside is verdant, the sun is not too hot and wild flowers abound. Many of the islands provide fine locations and scenery in which to walk, and the lack of too many organized trails gives a greater sense of freedom and discovery.

Trekking Hellas arranges walking holidays in the White Mountains of Crete, and on Andros and Tínos in the Cyclades. **Sherpa Expeditions** leads tours through the mountainous interior of western Crete, including the Samariá Gorge (see pp258–9), and **Ramblers Holidays** offers walking throughout the Greek islands, including some of the lesser-visited islands such as Nísyros and Ikaría.

Walkers climbing Mount Idi in central Crete

Pure Crete arranges walking and trekking tours in the coastal regions and inland to places such as the Samariá Gorge (see p258–9) on Crete, with a professional tour leader, while **Inntravel** features walking tours of Crete, Lésvos and Sámos. Walking tours of Crete to see the spring flowers are available through **Freelance Holidays** and from **Simpson Travel. Explore** organizes walks along the Corfu Trail, a walking holiday in Crete, and other trips including visiting several of the Aegean and Cyclades islands, while **Travelsphere** has walking in Crete and **Walks Worldwide** operate walking tours in Corfu. Tours are designed for leisure walkers, right through to those who have a high level of fitness and prefer more challenging terrains.

For the independent trekker, guides such as The Mountains of Greece: A Walker's Guide (Cicerone Press), and the various Sunflower Guides dealing with the Greek islands, are invaluable sources of information. If you are not one for the hardy mountain hike, there are plenty of less strenuous options too.

Trails in Greece tend not to be marked as well as in many other countries in Europe, with exceptions such as the excellent Corfu Trail. The Greek way is much simpler than signposts: a blob of red paint on rocks and walls indicates the path. Needless to say, these do not always work as well as intended.

On the positive side, many of the islands have locally published booklets or leaflets containing walks, which can be bought in shops, though some of them are available for free, supported by local walking groups and organizations.

Cruises and Boat Trips

Greece's unique combination of natural beauty and fascinating history makes a cruising holiday both relaxing and stimulating. Greek cruises run between April and October, and there are a variety of options available, ranging from a full luxury cruise to short boat trips.

Odyssey Sailing Greece provides information on a wide range of available options, from economy cabin cruises to fully crewed VIP motor yachts. Operators such as **Swan Hellenic Cruises, Travelsphere**

Daytrip boats in Mandráki Harbour, Rhodes

and **Voyages of Discovery** in the UK, **Metro Tours** and **Hellenic Holidays** in the US, offer all-inclusive holidays on board large luxury liners, with guest speakers versed on a range of subjects from archaeology to marine biology. Such cruises tend to incorporate the Greek islands into extensive routes from Italy to the Middle East, or to the Black Sea.

Explore runs week- and fortnight-long cruises aboard a traditional Greek caïque. A more informal option is to take a trip on one of the graceful tall ships operated by **Star Clippers**, which has various routes linking Athens with Venice or Istanbul, or through the Cycladic islands.

There are also less extensive boat trips to nearby islands and places of interest. Organized locally, these trips are best booked on the spot.

Cycling and Mopeds

Freewheeling cyclists can hire bikes at most holiday resorts, including the latest mountain bikes, but more organized options are available on Crete from **Simpson Travel** and on Kefalloniá with **Explore**. Even the smallest resorts will also have moped, scooter and perhaps motorbike rental agencies.

Mopeds are a cheap and easy way of getting about, but holiday-makers are advised to use them with caution, especially if you do not normally drive one when at home. In fact, some tour operators discourage their clients from renting them. Island roads can have many rough patches, with sudden potholes or patches of loose gravel, causing mopeds and scooters to skid and frequently come off the road. Greek car drivers also drive aggressively, some with little regard for vulnerable moped users. Accidents are so commonplace that anyone who rents a scooter or moped does so at their own risk.

Horse Riding

Those who prefer horse riding are also well catered for, with **Unicorn Trails** organizing trips to Kefalloniá, including the chance to swim with your horse in the sea, and to Crete, with its mountainous terrain. Riding in Corfu features in the programme of **Equitours**, based at the Vassilika Stables in the Rópa Plain to the south *(see p86)*.

Golf and Tennis

Tucked away on the Rópa Plain on Corfu is one of the best courses in Europe, where you can play as a guest if you happen to be there on holiday, or you can organize a special tour out there with golfing specialists such as **Bill Goff Golf Tours** and **3D Golf**. Golf courses on the islands are not widespread – apart from mini-golf and crazy golf! But **Golf Afandou** at Afántou on Rhodes has an 18-hole course, and there are two courses on Crete: the **Crete Golf Club** in Chersónisos and the Porto Elounda Golf Course located within the **Porto Elounda Resort**. For information on these and other golf courses on the mainland, contact the **Hellenic Golf Federation**.

Tennis players would be advised to book a holiday at one of the bigger hotels, many of which have their own tennis courts. Municipal courts and private clubs do exist but tend not to be as good. The Portomyrina Hotel on Límnos has three courts and two tennis coaches available, with special tennis holidays bookable through **Neilsen Active Holidays**.

Naturism

Nude sunbathing is only allowed in Greece on designated nudist beaches, but in practice people strip off on quiet beaches all over the islands. As long as the beaches are reasonably private and you do not offend local people, there is seldom a problem.

There is one licensed naturist hotel in the Greek islands: the

Vritomartis Hotel near Sfakiá on the south coast of Crete. It is a delightful hotel, and also welcomes non-naturist guests, as naturism is only practised around the swimming pool and at the beach, and not in any indoor areas.

Spas

Greece is well endowed with natural hot springs – a result of volcanic activity – and several islands have developed these as spas, offering such treatments as hydrotherapy, physiotherapy and hydromassage.

The main centres are listed on the EOT's (Greek Tourist Offices) information sheet *Spas in Greece*, and include Kos and Nísyros in the Dodecanese, Ikaría, Lésvos and Límnos in the Northeast Aegean group, Zákynthos in the Ionians and Kýthnos in the Cyclades.

Some of the large resort hotels also have excellent spa facilities, most notably around Eloúnta *(see p282)* on Crete and on upmarket islands such as Mýkonos and Santoríni.

Food and Wine

In medieval times Greece produced the best wine in Europe, and after a long lull, when a lot of Greek wine was barely drinkable, today's wine-makers have rediscovered their skills. A cruise which visits several of the country's leading vineyards can be booked through UK wine tour specialists, **Arblaster and Clarke**.

There is a growing interest in Greek cuisine too, and cookery holidays on the island of Sými are available with **Pure Crete**. Also offering holidays that combine the culture, food and wine of Crete are **Simpson Travel**. Their specialized tours take you into the homes of ordinary Cretans, where you can savour traditional home-cooked food and local wines. The tours are organized thematically and look at local activities, such as organic olive farming and wine-making.

DIRECTORY

Archaeological Tours

Andante Travels
The Clock Tower, Unit 4 Oakridge Office Park, Southampton Road, Whaddon, SP5 3HT, UK.
Tel 01722 713800.
W andantetravels.co.uk

Martin Randall Travel
Voysey House, Barley Mow Passage, London W4 4GF, UK. Tel 020 8742 3355.
W martinrandall.com

Creative Holidays

Aegean Scapes
Larnakos 35, Papagou 155669, Athens.
Tel 21064 10972.
W aegeanscapes.com

Simpson Travel
Boat Race House, 61–67 Mortlake High Street, London, SW14 8HL, UK.
Tel 0845 508 5175.
W simpsontravel.com

Travel à la Carte
Unit 4, 36 Queens Road, Newbury, Berkshire RG14 7NE, UK.
Tel 020 7286 9255.
W travelalacarte.co.uk

Greek Language Courses

Omilo
Pan. Tsaldari 13, 15122 Maroussi, Athens.
Tel 21061 22896.
W omilo.com

Nature Holidays

Hellenic Ornithological Society
Themistokleous 80, 10681 Athens.
Tel 21082 27937.
W ornithologiki.gr

Honeyguide Wildlife Holidays
36 Thunder Lane, Thorpe St Andrew, Norwich NR7 0PX, UK. Tel 01603 300552.
W honeyguide.co.uk

Limosa Holidays
West End Farmhouse, Chapelfield, Stalham, Norfolk, NR12 9EJ, UK.
Tel 01263 578143.
W limosaholidays.co.uk

Simpson Travel
(See Creative Holidays.)

The Travelling Naturalist
PO Box 3141, Dorchester, Dorset, DT1 2XD, UK.
Tel 01305 267994.
W naturalist.co.uk

Walking and Trekking

Explore
55 Victoria Road, Farnborough, Hants, GU14 7PA, UK.
Tel 01252 884223.
W explore.co.uk

Freelance Holidays
Falstaff House, Birmingham Road, Stratford Upon Avon CV37 0AA, UK. Tel 01789 297705. W freelance-holidays.co.uk

Inntravel
Whitwell Grange, nr Castle Howard, York, YO60 7JU, UK. Tel 01653 617001.
W inntravel.co.uk

Pure Crete
Bolney Place, Cowfold Road, Haywards Heath, W. Sussex RH17 5QT, UK.
Tel 01444 880404.
W purecrete.com

Ramblers Holidays
Lemsford Mill, Lemsford Village, Welwyn Garden City, Herts, AL8 7TR, UK.
Tel 01707 331133.
W ramblersholidays. co.uk

Sherpa Expeditions
81 Craven Gardens, London SW19 8LU, UK. Tel 020 8577 2717. W sherpa expeditions.com

Travelsphere
Compass House, Rockingham Road, Market Harborough, Leicestershire, LE16 7QD, UK. Tel 0844 567 9961.
W travelsphere.co.uk

Trekking Hellas
Saripolou 10, 10682 Athens. Tel 210 331 0323.
W trekking.gr

Walks Worldwide
Long Barn South, Sutton Manor Farm, Bishop's Sutton, Alresford

SO24 0AA, UK. Tel 01962 737565. W walksworld wide.com

Cruises and Boat Trips

Explore
(See Walking & Trekking.)

Hellenic Holidays
1501 Broadway, Suite 2004, New York, NY 10036, USA.
Tel 212 944 8288.
W hellenicholidays.com

Metro Tours
484 Lowell St, Peabody, MA 01960, USA. Tel 800 221 2810. W metrotours.com

Odyssey Sailing Greece
Antonopoulo 158D, 38221 Volos, Greece.
Tel 24210 36676.
W odysseysailing.gr

Star Clippers
Olympus House, 2 Olympus Close, Ipswich IP1 5LN, UK. Tel 0845 200 6145.
W starclippers.co.uk

Swan Hellenic Cruises
Compass House, Rockingham Road, Market Harborough, Leics LE16 7QD, UK.
Tel 0844 871 4603.
W swanhellenic.com

Travelsphere
(See Walking & Trekking.)

Voyages of Discovery
Compass House, Rockingham Road, Market Harborough, Leics LE16 7QD, UK. Tel 0844 822 0802. W voyagesof discovery.com

Cycling and Mopeds

Explore
(See Walking & Trekking.)

Simpson Travel
(See Creative Holidays.)

Horse Riding

Equitours
10 Stalnaker Street, Dubois, Wyoming 82513, USA. Tel 800 545 0019.
W equitours.com

Unicorn Trails
7 Baystrait House, Station Road, Biggleswade, Beds, SG18 8AL, UK.
Tel 01767 600606.
W unicorntrails.com

Golf and Tennis

3D Golf
Clerks Court, 18–20 Farringdon Lane, London EC1R 3AU, UK. Tel 020 7336 5349. W 3dgolf.com

Bill Goff Golf Tours
Clerks Court, 18–20 Farringdon Lane, London EC1R 3AU, UK.
Tel 0800 193 6624.
W billgoff.com

Crete Golf Club
PO Box 106, 70014 Hersonissos, Crete.
Tel 28970 26000.
W crete-golf.com

Golf Afandou
Afántou Bay, Rhodes.
Tel 22410 51451.
W afandougolf course.com

Hellenic Golf Federation
PO Box 70003, Glyfada, 16610 Athens. Tel 21089 45727. W hgf.gr

Neilson Active Holidays
Locksview, Brighton Marina, Brighton BN2 5HA. Tel 0333 014 3351.
W neilson.co.uk

Porto Elounda Resort
Elounda, Crete 72053.
Tel 28410 68000.
W portoelounda.com

Naturism

Vritomartis Hotel
Chora Sfakion, Crete.
Tel 282 509 1112.
W vritomartis.gr

Food and Wine

Arblaster and Clarke
Cedar Court, 5 College St, Petersfield, Hants, GU31 4AE, UK.
Tel 01730 263111.
W winetours.co.uk

Simpson Travel
(See Creative Holidays.)

BEACHES AND WATERSPORTS

With hundreds of islands, crystal-clear seas and beaches of every kind, it is not surprising that so many water-lovers are attracted to Greece. Although people swim most of the year round, the main season for watersports is from late May to early November. All kinds of watersports can be enjoyed, especially in the larger and more developed resorts, and rental fees are still quite reasonable compared with other Mediterranean destinations. But if you prefer a more leisurely vacation, you can always choose from the many beautiful and tranquil beaches to be found on the islands.

Holiday company flags flying on "Golden Beach", Páros

Beaches

Beaches vary greatly in the Greek islands, offering everything from shingle and volcanic rock to gravel and fine sand. The Cyclades and Ionian Islands are where the sandy beaches tend to be, and of these the best are usually on the south of the islands. Crete's beaches are also mostly sandy, but not exclusively. The Northeast Aegean, Dodecanese and Sporades are a mixture of sandy and pebbly beaches. Some islets, such as Chálki and Kastellórizo, have few or no beaches at all. But, in compensation, they often have very clear seas, which can be good for snorkelling.

Any beach with a Blue Flag (awarded annually by the Hellenic Society for the Protection of Nature, in conjunction with the European Union) is guaranteed to have its water tested every 15 days for cleanliness and purity, as well as meeting over a dozen other environmental criteria. These beaches tend to be among the best, and safest for children, though they can be very crowded.

Also worth trying out are beaches recommended in the headings for each entry in this guide. Occasionally the main beach near the port of an island is run by the EOT (Greek Tourist Office). There will be a charge for its use, but it will be kept clean and often have the added benefit of showers. Topless bathing is widespread, though nude bathing is still officially forbidden, except on a few designated beaches; it is never allowed within sight of a church.

The Greek seas are generally safe and delightful to swim in, though lifeguards are almost non-existent in Greece. Every year there are at least a few casualties, especially on windy days when the sea is rough and there are underwater currents. Sharks and stingrays are rare around beaches, but more common are sea urchins and jellyfish. Both can be painful, but are not particularly dangerous.

Watersports

With so much coastline, facilities catering for watersports are numerous. Windsurfing has become very popular, and waters recommended for this include those around Corfu, Lefkáda and Zákynthos in the Ionian islands, Lésvos and Sámos in the Northeast Aegean, Kos in the Dodecanese, Náxos in the Cyclades and the coast around Crete. The **Hellenic Water-ski Federation** can offer the best advice. For a little more money you could take up water-skiing or jet-skiing; and at the larger resorts parasailing is also available. If you need instruction, you will find that many of the places that rent equipment also provide tuition.

Swimmers diving off the boards at a pool by the beach on Rhodes

Holiday-makers learning the skills of windsurfing in coastal waters

Hire centre for watersports equipment, Rhodes

Scuba and Snorkelling

The amazingly clear waters of the Mediterranean and Aegean reveal a world of submarine life and archaeological remains. Snorkelling *(see pp28–9)* can be enjoyed almost anywhere along the coasts, though scuba diving is severely restricted. Designated areas for diving are around Crete, Rhodes, Kálymnos and Mýkonos, and also around most of the Ionian Islands. A complete list of places where it is permissible to dive with oxygen equipment can be obtained from the EOT, or by mail from the **Department of Underwater Archaeology** in Athens. Wherever you go snorkelling or diving, it is strictly forbidden to remove any antiquities you see, or even to photograph them.

Sailing Holidays

Sailing vacations can be booked through yacht charter companies in Greece or abroad. The season runs from April to the end of October or early November, and itineraries are flexible.

Charters fall into four main categories. Bareboat charter is without a skipper or crew and is available to those with previous sailing experience (contact the **Greek Professional and Bareboat Yacht Owners' Association**).

Crewed charters range from the modest services of a skipper, assistant or cook to a yacht with a full crew. Sailing within a flotilla, typically in a group of around 6 to 12 yachts, provides the opportunity of independent sailing with the support of a lead boat, contactable by radio. **Thomas Cook** and **Sunsail** both offer sailing holidays in a flotilla. They also offer the popular "combined vacation". This type of vacation mixes cruiser sailing with the added interest of coastal pursuits, such as shore-based dinghy sailing and windsurfing.

Learning the techniques of sailing

Sailing aboard a yacht in the Greek seas

DIRECTORY

Useful Organizations

Department of Underwater Archaeology
Areopagitou 59 & Erehthiou, 11742 Athens.
Tel 21092 35105.

Greek Professional and Bareboat Yacht Owners' Association
Alimos Marina, Alimos, 17455 Athens.
Tel 21098 41531.
Ⓦ sitesap.gr

Greek Yacht Brokers' and Consultants' Association
Marina Zeas, 18536 Piraeus.
Tel 21045 33134.

Hellenic Sailing Federation
Marina Dimou Kallitheas, 17602 Athens.
Tel 21094 04825.

Hellenic Water-ski Federation
Alexandrias 19, 16342 Athens. **Tel** 21099 44014.
Ⓦ waterski.gr

Yacht Charter Companies

Sunsail
Port House, Marina Keep, Portsmouth PO6 4TH, England. **Tel** 02392 222222.
Ⓦ sunsail.com

Tenrag Yacht Charters
Tenrag House, Freepost CU986, Preston, Canterbury, Kent CT3 1EB, England.
Tel 01227 721874.
Ⓦ tenrag.com

Thomas Cook Holidays
Tel 01733 224800.

Variety Cruises
Syngrou 214–216, 17672 Athens.
Tel 21069 19191.
Ⓦ varietycruises.com

SURVIVAL
GUIDE

PRACTICAL INFORMATION

Greece's appeal is both cultural and hedonistic. The islands' physical beauty, hot climate and warm seas, together with the easy-going outlook of the Greek people, are all conducive to a relaxing holiday. It does pay, however, to know something about the nuts and bolts of Greek life – when to visit, what to bring, how to get around and what to do if things go wrong – to avoid unnecessary frustrations. Greece is no longer the cheap holiday destination it once was, though public transport, vehicle hire, eating out and hotel accommodation are still fairly inexpensive compared to most Western European countries. The many tourist offices (see p347) offer information on all the practical aspects of your stay.

Boats sailing in a bay in high summer

When to Go

High season – from late June to early September – is the hottest and most expensive time to visit the Greek islands, as well as being very crowded. December to March are the coldest and wettest months, with reduced public transport and many hotels and restaurants closed throughout the winter.

Spring (from late April to May) is one of the loveliest times to visit the islands – the weather is sunny but not debilitatingly hot, there are relatively few tourists about, and the countryside is ablaze with brightly coloured wild flowers.

Visas and Passports

Visitors from EU countries need a valid passport or ID card to enter Greece, but do not need a visa. UK citizens need a passport. Some non-EU citizens such as those from the US, Canada, Australia and New Zealand, do not need a visa, but do need a valid passport for a stay of up to 90 days (there is no maximum stay for EU visitors). For longer stays, a resident's permit must be obtained from the **Aliens' Bureau** in Athens or the local police in the islands' main towns. Visitors should check visa requirements with their local Greek embassy before travelling.

Any non-EU citizen planning to work or study in Greece should contact their local Greek consulate about visas and work permits.

Customs Information

EU residents can import unlimited alcohol, perfumes and tobacco so long as they are for personal use. Visitors entering Greece from non-EU countries should check the following website for details of quantities that they can import free of charge: http://greece.visahq.com/customs.

Note that the unauthorized export of antiquities and archaeological artifacts from Greece is a serious offence, with strong penalties ranging from hefty fines to prison sentences. Any prescription drugs that are brought into the country should be accompanied by a copy of the prescription (see pp348–9).

Tourist Information

Tourist information is available in many towns and villages in the form of government-run EOT offices (Ellinikós Organismós Tourismoú, also often referred to as the **Greek National Tourism Organization, GNTO**), municipal tourist offices, the local tourist police (see p348) or travel agencies. Many of these offices operate only in summer. The GNTO publishes an array of tourist literature and brochures, but be aware that not all of this information is always up to date or reliable.

The addresses and phone numbers of the GNTO and municipal tourist offices are listed throughout this guide. A list of major Greek festivals and cultural events is given on pages 50–54, but it is also worth asking your nearest tourist office about what is happening locally.

Tourists exploring medieval Rhodes old town in summer

◄ Cyclists at the old harbour of Kos, on the island of the same name

The Minoan Palace of Knosós, Crete, a major archaeological site

Admission Prices

Most state-run museums and archaeological sites charge an entrance fee of €3–€12. However, visitors aged 18 or under from EU countries are entitled to free admission, as are EU travellers carrying an **International Student Identity Card** (ISIC) *(see pp346–7)*. Reductions of around 25 per cent are granted to EU citizens aged 65 and over (use your passport as proof of age), and reductions of 50 per cent to non-EU students armed with an ISIC card.

Though most museums and sites are closed on public holidays *(see p54)*, the ones that do stay open are free of charge. Admission to all state-run museums and archaeological sites is free on Sundays between November and April.

Opening Hours

Opening hours tend to be vague in Greece, varying from day to day, season to season and place to place. In addition to this, the financial crisis that the country has been experiencing since 2010 is having a significant impact on many attractions, causing staff budget cuts and reduced opening hours. Although the opening times in this book have been checked at the time of going to print, they are likely to keep changing. It is advisable to use the times in this book as a rough guideline only and to check with local information centres before visiting a sight.

Most attractions usually close on Mondays and the main public holidays *(see p54)*. Small and private museums may additionally be closed on local festival days.

Monasteries and convents are open during daylight hours, but will close for a few hours in the afternoon.

Most shops and offices are also closed on public holidays and local festival days, with the exception of some shops in tourist resorts. The dates of major local festivals are included in the Visitors' Check-lists in each main town entry in this guide.

Social Customs and Etiquette

For a carefree holiday in Greece, it is best to adopt the local philosophy: *sigá, sigá* ("slowly, slowly"). Within this principle is the ritual of the afternoon siesta, a practice that should be taken seriously, particularly during the hottest months, when it is almost a physiological necessity.

Like anywhere else, common courtesy and respect are appreciated in Greece, so try speaking a few words of the language, even if your vocabulary covers only the basics *(see pp396–400)*.

Though formal attire is rarely needed, modest clothing (trousers for men and skirts for women) is de rigueur for churches and monasteries. Topless sunbathing is generally tolerated, but nude bathing is officially restricted only to a few designated beaches.

In restaurants, the service charge is always included in the bill, but tips are still appreciated – the custom is to leave around 10 per cent if you were satisfied with the service. Public toilet attendants should also be tipped. Taxi drivers do not expect tips, but they are not averse to them either; likewise, hotel porters and chambermaids.

In 2010, Greece introduced a law officially banning smoking in enclosed public spaces, including in restaurants, bars and cafés. Around 40 per cent of Greeks smoke, and many ignore this law. Restaurant and café owners prefer to turn a blind eye to this, for fear of losing custom. Many bar staff smoke, too. All the same, visitors should avoid smoking in enclosed public spaces; smoking in outdoor areas such as café terraces is permitted.

Greek police will not tolerate rowdy or indecent behaviour, especially when fuelled by excessive alcohol consumption; Greek courts impose heavy fines or even prison sentences on people who behave indecently.

Religion

About 97 per cent of the population is Greek Orthodox. The symbols and rituals of the religion are deeply rooted in Greek culture, and they are visible everywhere. Saints' days are celebrated throughout Greece *(see p54)*, sometimes on a local scale and sometimes across the entire country. Greek Orthodox monasteries and churches, many dating back centuries, are among the country's top cultural attractions. Visitors to these sacred places should dress respectably (shoulders and legs covered for both men and women) and refrain from taking photographs (this is officially forbidden, though rules do vary).

A Greek priest

The largest religious minority are the Muslims of Thrace, who constitute only about 1.2 per cent of the country's total population. Many immigrants from Muslim countries such as Bangladesh, Pakistan, Somalia and Afghanistan, as well as Albania, live in Athens; small communities are also being formed on some islands close to the mainland. There are also sizeable communities of Roman Catholics, including ethnic Greeks and immigrants from Poland and the Philippines, who live mainly in Athens, the Cyclades and, to a lesser extent, the Dodecanese.

Sign for disabled parking

Travellers with Special Needs

There are few facilities for the disabled in Greece, so careful advance planning is essential; sights that have wheelchair access will have a wheelchair symbol at the start of their entry in this guide. Organizations such as **Disability Rights UK** and **Tourism for All** are invaluable sources of information. Specialized travel agencies including **Accessible Travel and Leisure** and **Responsible Travel** will arrange holidays for disabled travellers.

Note that disabled visitors (with a person assisting them) are entitled to free entry to state-run museums and archaeological sites.

Travelling with Children

Children are much loved by the Greeks and welcomed just about everywhere, including restaurants, where waiters will be happy to suggest special dishes for them. Babysitting facilities are provided by some hotels on request, but check before booking. Some coastal resorts have special amenities such as playgrounds, children's pools and even kids' clubs with organized activities.

Those aged 18 or under from EU countries enjoy free admission to state-run museums and archaeological sites, as do children aged 5 and under from non-EU countries. Concessions of up to 50 per cent are offered on most forms of public transport for children aged 10 and under (in some cases, 8 and under).

Swimming in the sea is generally safe for kids, but keep a close eye on them, as lifeguards are rare in Greece. Choose sandy beaches in sheltered bays with shallow water. Be aware of the hazards of overexposure to the sun and dehydration *(see p349)*.

Student and Youth Travellers

Concessions are offered on train, metro and bus travel in Greece to students below the age of 25 with a valid **International Student Identity Card (ISIC)**. They may also need to show their passport. There are plenty of deals to be had getting to Greece, especially during the low season. Agencies for student and youth travel include **STA Travel**. Before setting off, it is worth joining **Hostelling International** to enjoy discounts in Greek hostels.
Most state-run museums and archaeological sites are free to EU students with a valid ISIC card; non-EU students with an

An International Student Identity Card

ISIC card are usually entitled to a 50 per cent reduction. There are no youth concessions available for these entrance fees, but occasional discounts for museums and archaeological sites are possible with an International Youth Travel Card (IYTC), which can be obtained from any STA office by travellers under the age of 26.

Women Travellers

Greece is by and large a very safe country, and local communities are generally welcoming. Foreign women travelling alone are usually treated with respect, especially if they are dressed modestly. Although local men openly display their interest in women, making it clear that you are not interested in them is

usually enough to curtail any flirtation. Like elsewhere, hitch-hiking alone in Greece carries risks and is not advisable.

Time

Greece is 2 hours ahead of Britain, 1 hour ahead of countries on Central European Time (such as France and Italy), 7 hours ahead of New York, 10 hours ahead of Los Angeles and 8 hours behind Sydney.

Greece puts the clock forward to summertime, and back again to wintertime, on the same days as other EU countries, in order to avoid any confusion when travelling around Europe.

Electricity

Greece, like other European countries, runs on 220 volts/50 Hz AC. Plugs have two round pins (those for appliances that need to be earthed have three). The adaptors needed for British electrical appliances and the transformers for North American equipment are difficult to find in Greece, so bring one with you.

Responsible Tourism

Greece is lagging behind most EU countries in environmental awareness – recycling is scarcely practised, illegal dumping in rural areas is the norm, and waste management is a major problem. However, there is much interest in renewable energy sources.

Children enjoying the shallow waters at a sandy beach

Greece has great potential for developing solar energy – many families already have solar panels for heating water (they can sell the surplus to the National Grid), and there have been talks about producing solar power on a far larger scale. Wind energy is already used to some extent on the islands, but here too there is potential for further exploitation.

Agrotourism (working farms that offer accommodation and meals to visitors) has become well established on Crete, and is becoming increasingly so on the Ionian islands, especially Corfu, but is slow to develop on the other islands. Contact the **Hellenic Agrotourism Federation** or **Guest Inn** for a list of agro-tourism establishments.

Visitors can support local communities by shopping for local produce at the local markets held in all major towns throughout the islands (see p334). In some areas, you can buy local specialities directly from the producers – for example, Omalós honey in Irákleio, Crete and Kumquat liqueur in Corfu town – and visit vineyards for wine tastings and direct purchases.

Ethical tour operators include the UK-based **Responsible Travel** and the Athens-based **Trekking Hellas**; both run adventure sports packages including activities such as hiking, mountain biking, sea kayaking and rafting. The Athens-based **Ecotourism Greece** is a useful source of ideas for rural destinations, activities and small family-run hotels.

DIRECTORY

Visas and Passports

Aliens' Bureau
Petrou Ralli 24,
Tavros, 17778 Athens.
Tel 21034 05828.

Embassies

Australia
Level 6, Thon Building,
Kifissias & Alexandrias
Avenue, Ambelokipi,
11510 Athens.
Tel 21087 04000.

Canada
Ioannou Gennadíou 4,
11521 Athens.
Tel 21072 73400.

Ireland
Vassiléos Konstantínou 7,
10674 Athens.
Tel 21072 32771.

New Zealand (Consulate)
Kifissias 76, Ambelokipi,
11526 Athens.
Tel 21069 24136.

United Kingdom
Ploutárchou 1, 10675
Athens.
Tel 21072 72600.
Crete: Papalexándrou 16,
Irákleio, Crete.
Tel 28102 24012.
Corfu: Mantzaróu
18, Corfu.
Tel 26610 30055.

United States
Vasilíssis Sofías 91,
10160 Athens.
Tel 21072 12951.

Tourist Information Offices

Greek National Tourism Organization (GNTO)
Head office: Tsoha 7,
11521 Athens.
Tel 21087 07000.
W visitgreece.gr
Information centre:
Dionysiou Areopagitou
18–20, 11742 Athens.
Tel 21033 10529.

GNTO Chaniá
Kriari street 40, Pantheon
Mansion, 73100 Chaniá.
Tel 28210 92943.

GNTO Chíos
Kanari 18, Chíos.
Tel 22710 24442.

GNTO Corfu
Evangelistrías 4, Corfu.
Tel 26610 37520.

GNTO Crete
Xanthoudidou 1, 71202
Irákleio.
Tel 28102 28225.

GNTO Kos
Artemisias 2, Kos.
Tel 22420 29910.

GNTO Lefkáda
Lefkáda Marina,
Lefkáda.
Tel 26450 25292.

GNTO Réthymno
Eleftheriou Venizelou
Paralia Rethymnou, 74100
Réthymno.
Tel 28310 29148.

GNTO Rhodes
Papagou and
Archiepiskopou Makariou,
85100 Rhodes.
Tel 22410 21291.

GNTO Santoríni
Fira, Santoríni.
Tel 22860 27199.

GNTO Sýros
Thimaton Sperheióu 11,
Sýros.
Tel 22810 86725.

Travellers with Special Needs

Accessible Travel and Leisure
Tel 01452 729739.
W accessibletravel.
co.uk

Disability Rights UK
Tel 020 7250 8181.
W radar.org.uk

Responsible Travel
Tel 01273 823700.
W responsibletravel.
com

Tourism for All
Tel 01539 726111.
W tourismforall.org.uk

Student Travellers

Hostelling International
2nd Floor, Gate House,
Fretherne Road, Welwyn
Garden City, Herts AL8
6RD, UK.
Tel 01707 324170.
W hihostels.com

International Student Identity Card (ISIC)
W isic.org

STA Travel
52 Grosvenor Gardens,
London SW1W 0AG, UK.
Tel 0871 702 9849.
W statravel.co.uk

Responsible Tourism

Ecotourism Greece
Tel 21171 00050.
W ecotourism-greece.
com

Guest Inn
Tel 21096 07100.
W guestinn.com

Hellenic Agrotourism Federation (SEAGE)
Tel 69365 00670.
W agroxenia.net

Responsible Travel
(See Travellers with
Special Needs.)

Trekking Hellas
Saripolou 10, 10682
Athens.
Tel 21033 10323.
W trekking.gr

Personal Health and Security

Strikes and protest marches have always been regular features of Greek life. However, the rise in unemployment and in the cost of living caused by the ongoing economic crisis have led to higher levels of public unrest. Despite this, Greece remains a safe country to visit, although it is best to avoid protest marches and demonstrations, which can turn violent, especially in Athens. The biggest danger is the road: Greece has one of the highest accident rates in Europe. Considerable caution is recommended, for both drivers and pedestrians.

Police

Regular Greek police officers wear blue uniforms and keep a relatively low profile. However, there are several special units, the most conspicuous being the riot police (MAT), who wear a khaki military-type uniform and a helmet with a visor. The MAT are usually only seen at unruly demonstrations.

In addition, there are the tourist police, who combine normal police duties with dispensing advice to tourists. Tourist police wear a cap with a white band, a white belt and white gloves; as well as a badge saying "Tourist Police" on their shirt. If you suffer a theft, lose your passport or have cause to complain about restaurants, shops, taxi drivers or tour guides, your case should first be made to them. Every tourist police officer speaks several languages and each office claims to have at least one English speaker, so they can also act as interpreters if the case needs to involve the local police.

What to be Aware of

Most crime-related problems centre on major tourist destinations, such as Mýkonos, Rhodes and Corfu, and are usually relatively minor. Crime levels outside large towns are low. Visitors are advised to avoid public demonstrations. These most frequently occur in Athens on Sýntagma Square, in front of the Greek parliament and have become increasingly violent due to widespread public discontent.

Although still fairly rare, street muggings and burglaries have become more common than they once were. Visitors should take sensible precautions, such as keeping an eye on their bags in public, especially in crowded places, and keeping important documents and valuables in the hotel safe. If you do have anything stolen, contact the police or tourist police.

In an Emergency

In case of emergencies, the appropriate services to call are listed in the Directory on the opposite page. For accidents and other medical emergencies, a 24-hour ambulance service operates within major towns. In rural towns it is unlikely that ambulances will be on 24-hour call. Phone 166 for a medical emergency. If necessary, patients can be transferred from local ESY (Greek National Health Service) hospitals or surgeries to a main ESY hospital in Athens by ambulance or helicopter.

A complete list of ESY hospitals, private hospitals and clinics is available from the tourist police, so either call the tourist police number or visit the nearest office.

Hospitals and Pharmacies

Emergency medical care in Greece is free for all EU citizens in possession of a European Health Insurance Card (EHIC), available from main post offices. There are small hospitals in the main towns on the islands, but if you have a serious medical condition or an emergency, you will probably need to be taken by plane to one of the major hospitals in Athens.

Public hospitals are often understaffed, and it is not unusual for relatives to help feed and provide basic nursing care for patients. Corruption is rife within the Greek healthcare system, and it is considered perfectly normal to offer doctors under-the-table payments for priority treatment.

SOS Doctors is a service that provides free basic medical consultation by telephone or, should a condition require a diagnosis, will arrange for an immediate home visit. A fee is chargeable for a visit and for any prescription. The doctor will provide a receipt for claiming

Greek police officers wearing typical blue uniforms

Ambulance

Police car

Minor Hazards

The most obvious thing to avoid is over-exposure to the sun, particularly for the fair-skinned: always wear a hat and good-quality sunglasses, as well as a high-factor suntan lotion. Heatstroke is a real hazard for which medical attention should be sought immediately; heat exhaustion and dehydration are also serious.

Be sure to drink plenty of water, even if you don't feel thirsty; if in any doubt, invest in a packet of electrolyte tablets (a mixture of potassium salts and glucose) to replace lost minerals. These are available at any Greek pharmacy.

Tap water in Greece is generally safe to drink, but in remote communities it is a good precaution to check with the locals. Bottled spring water, for sale in shops and kiosks, is reasonably priced and often has the advantage of being chilled.

When swimming in the sea, hazards to be aware of are weaver fish, jellyfish and sea urchins. The latter are not uncommon and are extremely unpleasant if trodden on. If you do tread on one, the spine will need to be extracted using olive oil and a sterilized needle. Jellyfish stings can be relieved by applying vinegar, baking soda or various remedies sold at Greek pharmacies to the affected area. The sand-dwelling weaver fish has a powerful sting, its poison causing extreme pain. The immediate treatment is to immerse the affected area in very hot water to dilute the venom's strength.

No inoculations are required for visitors to Greece, though tetanus and typhoid boosters may be recommended by your doctor.

money back from your insurance. The service operates from Athens and has a data-base of doctors based on the islands. Telephone 1016 to speak directly with a doctor.

Greek pharmacists are highly qualified and can not only advise on minor ailments, but also dispense medication not usually available over the counter back home. Their premises, *farmakeía*, are identified by a green cross on a white background. Pharmacies are open from 8:30am to 2pm Monday to Friday, and they are usually closed at weekends. However, in larger towns, a rota system is usually in place to maintain a daily service from morning to night. Details of on-duty pharmacies are posted in pharmacy windows.

Pharmacy sign

Be sure to bring an adequate supply of any medication you may need while away, as well as a copy of the prescription with the generic name of the drug – this is useful not only in case you run out, but also for the purposes of customs when you enter the country.

Several international pharmaceutical companies have stopped selling to Greece due to delayed payments, so some drugs are now in short supply. Visitors should also be aware that codeine, a painkiller commonly found in headache tablets, is illegal in Greece.

Travel and Health Insurance

EU citizens should carry a European Health Insurance Card (EHIC) to receive free emergency medical care. Private medical insurance is needed for all other types of treatment. Visitors are strongly advised to take out comprehensive travel insurance – available from travel agents, banks and insurance brokers – covering both private medical treatment and loss or theft of personal possessions. Be sure to read the small print – not all standard policies, for instance, will cover you for activities of a "dangerous" nature, such as motorcycling and trekking, although some companies will cover such activities for an additional fee when you take out the policy. Not all policies will pay for doctors' or hospital fees direct, and only some will cover you for ambulances and emergency flights home. Paying for your flight with a credit card such as VISA or American Express will provide limited travel insurance, including reimbursement of your air fare if the agent happens to go bankrupt.

DIRECTORY

Emergency Numbers

Ambulance
Tel 166.

Coastguard patrol
Tel 108.

Emergencies
Tel 112.
🆆 sos112.info

Fire
Tel 199.

Police
Tel 100.

Road assistance
Tel 10400.

SOS Doctors
Tel 1016.
🆆 sosiatroi.gr

Tourist police
Tel 1571.

Banking and Currency

Greece replaced the drachma, a currency it had used for around ten centuries, in favour of the euro in 2002. Hit by the economic crisis of 2010, the country procured massive loans, but it soon emerged that it might be unable to repay them and therefore be forced to declare bankruptcy. In 2011, Greece's future within the Eurozone began to look uncertain, but by 2015 the economy had shown signs of stabilizing thanks to yet more loans, austerity measures and the sale of government bonds. Prices have risen, but the government introduced tax exemptions in the tourism sector in a bid to encourage holiday-makers.

Visitors changing money at a bureau de change

Banks and Bureaux de Change

The islands' larger towns, as well as tourist resorts, have the usual banking facilities, including 24-hour cash machines (ATMs). Alternatively, you can change foreign currency and travellers' cheques into euros at a bureau de change found in larger towns or, in more remote areas, at the local post office. Some travel agents, hotels, tourist offices and car-hire agencies can also change foreign currency. Always take your passport with you when cashing travellers' cheques, and check exchange rates and commission charges beforehand, since these vary greatly.

The main banks are Ethniki Trapeza tis Ellados (National Bank of Greece), Alpha Bank, ATE Bank, Piraeus Bank, Emporiki Bank and Eurobank. Banks are open from 8am to 2pm Monday to Thursday, and 8am to 1:30pm on Friday. They are closed on public holidays (see p54) and may also close on local festival days.

ATMs

Easily found in all major towns and resorts, ATMs can be used to withdraw cash using internationally recognized credit and debit cards. There has been a rise in ATM crime the world over, so exercise caution when using one, and always shield your PIN from passers-by.

Credit and Debit Cards

VISA, **MasterCard**, **American Express** and **Diners Club** are the most widely accepted credit cards in Greece. A credit card is the most convenient way to pay for air tickets, international ferry journeys, car hire, some hotels and large purchases. However, some small tavernas, shops and hotels do not take credit cards, so be sure to have cash with you when visiting these establishments.

A credit card can be used for drawing local currency at a cash machine. A 1.5 per cent processing charge is usually

levied for VISA at banks and ATMs, but this does not apply to other cards. Cirrus and Plus debit card systems operate in Greece. Cash can be obtained using Cirrus at National Bank of Greece ATMs, and Plus at Commercial Bank ATMs. Be sure to tell your bank that you are travelling to Greece, so that your card is not blocked while you are away.

Travellers' cheques are the safest way to carry large sums of money. They are refundable if lost or stolen, though the process can be time-consuming. American Express and VISA are the best-known brands of travellers' cheques in Greece.

DIRECTORY

Lost Credit Cards

American Express
Tel 00 44 1273 696 933.

Diners Club
Tel 00 44 1244 470 910.

MasterCard
Tel 0800 964 767.

VISA
Tel 00 800 891 725.

Lost Travellers' Cheques

American Express
Tel 00 800 44 127569 (toll-free).

VISA
Tel 00 800 44 131410 (toll-free).

Queueing up to use an ATM

The Euro

The euro (€) is the common currency of the European Union. It went into general circulation on 1 January 2002, initially for 12 participating countries, including Greece. The Greek drachma was phased out in March 2002. EU members using the euro as sole official currency are known as the Eurozone. Several EU members have opted out of joining this common currency. Euro notes are identical throughout the Eurozone countries, each one including designs of fictional architectural structures and monuments. The coins have one side identical (the value side), and one side with an image unique to each country. Both notes and coins are exchangeable in each of the Eurozone countries.

Bank notes

Euro bank notes have seven denominations. The €5 note (grey in colour) is the smallest, followed by the €10 note (pink), €20 note (blue), €50 note (orange), €100 note (green), €200 note (yellow) and €500 note (purple). All notes show the stars of the European Union.

5 euros

10 euros

20 euros

50 euros

100 euros

200 euros

500 euros

2 euros

1 euro

50 cents

20 cents

10 cents

Coins

The euro has eight coin denominations: €2 and €1; 50 cents, 20 cents, 10 cents, 5 cents, 2 cents and 1 cent. The €2 and €1 coins are both silver and gold in colour. The 50-, 20- and 10-cent coins are gold. The 5-, 2- and 1-cent coins are bronze.

5 cents

2 cents

1 cent

Communications and Media

The Greek national telephone company is OTE (Organismós Tilepikoinonión Elládos). Telecommunications in Greece and its islands are generally good, and there are direct lines to all major countries, which are often better than local lines. Mobile phone coverage is widespread. Internet access is available in most hotels and resorts, and at a growing number of cafés. Greek post is reasonably reliable and efficient, especially from the larger towns and resorts. The Greeks are avid newspaper readers, and in addition to a vast array of Greek publications there are a few good English-language papers and magazines. Foreign newspapers are also available.

International Telephone Calls

Public telephones have become increasingly rare on the streets of Greece as more and more people now have mobile (cell) phones. However, they still tend to be available in hotel foyers and at local OTE offices. Making long-distance calls from a hotel room can be very expensive.

The best deals on long-distance calls are to be found at privately run call centres, which have sprung up in all the larger cities (often close to the train or bus station) to serve Greece's immigrant communities. Each call centre displays specific rates, as well as information about peak and cheap times, which vary depending on the country you are phoning.

Mobile Phones

The main mobile phone network providers in mainland Greece and on the islands are **Cosmote**, **Vodafone Greece** and **WIND Hellas**. To reduce the cost of calls made while in Greece, it might be a good idea to purchase a Greek SIM card from one of these companies; however, this will work only if your phone has been unlocked. Alternatively, you can get excellent coverage by using your network's roaming facility, but this can work out expensive. All Greek mobile phone numbers begin with the digit "6".

Internet

There are many Internet cafés in all the main towns and resorts, as well as in some of the islands's more remote towns and villages. Internet cafés in some main towns are listed in the Directory opposite. An increasing number of hotels also offer free Wi-Fi.

A sign advertising the services of an Internet café

Postal Services

Greek post offices (tachy-dromeía) generally open from 7:30am to 2pm Monday to Friday, with some main branches, especially in larger towns or cities, staying open as late as 8pm and possibly for a few hours at weekends. All post offices are closed on public holidays (see p54).

Postboxes are usually bright yellow; some have two slots, marked esoterikó (domestic) and exoterikó (overseas). Bright-red postboxes are reserved for express mail,

both domestic and overseas. Express is a little more expensive, but it cuts delivery time by a few days. Airmail letters take three to six days to most European countries, and anywhere from five days to a week or more to North America, Australia and New Zealand.

Stamps (grammatósima) can be bought at post offices and sometimes from vending machines that are situated inside post offices.

The poste restante system – whereby mail can be sent to, and picked up from, a post office – is widely used in Greece, especially in more remote regions. Mail should be clearly marked "Poste Restante", with the recipient's surname underlined so that it gets filed in the right place. Proof of identity is needed when collecting the post, which is kept for a maximum of 30 days before being returned to the sender.

If you are sending a parcel to a non-EU country, do not seal

A standard bright-yellow Greek postbox

it before heading to the post office – its contents will need to be inspected by security before it is sent.

All of the main towns on the islands have a central post office. The main post offices on Crete are in Irákleio, off Plateía Venizélou, and in Chaniá, on Chálidon (near the cathedral). On the island of Corfu, the main post office is in Corfu town on Alexandras Street.

International courier services such as **ACS** and **DHL**, both of which have offices in Athens through which parcels are sent to handling agents on the islands, offer the best solution for express deliveries.

Newspapers and Magazines

The trusty corner *períptera* (newspaper kiosks), bookshops in larger towns and tourist shops in the resorts often sell day-old foreign newspapers and magazines, though the mark-up to the price can be substantial. Monthlies such as *The Ionian*

A street kiosk, selling a vast range of newspapers and magazines

for Corfu, Lefkáda, Kefalloniá, Zákynthos, Paxós and Antípaxos, *Crete Gazette* and the glossy bi-monthly *Odyssey* for all the islands are all published in English. They are good sources of information on local entertainment, festivals and other cultural goings-on, while also providing decent coverage of domestic and international news.

The most popular Greek-language newspapers are *Kathimeriní, Eleftherotypía* and *Ta Néa*.

Television and Radio

There are state-run and private TV channels, plus a host of cable and satellite stations from across Europe. The state-owned TV and radio broadcasting corporation is **New Hellenic Radio, Internet and TV (NERIT)**.

Most Greek stations cater to popular taste, with a mix of dubbed foreign soap operas, game shows, sport and films. Foreign-language films tend to be subtitled rather than dubbed. Satellite stations CNN and Euronews have international news in English around the clock. Guides detailing the coming week's television programmes are published in all the English-language papers.

With state-owned radio and a plethora of local stations, the airwaves are positively jammed in Greece, and reception is not always dependable. Many stations are devoted exclusively to Greek music, either traditional or contemporary. There are also classical music stations, such as ERA-TRITO (95.6 FM), and modern music stations such as Rock FM (96.9 FM). For the daily news in English, you can pick up the BBC World Service (frequency varies throughout Greece). The BBC and other English and American channels can also be received over the Internet.

DIRECTORY

Mobile Phones

Cosmote
Tel 13838.
W cosmote.gr

Vodafone Greece
Tel 13830.
W vodafone.gr

WIND Hellas
Tel 13800.
W wind.com.gr

Internet

3w
Chalidon, Chaniá, Crete.
Tel 28210 93478.

Aktaion Café Corfu
Corfu Town, Corfu
Tel 26610 37894.

Coffee Corner
Mýkonos Town, Mýkonos.
Tel 22890 22969.

Internet Café
Parthenon City, Anthoula Zervou 4, Rhodes Town, Rhodes.
Tel 22410 22351.

Postal Services

ACS
Tel 21081 90000.
W acscourier.gr

DHL
Tel 21098 90000.
W dhl.gr

Newspapers and Magazines

Crete Gazette
W cretegazette.com

The Ionian
W theionian.com

Kathimeriní
W ekathimerini.com

Odyssey
W odyssey.gr

Television and Radio

NERIT
W nerit.gr

TRAVEL INFORMATION

Reliably hot, sunny weather makes Greece a popular holiday destination. From mid-May to early October, countless visitors flock to the Greek islands. While many of the larger islands have their own airports and are easily accessible by plane from major European destinations, the ferry network ensures even the remotest islands are easy to reach. The extensive bus network has frequent services on major routes and also serves rural communities. It is also possible to reach Greece by car, rail and coach, or to fly into Athens, travelling on to the islands by ferry.

Most ferries depart from the port of Piraeus. Travelling around by car or motorcycle offers the most flexibility on larger islands, allowing travellers to reach places that are inaccessible by public transport. However, road conditions are variable, and in remoter parts can be rough and dangerous. Taxis provide another inexpensive option, and on many islands taxi boats sail around the coast offering various pick-up and drop-off points. Strikes (a regular occurrence in Greece) can cause disruption to public transport both to and within the country.

Green Travel

The concept of green travel in Greece is being pioneered in Athens, but has yet to have any real impact on the islands. Whereas Athens has driving restrictions to limit smog and traffic congestion, as well as a fleet of buses that run on natural gas and electric trolleybuses, introducing ecologically friendly public transport to the islands is still in the planning stages. Crete was to be the first island to have eco-friendly buses, but the initiative has been slow to develop.

A government incentive to encourage the withdrawal of old vehicles throughout Greece in order to reduce emissions is ongoing but has been criticized because of the programme's complexity. The upside is that jobs have been secured in the country's automobile industry.

Greece has been slow to embrace the introduction of cycle lanes and schemes to encourage the use of bicycles as a mode of transport. Other than in some areas of Athens and a handful of mainland cities like Thessaloníki and the Peloponnese city of Náfplio, cycle lanes are all but non-existent. However, cycling and hiking holidays on the islands are an increasingly popular option. Visitors relish the glorious unspoiled landscapes and the chance to explore what are often otherwise inaccessible areas. For details of companies that run walking, hiking and cycling holidays, see pages 336 to 339.

Arriving by Air

The main airlines operating direct scheduled flights from London to Athens are **Aegean Airlines** and **British Airways**. In addition, several budget airlines – including **easyJet** (from London Gatwick, Manchester and Edinburgh to Athens; and from London Gatwick to Thessaloníki) and **Ryanair** (from London Stansted to Thessaloníki and Athens) – also connect the UK to the Greek mainland. Irish airline **Aer Lingus** runs scheduled flights from Dublin to Athens and Corfu, while major European carriers such as **Air France** and **Alitalia** also operate scheduled flights.

In addition to regular flights to Athens, easyJet operates direct scheduled flights to Corfu, Crete, Kefalloniá, Zákythos, Kos, Rhodes, Mýkonos and Santoríni from the UK and numerous other European countries. Ryanair also flies direct to Corfu, Kefalloniá, Zákynthos, Crete, Rhodes and Kos from the UK and from a number of other European countries. Aegean Airlines operates flights from several European cities to Crete, Corfu, Rhodes, Chíos, Mýkonos, Santoríni, Skýros and Kos.

All scheduled long-haul flights to Greece land in Athens although many are not direct and will require changing at a connecting European city. There are direct flights to Athens from New York with **Delta**, **KLM** and Air France, and from Philadelphia with **US Airways**, British Airways and **Finnair**, while **Air Canada** and **Air Transat** fly from Montreal and Toronto. **Air China** flies directly from Beijing to Athens. Although there are no direct flights to Athens from Australia or New Zealand, there are more

Elefthérios Venizélos – Athens International Airport

Light and spacious check-in area at Athens International Airport

than five flight routes daily from that part of the world that involve changing to a connecting flight at hubs in the Middle East.

Charter Flights and Package Deals

Charter flights to Greece are nearly all from within Europe, and mostly operate between the months of May and October. They are usually the cheapest option during peak season (July to August), when air fares rise steeply, though discounted scheduled flights are worth considering in low season, when there are few charters available.

Tickets are sold through airline websites and, to a lesser extent, by travel agencies either as part of an all-inclusive package holiday or as a flight-only deal. Companies operating charter flights to the islands, especially to Crete, Corfu, Kefalloniá, Mykonos, Santoríni, Zákynthos and Rhodes, from the UK throughout the summer, are **Thomson Airways** and **Thomas Cook**.

Some real bargains can be found by buying tickets through price comparison websites such as **Kayak**, **Momondo** and **Skyscanner**.

Athens Airport

Greece's largest, most prestigious infrastructure development project for the millennium, **Elefthérios Venizélos – Athens International Airport** opened to air traffic in 2001. Located at Spata, 27 km (17 miles) southeast

of the city centre, the airport handles the majority of Greece's international and domestic flights, as well as all of Athens' passenger and cargo flights. It has two runways, designed for simultaneous, round-the-clock operation, and a Main Terminal Building for all arrivals and departures. Arrivals are located on the ground floor (Level 0) and departures on the first floor (Level 1). Passengers are advised to check in as early as possible and to contact their airline in advance to find out the recommended time to arrive at the airport for their flight.

Service facilities include a shopping mall, restaurants and cafés in the Main Terminal Building and a five-star Sofitel hotel in the airport complex. Car-rental firms, bureaux de change, banks and travel agencies are all in the arrivals area. There is also a small museum in the Main Terminal Building (departures) with archaeological findings from digs in the airport area.

Transport from Athens Airport

Metro line 3 (blue line) links the airport to Sýntagma and Monastiráki in the city centre every 30 minutes from 6:33am until 11:33pm, while the Proastiakos suburban rail service runs from 5:26am until 9:44pm from the airport to Ano Liosia just north of Athens (every 20 minutes), and to Kiato (every hour). From these it is possible to connect to the rest of the suburban or intercity rail network. Tickets for both metro and suburban rail journeys from the airport to the city centre cost €8 (single) and €14 (return).

Visitors who prefer to use road transport (for those arriving or departing between 11:30pm and 5:30am this is the only option), can take a bus, a taxi or a hired car. The X95 bus runs from the airport to Sýntagma Square, in the city centre, every 10–15 minutes (journey time: about 70 minutes). Bus X96 runs to Piraeus port every 20–25 minutes (journey time: about 90 minutes). Bus X93 runs to Kifisos and Liosion intercity bus stations in Athens every 25–30 minutes (journey time: about 65 minutes). Bus X97 runs to Dáfni metro station every 45–60 minutes (journey time: about 70 minutes). All four buses run 24 hours a day, and a single ticket costs €5. A taxi ride to the centre of Athens costs €35 by day and €50 by night (fixed prices). A six-lane toll motorway links the airport to the Athens ring road. Several car hire companies are also based at the airport.

A typical bus serving Athens International Airport

Island Airports

Crete has two international airports – Nikos Kazantzakis at Irákleio, located just 5 km (3 miles) from the city centre, and Ioannis Daskalogiannis Airport in Chaniá, which is situated around 15 km (9 miles) northwest of the city. KTEL buses and taxis run from outside the terminal to Chaniá's main square, with the journey taking about 45 minutes. Siteía also has a smaller domestic airport. Corfu is served by Ioannis Kapodistrias International

Airport, which is located less than 2 km (1 mile) from Corfu town centre. It is served by regular buses and taxis from the rank right outside the terminal. Corfu's fellow Ionian island Kefalloniá has its own international airport, which is situated 8 km (5 miles) from the centre of the principal town of Argostóli. The best way to reach central Argostóli is by taxi. The fare should cost around €15. Zákynthos is served by Dionýsios Solomós International Airport,

which is less than 5 km (3 miles) from Zákynthos town. Taxis cost around €6.

In the Dodecanese, Ippokratis International Airport serves the island of Kos. A compact airport, it lies 27 km (17 miles) from Kos town. KTEL buses from outside the arrivals exit make the 40-minute journey into Kos town centre, with a ticket costing €3.20. Taxis charge around €25–€27 for the same journey. Rhodes is served by Rodos Diagoras International Airport.

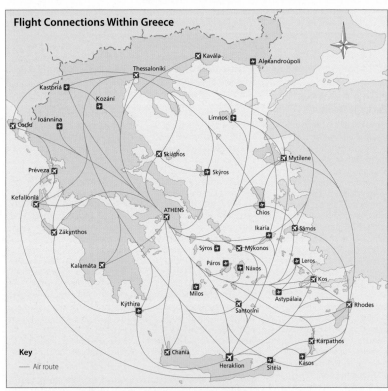

Flight Connections Within Greece

Key
— Air route

Island	Distance	Flying Time	Island	Distance	Flying Time
Corfu	381 km (237 miles)	40 minutes	Crete (Chaniá)	318 km (198 miles)	45 minutes
Rhodes	426 km (265 miles)	45 minutes	Santoríni	228 km (142 miles)	40 minutes
Skýros	128 km (80 miles)	40 minutes	Kos	324 km (201 miles)	45 minutes
Skiáthos	135 km (84 miles)	30 minutes	Mýkonos	153 km (95 miles)	30 minutes
Límnos	252 km (157 miles)	45 minutes	Páros	157 km (98 miles)	35 minutes

For keys to symbols *see back flap*

Kefalloniá International Airport, near Argostóli

Located 15 km (9 miles) west of Rhodes Town, KTEL airport buses link it to the city centre. The journey time is around 45 minutes, and the fare is less than €1. The journey by taxi is more expensive at €16.

Lésvos is served by Odyseas Elitís Mytilíni International Airport, which is located 7 km (4 miles) from the centre of the main town of Mytilíni. KTEL buses (fare €1.90) and taxis

(€10–€12) link the airport with the town. On the nearby island of Límnos, Ifestos International Airport is 18 km (11 miles) from the centre of Mýrina, its main town. It is served by taxis and primarily tourist buses. Sámos is served by Aristarchós of Sámos International Airport, which lies 14 km (9 miles) southwest of Vathý in Pythagório. Buses link the airport with many of the towns and villages on Sámos, and taxis are available to Vathý (fare €15).

Mýkonos and Santoríni have airports that receive international flights operated by Aegean Airlines from within Europe in the summer months.

Domestic and Connecting Flights

Greece's domestic airline network is fairly extensive. Most internal flights are operated by **Olympic Air** and **Aegean Airlines**, though a number of small private companies, including **Astra**

Airlines, provide connections between Thessaloníki and the rest of the country.

Elefthérios Venizélos International Airport in Athens and **Makedonia-Thessaloníki Airport** in Thessaloníki are hubs for onward air travel to around a dozen islands. Olympic Air and Aegean Airlines operate flights from these airports to and between Crete, Corfu, Zákynthos and Rhodes, as well as Lésvos and Límnos.

Astra airlines flies from Makedonia-Thessaloníki Airport to Chíos, Samos, Kos, Lésvos and Corfu.

Fares

Fares for domestic flights are often at least double those of a bus journey or deck-class ferry trip. Tickets and timetables for Olympic Air and Aegean Airlines flights are available from the respective websites, as well as from most major travel agencies.

Travelling by Sea

Greece has always been a nation of seafarers and, with its hundreds of islands and thousands of miles of coastline, the sea has played an important part in the history of the country and continues to do so today. It is a major source of revenue for Greece, with millions of holiday-makers choosing the idyllic Greek islands for their vacation. The network of ferries is a lifeline for islanders, and for tourists an enjoyable and relaxing way of island-hopping or reaching a single destination.

Arriving by Sea

There are regular year-round ferry crossings from the Italian ports of Ancona, Bari and Brindisi to the Greek mainland ports of Igoumenítsa in Epirus and Pátra in the Peloponnese. During the summer, there are additional sailings from Venice and Trieste. **Minoan** and **Superfast** are the main Greek companies covering these routes. Journey times and fares vary considerably, depending on the time of year, point of embarkation, ferry company and type of ticket. Fare reductions are possible for students, travellers under the age of 26 who have an ISIC card (see p346) and railcard holders. Booking in advance is recommended in high season, especially if you are travelling by car or would like a cabin.

The Ionian islands have regular ferry services from Igoumenítsa and Pátra ports. Other ferry services include the route from Turkey's Aegean coast between Kusadasi and Sámos, and

Car ferry leaving from Mandráki harbour, on Nísyros

Çeşme and Chíos, with additional summer sailings between Bodrum and Kos, Kálymnos and Rhodes, and Ayvalik and Lésvos.

If you are transporting your car into Greece by ferry, you will require a vehicle registration document and, in summer, will need to reserve a berth in advance. Addresses and telephone numbers of agents for advance bookings are given on page 361.

Greek Ferry Services

The smaller ports have limited services, so check the timetable on arrival to see if you can get a ferry on the day and for the destination you want. The larger ports, such as Piraeus and Thessaloníki, have many more services. Piraeus, the port of Athens, is Greece's busiest port and has many routes emanating from its harbour. The hub of activity is at Plateía Karaïskáki, where most ticket agents reside, as well as the port police (limenarcheío). A number of competing companies, such as **Blue Star Ferries**, run the ferry services, each with its own agents handling bookings and enquiries. This makes the task of finding out when ferries sail, and from which dock, a more challenging one. The ferries are approximately grouped by destination, but when the port is busy ferries dock wherever space permits. So, finding your ferry usually involves studying the agency's information board or asking the port police.

In this guide, the direct ferry routes in high season are shown on the individual island maps, pictorial maps for each island group, and the back endpaper for the country-wide network; high season is from June to August. In low season, expect all services to be greatly reduced and some routes to be suspended altogether. The routes on these maps should be taken as guidelines only – check local sources for the latest information.

The Greek Tourist Office's weekly schedules can serve as a useful guideline to departure times. Visit the Greek Travel Pages website (www.gtp.gr) for information. Alternatively, ask at a local travel agency. Some of the English-language papers also print summer ferry schedules. Hydrofoils, catamarans, caïques and taxi boats supplement the ferry services (see pp360–61).

Ferry Tickets

Tickets for all ferry journeys can be purchased from the shipping line office, any

A Minoan Lines ferry leaving a port

Cruise ship sailing towards the harbour at Zákynthos

authorized travel agency, on the quayside or on the ferry itself. All fares except first class are set by the Ministry of Transport, so a journey should cost the same amount regardless of which shipping line you choose. For motorbikes and cars, a supplement is also payable.

Cars can cost as much as three or four times the passenger fare.

Children under two travel free, those aged from two to nine pay half fare, and once over the age of ten, children must pay the full adult fare.

On major routes, ferries have essentially three classes,

ranging from deck class to deluxe – the latter costing almost as much as flying.

First class usually entitles you to a two-bunk exterior cabin with bathroom facilities. A second-class ticket costs around 25 per cent less and gives you a three- or four-bunk cabin with washing facilities, such as a basin. Second-class cabins are usually within the interior of the vessel.

A deck-class ticket gives you access to most of the boat, including a lounge with reclining seats. During the summer, on a warm, starry night, the deck is often the best place to be.

Ferry Company Funnels

The funnels of each company's fleet are bold and brightly coloured, and serve as beacons for travellers searching the harbour for their ferry. In the busiest port, Piraeus, ferries often dock wherever there is space, and even in high season each company is unlikely to have more than two or three boats in dock at a time. Targeting the funnel, therefore, is often the easiest way to find your ferry.

ANEK Lines

GA Ferries

Ventouris Sea Lines

NEL Lines

Piraeus Port Map

This shows where you are likely to find ferries to various destinations.

Piraeus Port Authority
Tel 210 455 0000.

Coastal Services Timetables
Tel 14541.

Key to Departure Points

- Argo-Saronic Islands
- Northeast Aegean Islands
- Dodecanese
- Cyclades
- Crete
- International ferries
- Hydrofoils and catamarans

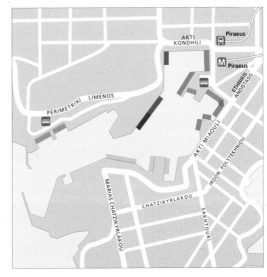

For keys to symbols *see back flap*

Hydrofoil, known as a "Flying Dolphin"

Hydrofoils and Catamarans

Some of the islands can be reached by Greece's 60 or so hydrofoils. The main operators are **Hellenic Seaways** and **Dodekanisos Seaways**, though there are many smaller companies running "Flying Dolphins", as they are known locally. They are twice as fast as a ferry but, as a consequence, are double the price.

The major drawback of hydrofoils is that most vessels only run in the summer and are often cancelled if weather conditions are poor. In fact, on seas that are anything other than calm, hydrofoils are quite slow, and can prove a bad idea for those prone to seasickness.

Hydrofoils can accommodate around 140 passengers, but have no room for cars or motorcycles. Advance booking is often essential, and it is as well to book as early as possible during high season. Tickets are bought from an agent or on the quayside, but rarely on board the vessel itself. Routes are around the mainland and Peloponnese coasts, and to island groups close to the mainland – the Argo-Saronic group, Evvoia and the Sporades, and to several islands within the Cyclades. There are also routes between Rhodes, in the Dodecanese, and Sámos, at the southern end of the Northeast Aegean.

Catamarans offer an airline-type service in terms of seating, bar facilities and on-board television. They are also better designed for disabled passengers. There are services around the Ionian Islands, and about half a dozen catamarans operate in the Aegean, mostly between the mainland port of Rafína and the islands of Andros, Tínos and Mýkonos. Costs are on a par with hydrofoils, and tickets should be bought from a travel agency a few days ahead.

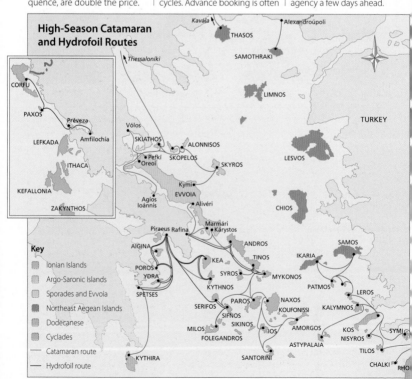

High-Season Catamaran and Hydrofoil Routes

Kavála
Alexandroúpoli
THASOS
SAMOTHRAKI
Thessaloníki
LIMNOS
TURKEY
CORFU
PAXOS
Préveza
Vólos
LEFKADA
Amfilochía
SKIATHOS
ALONNISOS
LESVOS
Pefki
SKOPELOS
ITHACA
Oreoí
SKYROS
Kými
KEFALLONIA
EVVOIA
CHIOS
Agios
Ioánnis
Alivéri
ZAKYNTHOS
Marmári
Kárystos
Piraeus Rafína
SAMOS
AIGINA
ANDROS
IKARIA
POROS
KEA
TINOS
YDRA
SYROS
MYKONOS
PATMOS
SPETSES
KYTHNOS
LEROS
PAROS
NAXOS
SERIFOS
KOUFONISSI
KALYMNOS
SIFNOS
MILOS
SIKINOS
IOS
AMORGOS
KOS
SYMI
FOLEGANDROS
ASTYPALAIA
NISYROS
KYTHIRA
SANTORINI
TILOS
CHALKI
RHO

Key

- ☐ Ionian Islands
- ☐ Argo-Saronic Islands
- ☐ Sporades and Evvoia
- ▓ Northeast Aegean Islands
- ☐ Dodecanese
- ☐ Cyclades
- — Catamaran route
- — Hydrofoil route

An excursion caïque on the Dodecanese island of Sými

Tourist Excursions

Many holiday resorts put on small excursion boats that take groups of tourists to out-of-the-way beaches and caves, or on day cruises and picnics. Routes and times are dictated by local conditions. Up-to-date information and booking arrangements are available on arrival in the islands at the resort or any local travel agency or information centre.

Local Inter-Island Ferries

In addition to the large ferries that cover the main routes, there are smaller ferries making inter-island crossings in the summer. Local ferries, regardless of size, are subject to government price controls, but boats chartered by tourist agencies can charge what they like, and often prove expensive. It is therefore worth doing some research locally on the various options before purchasing a ticket. These boats do, however, provide direct connections to the islands, which shortcut circuitous routes via mainland ports, saving you valuable holiday time.

Taxi Boats

Taxi boats (or caïques) are even more ad hoc, sailing along coastlines and making short trips between adjacent islands. They are usually only available during high season and, as the smallest vessels, are most prone to cancellation in adverse sea conditions. They tend to be more expensive than ferries, given the short distances involved, but often provide a route where few or no others are available.

Routes and itineraries are at the discretion of the boat owners, and the only place to determine if one is going your way is at the quayside.

An inter-island ferry run by one of the main companies

Travelling by Road and Rail

There has been much upgrading of the roads on the islands but, particularly in remote areas, they can still be rough, and in some cases suitable only for four-wheel-drive vehicles. Cars and motorcycles are easily rented, though, and the extensive bus network is complemented by many taxis. Maps from local travel agents are less than reliable, however, so visitors should bring their own: GeoCenter and Freytag & Berndt are both good.

You have
priority

You have right
of way

Do not use car
horn

Wild animals
crossing

Hairpin bend
ahead

Roundabout
(traffic circle)

Arrving by Car

The most direct overland routes to Greece are through the countries that made up the former Yugoslavia. However, road quality is poor through some parts of this route. The alternative route is through France, Switzerland and Italy, and from there to Greece by ferry. Once on mainland Greece, you can drive your vehicle aboard one of the car ferries that travel between the islands. There is only one island that can be accessed by road: Lefkáda in the Ionian island group. It is possible to drive across the causeway and floating bridge from the mainland to Lefkáda town, where you can pick up the coastal road for onward travel around the island.

Motoring organizations, such as the **AA**, **RAC** or **ELPA** (the Automobile and Touring Club of

Greece), offer advice on routes and regulations. You need a full, valid national driving licence, and insurance cover (at least third party insurance is compulsory).

Rules of the Road

Driving is on the right in Greece and, with the exception of some rural backroads, road signs conform to European norms. The speed limit on national highways is 120 km/h (75 mph) for cars; on country roads it is 90 km/h (55 mph) and in towns 50 km/h (30 mph). Seat belts are required by law and children under ten are not allowed in the front seat.

Car Hire

Scores of car-rental agencies in all main resorts offer a range of vehicles from small cars to minibuses. International companies such as **Avis** and **Budget** tend to be considerably more expensive than their local counterparts, though the latter are generally as reliable. Third party is the minimum insurance required by law, but personal accident insurance is strongly recommended. A full licence that has been held for at least one year is needed, and the minimum age requirement ranges from 21 to 25 years.

Petrol

Petrol stations are plentiful in towns, though less so in rural areas – always set out with a full tank, to be on the safe side. Fuel is sold by the litre, and there are usually three or four grades available: super (95 octane), unleaded, super unleaded and diesel, which is confusingly

called *petrélaio*. Filling stations set their own working hours, but most are open seven days a week from 7am or 8am to between 7pm and 9pm.

Motorcycle, Moped and Bicycle Rental

Motorcycles and mopeds are readily available to rent on the islands. The latter are ideal for short distances on flattish terrain, but for anything mountainous a motorcycle is a must. Make sure that the vehicle is in good condition before you set out and that the price includes adequate insurance cover; also check that your own travel insurance covers motorcycle accidents, as many do not.

The speed limit on national highways is 70 km/h (45 mph) for bikes up to 100 cc, and 90 km/h (55 mph) for larger bikes; helmets are compulsory.

Bicycles can also be rented in some resorts, though the steep mountainous terrain and hot sun can be deterrents to even the toughest enthusiast. Bicycles can, however, be transported free on most Greek ferries and buses.

Rack of bikes for hire at the beach in Kos Town

Coaches and Buses

International buses connect Greece with the rest of Europe, though fares are not as cheap as charter flights during the holiday season.

Greece's domestic bus system is operated by **KTEL** (Koinó Tameío Eispráxeon Leoforeíon), a syndicate of privately run companies that provides almost every community with services

Passengers boarding a local bus at Irákleio bus station, Crete

of some sort. In remote rural villages the bus might call once or twice a day, usually at the local taverna or *kafeneío*, while services between the larger towns are frequent and efficient. You can also usually rely on there being a bus service between the port and main town of any island, if the latter is inland.

On many of the larger islands travel agents offer a wide range of excursions on air-conditioned coaches, accompanied by qualified guides. These include trips to major archaeological and historical sites, other towns and resorts, popular beaches, areas for established walks, such as the Samariá Gorge in Crete, and organized events, such as an evening out in a "typical Greek taverna".

Taxis

Taxis provide a very reasonably priced way of getting around on the islands. Although all taxis are metered, it is worth asking the driver to give you a rough idea of the price before setting out. Round up the fare to the nearest euro as a tip. For longer journeys, a price can usually be negotiated per day, or per trip. Drivers are generally amenable to dropping you off and picking you up a few hours later. Most rural villages have at least one taxi, and the best place to arrange for one is at the local *kafeneío* (café). Taxi trucks often take several passengers, each paying for their part of the journey.

Trains

Due to Greece's ailing economy, Greek train services to and from the country are limited. The route into neighbouring Bulgaria resumed in 2014; however, the long-established direct service to Turkey remains suspended until further notice.

Within Greece, the train network is run by the **OSE** (Organismós Sidirodrómon Elládos). The system is restricted to the mainland, but there is a useful route out of Athens to Vólos for ferries to Skiáthos and Skópelos.

DIRECTORY

Arriving by Car

AA
Tel 0800 072 3279 (UK).
w theaa.com

ELPA
Mesogeíon 395, 15343 Athens.
Tel 21060 68800.
Tel 10400 (road assistance).
w elpa.gr

RAC
Tel 0800 015 6000 (UK).
w rac.co.uk

Car Hire

Avis
Sygrou 23, 11742 Athens.
Tel 21032 24951.
w avis.gr

Budget
Metaxa 29, 16674 Athens.
Tel 21089 81444.
w budgetrentacar.gr

Coaches and Buses

Bus terminals in Athens
Terminal A: Kifisoú 100
Tel 21051 24910/32601.
Terminal B: Liosíon 260
Tel 21083 17153.

KTEL
Tel 14505.
w ktel.org

Trains

Athens train station
Laríssis station.
Tel 21052 98837/37741.
Tel 14511 (train timetable).

OSE (information & reservations)
Karólou 1, Athens.
Tel 21052 97865.
w ose.gr

Passengers aboard a taxi truck on the island of Lipsí

General Index

Acknowledgments

Dorling Kindersley would like to thank the following people whose contributions and assistance have made the preparation of this book possible.

Main Contributor

Marc Dubin is an American expatriate who divides his time between London and Sámos. Since 1978 he has travelled in every province of Greece. He has written or contributed to numerous guides to Greece, covering such diverse topics as trekking and contemporary Greek music.

Stephanie Ferguson, a freelance journalist and travel writer, has hopped around almost 50 Greek islands. She became bewitched by Greece after a holiday 20 years ago and since then has contributed to eight guide books and written travel features on Greece for several national publications.

Mike Gerrard is a travel writer and broadcaster who has written several guides to various parts of Greece, which he has been visiting annually since 1964.

Andy Harris is a travel and food journalist based in Athens. He is the author of *A Taste of the Aegean*.

Tanya Tsikas is a Canadian writer and travel guide editor. Married to a Greek, she has spent time in Crete and currently lives in Oxford.

Deputy Editorial Director Douglas Amrine
Deputy Art Director Gillian Allan
Managing Editor Georgina Matthews
Managing Art Editor Annette Jacobs

Additional Illustrations

Richard Bonson, Louise Boulton, Gary Cross, Kevin Goold, Roger Hutchins, Claire Littlejohn.

Revisions Team

Emma Anacootee, Claire Baranowski, Marta Bescos, Sonal Bhatt, Tessa Bindloss, Hilary Bird, Tony Clark, Elspeth Collier, Michelle Crane, Michele Crawford, Catherine Day, Mariana Evmolpidou, Jim Evoy, Carole French, Robin Gauldie, Emily Green, Lydia Halliday, Emily Hatchwell, Leanne Hogbin, Kim Inglis, Maria Kelesidi, Lorien Kite, Priya Kukadia, Esther Labi, Felicity Laughton, Nicola Malone, Paul Marsden, Andreas Michael, Ella Milroy, Lisa Minsky, Robert Mitchell, Adam Moore, Jennifer Mussett, Tamsin Pender, Eva Petrou, Marianne Petrou, Pollyanna Poulter, Jake Reimann, Ellen Root, Simon Ryder, Collette Sadler, Alice Saggers, Sands Publishing Solutions, Rita Selvaggio, Liz Sharp, Ellie Smith, Claire Stewart, Claire Tennant-Scull, Amanda Tomeh, Nikky Twyman, Conrad Van Dyk, Dora Whitaker, Andy Wilkinson.

Dorling Kindersley would also like to thank the following for their assistance: The Greek Wine Bureau, Odysea.

Additional Research

Anna Antoniou, Garifalia Boussiopoulou, Anastasia Caramanis, Michele Crawford, Magda Dimouti, Shirley Durant, Panos Gotsi, Zoi Groummouti, Peter Millett, Tasos Schizas, Garifalia Tsiola, Veronica Wood.

Artwork Reference

Ideal Photo S.A., The Image Bank, Melissa Publishing House, Tony Stone Worldwide.

Additional Photography

Jane Burton, Mariana Evmolpidou, Frank Greenaway, Derek Hall, Nigel Hicks, Dave King, Neil Lucas, National History Museum, Ian O'Leary, Stephen Oliver, Roger Philips, Rough Guides/Chris Chrstoforou, Kim Sayer, Tony Souter, Clive Steeter, Harry Taylor, Kim Taylor, Mathew Ward, Stuart West, Jerry Young.

Photography Permissions

Dorling Kindersley would like to thank the following for their assistance and kind permission to photograph at their establishments:

Museum of Greek Folk Art, Athens; Karpathos Museum; Markos Vamvakaris Museum, Syros; Kymi Folk Museum, Evvoia; Stavros Kois's House, Syros. Also all other cathedrals, churches, museums, hotels, restaurants, shops, galleries, and sights too numerous to thank individually.

Picture Credits

a = above; b = below/bottom; c = centre;
f = far; l = left; r = right; t = top.
Works of art have been reproduced with the
permission of the following copyright holders:
© ADAGP, Paris and DACS, London 2011 *The
Kiss* Constantin Brancusi 215br. The work of art
Three Standing Figures, Henry Moore (1947)
215bc is reproduced by permission of the
Henry Moore Foundation.

The publisher would like to thank the
following individuals, companies and
picture libraries for permission to reproduce
their photographs:

4Corners: Guido Cozzi 146-7; SIME/Johanna
Huber 108; SIME/Riccardo Spila 2-3.

Adrina Hotel Group, Skopelos: 305tr, 309br;
AISA Archivo Icongrafico, Barcelona:
Museo Archeologique, Bari 63cra; **AKG,
London:** 194bl; Antiquario Palatino 59bl;
British Museum 293bc; Erich Lessing
Akademie der Bildenden Künste, Vienna 60c;
Musée du Louvre 59tc; Naples Archaeological
Museum 149bc; National Archeological
Museum, Athens 32–3(d), 33tl; Staatliche
Kunstsammlungen, Albertinum, Dresden
37crb, Liebighaus, Frankfurt/Main 39crb;
Staatliche Antikensammlungen und
Glyptotek, München 58bl; Mykonos Museum
61tr; **Alamy Images:** Art Directors & TRIP/Bob
Turner 355br; David Crosby 346tl; Danita
Delimont 352cb; Peter Eastland 260-1; Werli
Francois 23c; funkyfood London - Paul
Williams 220-1; Terry Harris just greece photo
library 12bl,16bl; hemis.fr/Franck Guiziou
313br; Peter Horree 319tl; Mike Hughes 300-1;
imageBROKER/Joachim Hiltmann 312tr; IML
Image Group Ltd/George Detsis 330bc; Yadid
Levy 328tr Dennis MacDonald 286; Hercules
Milas 162, 204-5, 208, 248, 342-3; Jeff Morgan
01 357tl; Kostas Pikoulas 349tl; Massimo
Pizzocaro 329br; Robert Harding World
Imagery 319cb, /Tuul 72; Sebastian Rothe
177br; Peter Titmuss 349cla; Travelshots.com/
Peter Phipp 323tr; Konstantinos Tsakalidis
118t;
Ancient Art and Architecture: 35crb, 40ca,
41cb, 43tl, 62cb, 62b(d), 103bl;
**Antikenmuseum Basel Und Sammlung
Ludwig:** 66–7; **Aperion:** John Hios 51c;
Argyropoulos Photo Press: 53cr, 54tl, 54crb;

Athens International Airport: 354bl, 355tl;
Athens Urban Transport Organization
(OASA): 296cla, 297tl; **AWL Images:** Doug
Pearson 68-9; Travel Pix Collection 96

Benaki Museum: 31b, 42cla, 45tl, 45crb, 47cra,
295br; **Paul Bernard:** 39tc; **Bibliotheque
National,** Paris: Caoursin folio 175 4cr(d),
44–5(d), Caoursin folio 33 193bl, Caoursin folio
79 193br; **Bodleian Library,** Oxford: MS Canon
Misc 378 170v 40clb; **Bridgeman Art Library,**
London: Acropolis Museum, Athens, Greece
*Relief depicting hydria carriers from the North
Frieze of the Parthenon, c.447-432 BC* (marble)
56-7; Birmingham City Museums and Art
Galleries *Pheidias Completing the Parthenon
Frieze,* Sir Lawrence Alma-Tadema 64tr;
Bibliothèque Nationale, Paris *The Author
Guillaume Caoursin, Vice Chancellor of the Order
of St John of Jerusalem Dedicating his Book to
Pierre d'Aubusson, Grand Master of the Order of
St John of Jerusalem who is Seated Surrounded
by High Dignitaries of the Order* (1483),
illustrated by the Master of Cardinal of
Bourbon Lat 6067 f 3v 30(d); British Museum,
London *Cup, Tondo, with Scene of Huntsmen
Returning Home* 35cb, *Greek Vase Showing Diver
About to Enter the Sea in Search of Sponges*
(c.500 BC) 173br; Fitzwilliam Museum,
University of Cambridge *Figurine of
Demosthenes* Enoch Wood of Burslem (c.1790)
(lead glazed earthenware) 63tl, *Attic Red-
figured Pelike: Pigs, Swineherd and Odysseus,* Pig
Painter (470–60BC) 91clb; House of Masks,
Delos *Mosaic of Dionysus riding a Leopard* (c.AD
180) 41tc; Kunsthistorisches Museum, Vienna
*Elizabeth of Bavaria, Wife of Emperor Franz
Joseph I of Austria,* Franz Xavier Winterhalter
87c(d); Lauros-Giraudon/Louvre, Paris *Rhodes
Winged Victory of Samothrace* (early 2nd
century BC) 136c; National Archaeological
Museum, Athens *Bronze Statue of Poseidon*
(c.460–450 BC) photo Bernard Cox 58cl; Private
Collection *Two-tiered Icon of the Virgin and
Child and Two Saints,* Cretan School (15th
century) 44clb; Victoria and Albert Museum,
London *Corfu,* Edward Lear 85br; © **The British
Museum:** 32clb, 33crb, 36cb, 37clb, 59cra(d),
59br, 61cl(d), 67tl, 67ca, 143cra.

Camera Press, London: ANAG 49tl, 49bl;
Christopher Simon Sykes 81tl; Wim Swaan
223tl; **Bruce Coleman Ltd.:** Philip van de Berg
251cl; Luiz Claudio Marigo 95tl; Natalio Feneck

119cl; Gordon Langsbury 251bl; Andrew J Purcell 28cla; Kim Taylor 119bl; World Wildlife Fund for Nature 119tl; Konrad Wothe 251br; **Corbis**: EPA/Orestis Panagiotou 51tl; Hemis/ Tuul 223clb; John Heseltine 318cla.

C M Dixon Photo Resources: 33cr; Glyptotek, Munich 38tr; **Dreamstime.com**: Arsty 11bl; Dstamatelatos 331tr; Dudau 350bl; Enisu 316br, 324tr; Ivan Jelisavcic 15cla; Pavel Kalouš 344cla; Panagiotis Karapanagiotis 13br; Iancucristi 14bl; Leannevorrias 15br; Lornet 12tl; Lucasdm 13tc; Anna Lurye 332tr; Stephen Outram 317br; Konstantinos Papaioannou 358bl; Radist 198bl; George Tsartsianidis 325br; Xiaoma 10clb; **Marc Dubin**: 26tr, 27bca, 27bra, 52tl, 178t, 203tr, 203br, 227br, 237br, 239br, 256b.

ECB: 351 all; **Ecole Française d'Athènes**: 222tr; **Ecole Nationale Superieure Des Beaux Arts**, Paris: *Delphes Restauration du Sanctuaire Envoi*, Tournaire (1894) 36–7; **Ekdotiki Athinon**: 32crb, 168bc, 191br(d); **ELIA**: 113br; **Janice English**: 197t, 197bl, 198c, 198b; **Esperas**: 311tr; **ET Archive**: National Archaeology Museum, Naples 38cla; **Mary Evans Picture Library**: 1, 91cr, 91bc, 91br, 137bc, 160bl.

Ferens Art Gallery: Hull City Museums and Art Galleries and Archives *Electra at the Tomb of Agamemnon* (1869), Lord Frederick Leighton 61br; **Fotolia**: KaYann 345tl.

Giraudon, Paris: Chateau Ecouen *Retour d'Ulysse* Ecole Siennoise 91cl; Louvre Paris 66cl, *Scène de Massacres de Scio*, Eugene Delacroix 46ca(d), 151br(d); Musée Nationale Gustave Moreau *Hesiode et Les Muses*, Gustave Moreau 62cla; Musée d'Art Catalan, Barcelona 293br; **La Gorgona**: 327tr; **Nicholas P Goulandris Foundation Museum of Cycladic and Ancient Greek Art**: 215tl, 215tc, 215tr, 215cl, 215cr, 288bl, 295ca; **Ronald Grant Archive**: *Zorba the Greek*, 20th Century Fox 280bl.; **Grecotel Amirandes**: 303br, 304tr.

Robert Harding Picture Library: David Beatty 50br; Tony Gervis 50cla, 50bl; Adam Woolfitt 52c; **Helio Photo**: 102ca; **Hellenic Post Service**: 49cla, 353tl; **Hellenic War Museum, Athens**: 255tl; **Historical Museum of Crete, Irákleio**: *Landscape of the Gods-Trodden Mount Sinai*, El Greco 272tr; **Michael Holford**: British Museum

38cb, 58tr; **Hotel Andreas**: 310bc; **Hulton Getty Collection**: 47crb(d); Central Press Photo 48clb(d).

Ideal Photo SA: T Dassios 305clb; A Pappas 239tr; C Vergas 51clb, 51br, 52tr, 87bl, 121bc, 165cra; **Images Colour Library**: 98tr, 274bl; **Impact Photos**: Jeremy Nicholl 336bl; Caroline Penn 51bl; **ISIC**: 346ca.

Carol Kane: 202br; **Gulia Klimi**: 53tr; **Kostos Kontos**: 5clb, 32cra, 48cra, 48cla, 51crb, 54c, 167br, 168cl, 192br, 297br, 341cb, 349c.; **Kriti Restaurant, Athens**: 333br;

Frank Lane Pictures: Eric and David Hoskings 250bl; **Ilias Lalaounis**: 335cla.

Mani- Mani/conceptcom.gr: 314bl; **Mansell Collection**: 58–9; **Merchant's House**: 306bc;

National Gallery of Victoria, Melbourne: *Greek by the Inscriptions Painter Challidian* Felton Bequest (1956) 60bl; **National Historical Museum**: 46–7(d), 47tl, 48bl; **Natural Image**: Bob Gibbons 251tl; **Nature Photographers**: Brinsley Burbridge 251tr; Robin Bush 251cr; Michael J Hammett 28tr; Paul Sterry 119br, 251tc; **Antonis Nicolopoulos**: 340cla, 341bl.

Oronoz Archivo Fotografico: Biblioteca National Madrid *Invasions Bulgares Historia Matriksiscronica FIIIV* 42clb(d); Charlottenberg, Berlin 60tr; El Escorial, Madrid *Battle of Lepanto*, Cambiaso Luca 44cr(d); Museo Julia 59cr(d); Musée du Louvre 66br; Museo Vaticano 63b, 67bl; **Oxford Scientific Films**: Paul Kay 29tr.

Romylos Parisis: City of Athens Museum 46clb; **Patras Apartments**: 303tl; **Phaedra Hotel, Ydra**: 302bl, 307tr; **Photoshot**: Rob Wyatt 344br; **Pictor International**: 52bl; **Pictures**: 50crb, 340br; **Planet Earth Pictures**: Wendy Dennis 250clb; Jim Greenfield 29crb; Ken Lucas 119crb; Marty Snyderman 29tl; **Private Collection**: 285c; **Popperfoto**: 48cb, 49c, 49bc.

Rex Features: Sipa Press/C Brown 49ca; **Robert Harding Picture Library**: Phil Robinson 8-9.

Scala, Florence: Gallerie degli Uffizi 34clb; Museo Archeologico, Firenze 35tl; Museo

Mandralisca Cefalu 36cla; Museo Nationale Tarquinia 67bc; Museo de Villa Giulia 34–5, 66tr; **Spectrum Colour Library**: 258tr; **Maria Stefossi**: 24bl.

TAP (Service Archaeological Receipts Fund) Hellenic Republic Ministry of Culture: 1st Epharat of Antiquities 49tc, 64br, 292cl, 292br, 293tl, 293cb; Acropolis Museum 294cla, 294tr; Andros Archaeological Museum 212c; Agios Nikolaos Archaeological Museum 282c; 2nd Epharat of Antiquities 70bl, 102tl, 102clb, 102br, 103tc, 103cra; Chalkida Archaeological Museum 5tc, 124c; Chania Archaeological Museum 257c; Corfu Archaeological Museum 73t, 83cl; Eretreia Archaeological Museum 5t, 123bl, 125c; 5th Epharat of Byzantine Antiquities 43ca; 14th Epharat of Byzantine Antiquities 141tc, 148br; 4th Epharat of Byzantine Antiquities 22c, 169tc, 169cr, 169br, 203clb; 18th Epharat of Antiquities 132tr, 133tr, 133br; 19th Epharat of Antiquities 136bl, 137cla, 137crb; Irakleio Archaeological Museum 270bl, 274tr, 274cla, 274cl, 274cb, 275tl, 275cr, 276br; 20th Epharat of Antiquities 139bc; 21st Epharat of Antiquities 71tl, 160tl, 160c, 222clb, 223cra, 223cr, 223bl, 223br, 234cla, 244crb, 245crb, 336cla; 22nd Epharat of Antiquities 175b, 176cla, 184cla, 190tr, 190bl 190bc, 191tl, 191cr, 200cr; 23rd Epharat of Antiquities 253cra, 267cr, 268cr, 268bc, 269 all, 270tr, 270br, 271 all, 276tr, 276cra, 276bl, 277tl, 277crb, 277br, 278 all, 279tr, 279b, 225c; 24th Epharat of Antiquities 281cl; Kos Archaeological Museum 176br; Milos Archaeological Museum 240tr; Mykonos Archaeological Museum 218tr; National Archeological Museum, Athens 34tr, 245bc, 290c; Naxos Archaeological Museum 234tr; Nea Moni Archaeological Museum 71ca; Numismatic Museum of Athens 279cl; Rhodes Archaeological Museum 188cb; Vathy Archaeological Museum, Samos 158cra; 2nd Epharat of Byzantine Antiquities 217c, 225tc, 228tc, 231br, 255c; 7th Epharat of Byzantine Antiquities 113tl; 6th Epharat of Byzantine Antiquities 93c; Thessaloniki Archaeological Museum 37tl; 3rd Epharat of Byzantine Antiquities 156tl, 157cl, 159br; 13th Epharat of Byzantine Antiquities 154–5 all, 265cra, 281bl, 283tl; Tinos Archaeological Museum 216tr; 3rd Epharat of Antiquities 291bl; **Teriade Museum**: *Dafnis and Chloe*, Marc Chagall ©ADAGP, Paris and DACS, London 2011 142br; **Travel Library:** Faltaits Museum 120tr; **Yannis Tsarouchis Foundation:** Private Collection *Barber Shop in Marousi*, Yannis Tsarouchis (1947) 48tr; **Tsipouradiki:** 322bc.

Villa Melina: 308tr.

Lorraine Wilson: 70br; **Peter Wilson:** 24c, 65bl, 71br, 292cla; **Brian Woodyatt:** 22b.

Front Endpaper
4Corners: SIME/Johanna Huber Ltr; **Alamy Images:** Dennis MacDonald Ltc; Hercules Milas Lbc, Rcr, Rtc; Robert Harding World Imagery/ Tuul Lcl; **AWL Images:** Travel Pix Collection Lbl; **Getty Images:** J. Alemañ Rtr.

Jacket
Front and spine top - Getty Images: Cornelia Doerr.

All other images © Dorling Kindersley. For further information see: www.dkimages.com

Phrase Book

There is no universally accepted system for representing the modern Greek language in the Roman alphabet. The system of transliteration adopted in this guide is the one used by the Greek government. Though not yet fully applied throughout Greece, most of the street and place names have been transliterated according to this system. For Classical names this guide uses the k, os, on and f spelling, in keeping with the modern system of transliteration. In a few cases, such as Socrates, the more familiar Latin form has been used. Classical names do not have accents. Where a well-known English form of a name exists, such as Athens or Corfu, this has been used. Variations in transliteration are given in the index.

Guidelines for Pronunciation

The accent over Greek and transliterated words indicates the stressed syllable. In this guide the accent is not written over capital letters nor over monosyllables, except for question words and the conjunction ή (meaning "or"). In the right-hand "Pronunciation" column below, the syllable to stress is given in bold type.

On the following pages, the English is given in the left-hand column with the Greek and its transliteration in the middle column. The right-hand column provides a literal system of pronunciation and indicates the stressed syllable in bold.

The Greek Alphabet

Α α	A a	*arm*
Β β	V v	*vote*
Γ γ	G g	*year* (when followed by e and i sounds) **n**o (when followed by ξ or γ)
Δ δ	D d	*that*
Ε ε	E e	*egg*
Ζ ζ	Z z	*zoo*
Η η	I i	*believe*
Θ θ	Th th	*think*
Ι ι	I i	*believe*
Κ κ	K k	*kid*
Λ λ	L l	*land*
Μ μ	M m	*man*
Ν ν	N n	*no*
Ξ ξ	X x	*taxi*
Ο ο	O o	*fox*
Π π	P p	*port*
Ρ ρ	R r	*room*
Σ σ	S s	*sorry* (zero when followed by μ)
ς	s	(used at end of word)
Τ τ	T t	*tea*
Υ υ	Y y	*believe*
Φ φ	F f	*fish*
Χ χ	Ch ch	*loch* in most cases, but *he* when followed by a, e or i sounds
Ψ ψ	Ps ps	*maps*
Ω ω	O o	*fox*

Combinations of Letters

In Greek there are two-letter vowels that are pronounced as one sound:

Αι αι	Ai ai	*egg*
Ει ει	Ei ei	*believe*
Οι οι	Oi oi	*believe*
Ου ου	Ou ou	*lute*

There are also some two-letter consonants that are pronounced as one sound:

Μπ μπ	Mp mp	**b**ut, sometimes nu**mb**er in the middle of a word
Ντ ντ	Nt nt	**d**esk, sometimes u**nd**er in the middle of a word
Γκ γκ	Gk gk	**g**o, sometimes bi**ng**o in the middle of a word
Γξ γξ	nx	a**nx**iety
Τζ τζ	Tz tz	han**ds**
Τσ τσ	Ts ts	it'**s**
Γγ γγ	Gg gg	bi**ng**o

In an Emergency

Help!	Βοήθεια! *Voítheia*	vo-**ee**-theea
Stop!	Σταματήστε! *Stamatíste*	sta-ma-**tee**-steh

Call a doctor!	Φωνάξτε ένα γιατρό *Fonáxte éna giatró*	fo-**nak**-steh **e**-na ya-**tro**
Call an ambulance/ the police/the fire brigade!	Καλέστε το ασθενοφόρο/την αστυνομία/την πυροσβεστική *Kaléste to asthenofóro/tin astynomía/tin pyrosvestikí*	ka-**le**-steh to as-the-no-**fo**-ro/teen a-sti-no-**mia**/teen pee-ro-zve-stee-**kee**
Where is the nearest telephone/hospital/ pharmacy?	Πού είναι το πλησιέστερο τηλέφωνο/νοσοκο-μείο/φαρμακείο; *Poú eínai to plisiés-tero tiléfono/nosoko-meío/farmakeío?*	poo **ee**-ne to plee-see-**e**-ste-ro tee-**le**-pho-no/no-so-ko-**mee**-o/far-ma-**kee**-o?

Communication Essentials

Yes	Ναι *Nai*	neh
No	Όχι *Ochi*	**o**-chee
Please	Παρακαλώ *Parakaló*	pa-ra-ka-**lo**
Thank you	Ευχαριστώ *Efcharistó*	ef-cha-ree-**sto**
You are welcome	Παρακαλώ *Parakaló*	pa-ra-ka-**lo**
OK/alright	Εντάξει *Entáxei*	en-**dak**-zee
Excuse me	Με συγχωρείτε *Me synchoreíte*	me seen-cho-**ree**-teh
Hello	Γειά σας *Geiá sas*	yeea sas
Goodbye	Αντίο *Antío*	an-**dee**-o
Good morning	Καλημέρα *Kaliméra*	ka-lee-**me**-ra
Good night	Καληνύχτα *Kalinýchta*	ka-lee-**neech**-ta
Morning	Πρωί *Proí*	pro-**ee**
Afternoon	Απόγευμα *Apógevma*	a-**po**-yev-ma
Evening	Βράδυ *Vrádi*	vrath-i
This morning	Σήμερα το πρωί *Símera to proí*	see-me-ra to pro-**ee**
Yesterday	Χθές *Chthés*	chthes
Today	Σήμερα *Símera*	see-me-ra
Tomorrow	Αύριο *Avrio*	**av**-ree-o
Here	Εδώ *Edó*	ed-**o**
There	Εκεί *Ekeí*	e-**kee**
What?	Τι; *Tí?*	tee?
Why?	Γιατί; *Giatí?*	ya-tee?
Where?	Πού; *Poú?*	poo?
How?	Πώς; *Pós?*	pos?
Wait!	Περίμενε! *Perímene!*	pe-ree-me-neh

Useful Phrases

How are you?	Τί κάνεις;	tee ka-nees
	Tí kάneis?	
Very well, thank you	Πολύ καλά,	po-lee ka-la, ef-cha-ree-sto
	ευχαριστό	
	Poly kalá, efcharistó	
How do you do?	Πώς είστε;	pos ees-te?
	Pós eíste?	
Pleased to meet you	Χαίρω πολύ	che-ro po-lee
	Chaíro polý	
What is your name?	Πώς λέγεστε;	pos le-ye-ste?
	Pós légeste?	
Where is/are…?	Πού είναι;	poo ee-ne?
	Poú eínai?	
How far is it to…?	Πόσο απέχει…;	po-so a-pe-chee…?
	Póso apéchei…?	
How do I get to?	Πώς μπορώ να	pos bo-ro-na pa-o?
	πάω….;	
	Pós mporó na páo…?	
Do you speak English?	Μιλάτε Αγγλικά;	mee-la-te an-glee-ka?
	Miláte Angliká?	
I understand	Καταλαβαίνω	ka-ta-la-ve-no
	Katalavaíno	
I don't understand	Δεν καταλαβαίνω	then ka-ta-la-ve-no
	Den katalavaíno	
Could you speak slowly?	Μιλάτε λίγο πιο	mee-la-te lee-go pyo ar-ga pa-ra-ka-lo?
	αργά παρακαλώ;	
	Miláte ligo pio argá parakaló?	
I'm sorry	Με συγχωρείτε	me seen-cho-ree teh
	Me synchoreíte	
Does anyone have a key?	Έχει κανένας	e-chee ka-ne-nas klee-dee?
	κλειδί;	
	Echei kanénas kleidí?	

Useful Words

big	Μεγάλο	me-ga-lo
	Megálo	
small	Μικρό	mi-kro
	Mikró	
hot	Ζεστό	zes-to
	Zestó	
cold	Κρύο	kree-o
	Krýo	
good	Καλό	ka-lo
	Kaló	
bad	Κακό	ka-ko
	Kakó	
enough	Αρκετά	ar-ke-ta
	Arketá	
well	Καλά	ka-la
	Kalá	
open	Ανοιχτά	a-neech-ta
	Anoichtá	
closed	Κλειστά	klee-sta
	Kleistá	
left	Αριστερά	a-ree-ste-ra
	Aristerá	
right	Δεξιά	dek-see-a
	Dexiá	
straight on	Ευθεία	ef-thee-a
	Eftheía	
between	Ανάμεσα / Μεταξύ	a-na-me-sa/me-tak-see
	Anámesa / Metaxý	
on the corner of….	Στη γωνία του…	stee go-nee-a too
	Sti gonía tou…	
near	Κοντά	kon-da
	Kontá	
far	Μακριά	ma-kree-a
	Makriá	
up	Επάνω	e-pa-no
	Epáno	
down	Κάτω	ka-to
	Káto	
early	Νωρίς	no-rees
	Norís	
late	Αργά	ar-ga
	Argá	
entrance	Η είσοδος	ee ee-so-thos
	I eísodos	
exit	Η έξοδος	eee-kso-dos
	I éxodos	
toilet	Οι τουαλέτες /WC	ee-too-a-le-tes
	Oi toualétes / WC	
occupied/engaged	Κατειλημμένη	ka-tee-lee-me-nee
	Kateiliméni	
unoccupied/vacant	Ελεύθερη	e-lef-the-ree
	Eléftheri	

free/no charge	Δωρεάν	tho-re-an
	Doreán	
in/out	Μέσα/Έξω	me-sa/ek-so
	Mésa/ Exo	

Making a Telephone Call

Where is the nearest public telephone ?	Πού βρίσκεται ο πλησιέστερος τηλεφωνικός θάλαμος;	poo vrees-ke-teh o plee-see-e-ste-ros tee-le-fo-ni-kos tha-la-mos?
	Poú vrísketai o plisiésteros tilefonikós thálamos?	
I would like to place a long-distance call	Θα ήθελα να κάνω ένα υπεραστικό τηλεφώνημα	tha ee-the-la na ka-no e-na ee-pe-ra-sti-ko tee-le-fo-nee-ma
	Tha íthela na káno éna yperastikó tilefónima	
I would like to reverse the charges	Αα ήθελα να χρεώσω το τηλεφώνημα στον παραλήπτη	tha ee-the-la na chre-o-so to tee-le-fo-nee-ma ston pa-ra-lep-tee
	Tha íthela na chreóso to tilefónima ston paralípti	
I will try again later	Θα ξαναντηλε φωνήσω αργότερα	tha ksa-na-tee-le-fo-ni-so ar-go-te-ra
	Tha xanatilefoníso argótera	
Can I leave a message?	Μπορείτε να του αφήσετε ένα μήνυμα;	bo-ree-te na too a-fee-se-teh e-na mee-nee-ma?
	Mporeíte na tou afísete éna mínyma?	
Could you speak up a little please?	Μιλάτε δυνατότερα, παρακαλώ;	mee-la-teh dee-na-to-te-ra, pa-ra-ka-lo
	Miláte dynatótera, parakaló	
Local call	Τοπικό τηλεφώνημα	to-pi-ko tee-le-fo-nee-ma
	Topikó tilefónima	
Hold on	Περιμένετε	pe-ri-me-ne-teh
	Periménete	
OTE telephone office	Ο ΟΤΕ / Το τηλεφωνείο	o O-TE / To tee-le-fo-nee-o
	O OTE / To tilefoneío	
Phone box/kiosk	Ο τηλεφωνικός θάλαμος	o tee-le-fo-ni-kos tha-la-mos
	O tilefonikós thálamos	
Phone card	Η τηλεκάρτα	ee tee-le-kar-ta
	I tilekárta	

Shopping

How much does this cost?	Πόσο κάνει;	po-so ka-nee?
	Póso kánei?	
I would like….	Αα ήθελα…	tha ee-the-la…
	Tha íthela…	
Do you have….?	Έχετε…;	e-che-teh
	Echete…?	
I am just looking	Απλώς κοιτάω	a-plos kee-ta-o
	Aplós koitáo	
Do you take credit cards/travellers' cheques?	Δέχεστε πιστωτικές κάρτες/travellers' cheques;	the-ches-teh pee-sto-tee-kes kar-tes/ travellers' cheques?
	Décheste pistotikés kártes/travellers' cheques?	
What time do you open/close?	Ποτέ ανοίγετε/ κλείνετε;	po-teh a-nee-ye-teh/ klee-ne-teh?
	Póte anoígete/ kleínete?	
Can you ship this overseas?	Μπορείτε να το στείλετε στο εξωτερικό;	bo-ree-teh na to stee-le-teh sto e-xo-te-ree ko?
	Mporeíte na to steílete sto exoterikó?	
This one	Αυτό εδώ	af-to e-do
	Aftó edó	
That one	Εκείνο	e-kee-no
	Ekeíno	
expensive	Ακριβό	a-kree-vo
	Akrivó	
cheap	Φθηνό	fthee-no
	Fthinó	
size	Το μέγεθος	to me-ge-thos
	To mégethos	
white	Λευκό	lef-ko
	Lefkó	

black	Μαύρο **Μáνro**	*mav-ro*
red	Κόκκινο **Kókkino**	*ko-kee-no*
yellow	Κίτρινο **Kítrino**	*kee-tree-no*
green	Πράσινο **Prásino**	*pra-see-no*
blue	Μπλε **Mple**	*bleh*

Types of Shop

antique shop	Μαγαζί με αντίκες *Magazí me antíkes*	*ma-ga-zee me an-dee-kes*
bakery	Ο φούρνος *O foúrnos*	*o foor-nos*
bank	Η τράπεζα *I trápeza*	*ee tra-pe-za*
bazaar	Το παζάρι *To pazári*	*to pa-za-ree*
bookshop	Το βιβλιοπωλείο *To vivliopoleío*	*to vee-vlee-o-po-lee-o*
butcher	Το κρεοπωλείο *To kreopoleío*	*to kre-o-po-lee-o*
cake shop	Το ζαχαροπλαστείο *To zacharoplasteío*	*to za-cha-ro-plastee-o*
cheese shop	Μαγαζί με αλλαντικά *Magazí me allantiká*	*ma-ga-zee me a-lan-dee-ka*
department store	Πολυκάτάστημα *Polykátástima*	*Po-lee-ka-ta-stee-ma*
fishmarket	Το ιχθυοπωλείο/ ψαράδικο *To ichthyopoleío/ psarádiko*	*to eech-thee-o-po-lee-o/psa-ra-dee-ko*
greengrocer	Το μανάβικο *To manáviko*	*to ma-na-vee-ko*
hairdresser	Το κομμωτήριο *To kommotírio*	*to ko-mo-tee-ree-o*
kiosk	Το περίπτερο *To períptero*	*to pe-reep-te-ro*
leather shop	Μαγαζί με δερμάτινα είδη *Magazí me dermátina eídi*	*ma-ga-zee me ther-ma-tee-na ee-thee*
street market	Η λαϊκή αγορά *I laïkí agorá*	*ee la-ee-kee a-go-ra*
newsagent	Ο εφημεριδοπώλης *O efimeridopólis*	*O e-fee-me-ree-tho-po-lees*
pharmacy	Το φαρμακείο *To farmakeío*	*to far-ma-kee-o*
post office	Το ταχυδρομείο *To tachydromeío*	*to ta-chee-thro-mee-o*
shoe shop	Κατάστημα υποδημάτων *Katástima ypodimáton*	*ka-ta-stee-ma ee-po-dee-ma-ton*
souvenir shop	Μαγαζί με "souvenir" *Magazí me "souvenir"*	*ma-ga-zee meh "souvenir"*
supermarket	Σουπερμάρκετ/ Υπεραγορά *"Supermarket"/ Yperagorá*	*"Supermarket"/ ee-per-a-go-ra*
tobacconist	Είδη καπνιστού *Eídi kapnistoú*	*Ee-thee kap-nees-too*
travel agent	Το ταξειδιωτικό γραφείο *To taxeidiotikó grafeío*	*to tak-see-thy-o-tee-ko gra-fee-o*

Sightseeing

tourist information	Ο ΕΟΤ *O EOT*	*o E-OT*
tourist police	Η τουριστική αστυνομία *I touristikí astynomía*	*ee too-rees-tee-kee a-stee-no-mee-a*
archaeological	αρχαιολογικός *archaiologikós*	*ar-che-o-lo-yee-kos*
art gallery	Η γκαλερί *I gkalerí*	*ee ga-le-ree*
beach	Η παραλία *I paralía*	*ee pa-ra-lee-a*
Byzantine	βυζαντινός *vyzantinós*	*vee-zan-dee-nos*
castle	Το κάστρο *To kástro*	*to ka-stro*
cathedral	Η μητρόπολη *I mitrópoli*	*ee mee-tro-po-lee*

cave	Το σπήλαιο *To spílaio*	*to spee-le-o*
church	Η εκκλησία *I ekklisía*	*ee e-klee-see-a*
folk art	λαϊκή τέχνη *laïkí téchni*	*la-ee-kee tech-nee*
fountain	Το συντριβάνι *To syntriváni*	*to seen-dree-va-nee*
hill	Ο λόφος *O lófos*	*o lo-fos*
historical	ιστορικός *istorikós*	*ee-sto-ree-kos*
island	Το νησί *To nisí*	*to nee-see*
lake	Η λίμνη *I límni*	*ee leem-nee*
library	Η βιβλιοθήκη *I vivliothíki*	*ee veev-lee-o-thee-kee*
mansion	Η έπαυλις *I épavlis*	*eee-pav-lees*
monastery	Μονή *moní*	*mo-ni*
mountain	Το βουνό *To vounó*	*to voo-no*
municipal	δημοτικός *dimotikós*	*thee-mo-tee-kos*
museum	Το μουσείο *To mouseío*	*to moo-see-o*
national	εθνικός *ethnikós*	*eth-nee-kos*
park	Το πάρκο *To párko*	*to par-ko*
garden	Ο κήπος *O kípos*	*o kee-pos*
gorge	Το φαράγγι *To farángi*	*to fa-ran-gee*
grave of….	Ο τάφος του… *O táfos tou…*	*o ta-fos too*
river	Το ποτάμι *To potámi*	*to po-ta-mee*
road	Ο δρόμος *O drómos*	*o thro-mos*
saint	άγιος/άγιοι/αγία /αγίες *ágios/ágioi/agía/agies*	*a-yee-os/a-yee-ee/a-yee-a/a-yee-es*
spring	Η πηγή *I pigí*	*ee pee-yee*
square	Η πλατεία *I plateía*	*ee pla-tee-a*
stadium	Το στάδιο *To stádio*	*to sta-thee-o*
statue	Το άγαλμα *To ágalma*	*toa-gal-ma*
theatre	Το θέατρο *To théatro*	*to the-a-tro*
town hall	Το δημαρχείο *To dimarcheío*	*To thee-mar-chee-o*
closed on public holidays	κλειστό τις αργίες *kleistó tis argíes*	*klee-sto tees aryee-es*

Transport

When does the …. leave?	Πότε φεύγει το ….; *Póte févgei to…?*	*po-teh fev-yee to…?*
Where is the bus stop?	Πού είναι η στάση του λεωφορείου; *Poú eínai i stási tou leoforeíou?*	*poo ee-eneh ee sta-see too le-o-fo-ree-oo?*
Is there a bus to…?	Υπάρχει λεωφορείο για….; *Ypárchei leoforeío gia…?*	*ee-par-chee le-o-fo-ree-o yia…?*
ticket office	Εκδοτήρια εισιτηρίων *Ekdotíria eisitiríon*	*Ek-tho-tee-reea ee-see-tee-ree-on*
return ticket	Εισιτήριο με επιστροφή *Eisitírio me epistrofí*	*ee-see-tee-ree-o meh e-pee-stro-fee*
single journey	Απλό εισιτήριο *Apló eisitírio*	*a-plo ee-see-tee-reeo*
bus station	Ο σταθμός λεωφορείων *O stathmós leoforeíon*	*o stath-mos leo-fo-ree-on*
bus ticket	Εισιτήριο λεωφορείου *Eisitírio leoforeíou*	*ee-see-tee-ree-o leo-fo-ree-oo*
trolley bus	Το τρόλλεϋ *To trólley*	*to tro-le-ee*
port	Το λιμάνι *To limáni*	*to lee-ma-nee*

train/metro	Το τρένο *To tréno*	*to tre-no*
railway station	Ο σιδηροδρομικός σταθμός *sidirodromikós stathmós*	*see-thee-ro-thro-mee-kos stath-mos*
moped	Το μοτοποδήλατο / το μηχανάκι *To motopodílato / To michanáki*	*to mo-to-po-thee-la-to/to mee-cha-na-kee*
bicycle	Το ποδήλατο *To podílato*	*to po-thee-la-to*
taxi	Το ταξί *To taxí*	*to tak-see*
airport	Το αεροδρόμιο *To aerodrómio*	*to a-e-ro-thro-mee-o*
ferry	Το φερυμπώτ *To "ferry-boat"*	*to fe-ree-bot*
hydrofoil	Το δελφίνι / Το υδροπτέρυγο *To delfíni / To ydroptérygo*	*to del-fee-nee / To ee-throp-te-ree-go*
catamaran	Το καταμαράν *To katamarán*	*to catamaran*
for hire	Ενοικιάζονται *Enoikiázontai*	*e-nee-kya-zon-deh*

Staying in a Hotel

Do you have a vacant room?	Έχετε δωμάτια; *Echete domátia?*	*ee-che-teh tho-ma-tee-a?*
double room with double bed	Δίκλινο με διπλό κρεβάτι *Díklino me dipló kreváti*	*thee-klee-no meh thee-plo kre-va-tee*
twin room	Δίκλινο με μονά κρεβάτια *Díklino me moná krevátia*	*thee-klee-no meh mo-na kre-vat-ya*
single room	Μονόκλινο *Monóklino*	*mo-no-klee-no*
room with a bath	Δωμάτιο με μπάνιο *Domátio me mpánio*	*tho-ma-tee-o meh ban-yo*
shower	Το ντουζ *To douz*	*To dooz*
porter	Ο πορτιέρης *O portiéris*	*o por-tye-rees*
key	Το κλειδί *To kleidí*	*to klee-dee*
I have a reservation	Έχω κάνει κράτηση *Echo kánei krátisi*	*e-cho ka-nee kra-tee-see*
room with a sea view/balcony	Δωμάτιο με θέα στη θάλασσα/μπαλκόνι *Domátio me théa stí thálassa/balpkóni*	*tho-ma-tee-o meh the-a stee tha-la-sa/bal-ko-nee*
Does the price include breakfast?	Το πρωινό συμπεριλαμβάνεται στην τιμή; *To proïnó symperi-lamvánetai stin timí?*	*to pro-ee-no seem-be-ree-lam-va-ne-teh steen tee-mee?*

Eating Out

Have you got a table?	Έχετε τραπέζι; *Echete trapézi?*	*e-che-te tra-pe-zee?*
I want to reserve a table	Θέλω να κρατήσω ένα τραπέζι *Thélo na kratíso éna trapézi*	*the-lo na kra-tee-so e-na tra-pe-zee*
The bill, please	Τον λογαριασμό, παρακαλώ *Ton logariazmó parakaló*	*ton lo-gar-yas-mo pa-ra-ka-lo*
I am a vegetarian	Είμαι χορτοφάγος *Eímai chortofágos*	*ee-meh chor-to-fa-gos*
What is fresh today?	Τί φρέσκο έχετε σήμερα; *Ti frésko échete símera?*	*tee fres-ko e-che-teh see-me-ra?*
waiter/waitress	Κύριε / Γκαρσόν / Κυρία (female) *Kýrie/Garson'/Kyría*	*Kee-ree-eh/Gar-son/Kee-ree-a*
menu	Ο κατάλογος *O katálogos*	*o ka-ta-lo-gos*
cover charge	Το κουβέρ *To "couvert"*	*to koo-ver*
wine list	Ο κατάλογος με τα οινοπνευματώδη *O katálogos me ta oinopnevmatódi*	*o ka-ta-lo-gos meh ta ee-no-pnev-ma-to-thee*

glass	Το ποτήρι *To potíri*	*to po-tee-ree*
bottle	Το μπουκάλι *To mpoukáli*	*to bou-ka-lee*
knife	Το μαχαίρι *To machaíri*	*to ma-che-ree*
fork	Το πηρούνι *To piroúni*	*to pee-roo-nee*
spoon	Το κουτάλι *To koutáli*	*to koo-ta-lee*
breakfast	Το πρωινό *To proïnó*	*to pro-ee-no*
lunch	Το μεσημεριανό *To mesimerianó*	*to me-see-mer-ya-no*
dinner	Το δείπνο *To deípno*	*to theep-no*
main course	Το κυρίως γεύμα *To kyríos gévma*	*to kee-ree-os yev-ma*
starter/first course	Τα ορεκτικά *Ta orektiká*	*ta o-rek-tee-ka*
dessert	Το γλυκό *To glykó*	*to ylee-ko*
dish of the day	Το πιάτο της ημέρας *To piáto tis iméras*	*to pya-to tees ee-me-ras*
bar	Το μπαρ *To "bar"*	*To bar*
taverna	Η ταβέρνα *I tavérna*	*ee ta-ver-na*
café	Το καφενείο *To kafeneío*	*to ka-fe-nee-o*
fish taverna	Η ψαροταβέρνα *I psarotavérna*	*ee psa-ro-ta-ver-na*
grill house	Η ψησταριά *I psistariá*	*ee psee-sta-rya*
wine shop	Το οινοπωλείο *To oinopoleío*	*to ee-no-po-lee-o*
dairy shop	Το γαλακτοπωλείο *To galaktopoleío*	*to ga-lak-to-po-lee-o*
restaurant	Το εστιατόριο *To estiatório*	*to e-stee-a-to-ree-o*
ouzeri	Το ουζερί *To ouzerí*	*to oo-ze-ree*
meze shop	Το μεζεδοπωλείο *To mezedopoleío*	*To me-ze-do-po-lee-o*
take away kebabs	Το σουβλατζίδικο *To souvlatzídiko*	*To soo-vlat-zee-dee-ko*
rare	Ελάχιστα ψημένο *Eláchista psiméno*	*e-lach-ees-ta psee-me-no*
medium	Μέτρια ψημένο *Métria psiméno*	*met-ree-a psee-me-no*
well done	Καλοψημένο *Kalopsiméno*	*ka-lo-psee-me-no*

Basic Food and Drink

coffee	Ο καφές *O Kafés*	*o ka-fes*
with milk	με γάλα *me gála*	*me ga-la*
black coffee	σκέτος *skétos*	*ske-tos*
without sugar	χωρίς ζάχαρη *chorís záchari*	*cho-rees za-cha-ree*
medium sweet	μέτριος *métrios*	*me-tree-os*
very sweet	γλυκύς *glykýs*	*glee-kees*
tea	τσάι *tsái*	*tsa-ee*
hot chocolate	ζεστή σοκολάτα *zestí sokoláta*	*ze-stee so-ko-la-ta*
wine	κρασί *krasí*	*kra-see*
red	κόκκινο *kókkino*	*ko-kee-no*
white	λευκό *lefkó*	*lef-ko*
rosé	ροζέ *rozé*	*ro-ze*
raki	Το ρακί *To rakí*	*to ra-kee*
ouzo	Το ούζο *To oúzo*	*to oo-zo*
retsina	Η ρετσίνα *I retsína*	*ee ret-see-na*
water	Το νερό *To neró*	*to ne-ro*
octopus	Το χταπόδι *To chtapódi*	*to chta-po-dee*
fish	Το ψάρι *To psári*	*to psa-ree*

cheese	**Το τυρί**	*to tee-ree*
	To tyrí	
halloumi	**Το χαλούμι**	*to cha-loo-mee*
	To chaloúmi	
feta	**Η φέτα**	*ee fe-ta*
	I féta	
bread	**Το ψωμί**	*to pso-mee*
	To psomí	
bean soup	**Η φασολάδα**	*ee fa-so-la-da*
	I fasoláda	
houmous	**Το χούμους**	*to choo-moos*
	To houmous	
halva	**Ο χαλβάς**	*o chal-vas*
	O chalvás	
meat kebabs	**Ο γύρος**	*o yee-ros*
	O gýros	
Turkish delight	**Το λουκούμι**	*to loo-koo-mee*
	To loukoúmi	
baklava	**Ο μπακλαβάς**	*o bak-la-vas*
	O mpaklavás	
klephtiko	**Το κλέφτικο**	*to klef-tee-ko*
	To kléftiko	

Numbers

1	**ένα**	*e-na*
	éna	
2	**δύο**	*thee-o*
	dýo	
3	**τρία**	*tree-a*
	tría	
4	**τέσσερα**	*te-se-ra*
	téssera	
5	**πέντε**	*pen-deh*
	pénte	
6	**έξι**	*ek-si*
	éxi	
7	**επτά**	*ep-ta*
	eptá	
8	**οχτώ**	*och-to*
	ochtó	
9	**εννέα**	*e-ne-a*
	ennéa	
10	**δέκα**	*the-ka*
	déka	
11	**έντεκα**	*en-de-ka*
	énteka	
12	**δώδεκα**	*tho-the-ka*
	dódeka	
13	**δεκατρία**	*de-ka-tree-a*
	dekatría	
14	**δεκατέσσερα**	*the-ka-tes-se-ra*
	dekatéssera	
15	**δεκαπέντε**	*the-ka-pen-de*
	dekapénte	
16	**δεκαέξι**	*the-ka-ek-si*
	dekaéxi	
17	**δεκαεπτά**	*the-ka-ep-ta*
	dekaeptá	
18	**δεκαοχτώ**	*the-ka-och-to*
	dekaochtó	
19	**δεκαεννέα**	*the-ka-e-ne-a*
	dekaennéa	
20	**είκοσι**	*ee-ko-see*
	eíkosi	
21	**εικοσιένα**	*ee-ko-see-e-na*
	eikosiéna	
30	**τριάντα**	*tree-an-da*
	triánta	
40	**σαράντα**	*sa-ran-da*
	saránta	
50	**πενήντα**	*pe-neen-da*
	penínta	
60	**εξήντα**	*ek-seen-da*
	exínta	
70	**εβδομήντα**	*ev-tho-meen-da*
	evdomínta	
80	**ογδόντα**	*og-thon-da*
	ogdónta	
90	**ενενήντα**	*e-ne-neen-da*
	enenínta	

100	**εκατό**	*e-ka-to*
	ekató	
200	**διακόσια**	*thya-kos-ya*
	diakósia	
1,000	**χίλια**	*cheel-ya*
	chília	
2,000	**δύο χιλιάδες**	*thee-o cheel-ya-thes*
	dýo chiliádes	
1,000,000	**ένα εκατομμύριο**	*e-na e-ka-to-mee-ree-o*

Time, Days and Dates

one minute	**ένα λεπτό**	*e-na lep-to*
	éna leptó	
one hour	**μία ώρα**	*mee-a o-ra*
	mía óra	
half an hour	**μισή ώρα**	*mee-see o-ra*
	misí óra	
quarter of an hour	**ένα τέταρτο**	*e-na te-tar-to*
	éna tétarto	
half past one	**μία και μισή**	*mee-a keh mee-see*
	mía kai misí	
quarter past one	**μία και τέταρτο**	*mee-a keh te-tar-to*
	mía kai tétarto	
ten past one	**μία και δέκα**	*mee-a keh the-ka*
	mía kai déka	
quarter to two	**δύο παρά τέταρτο**	*thee-o pa-ra te-tar-to*
	dýo pará tétarto	
ten to two	**δύο παρά δέκα**	*thee-o pa-ra the-ka*
	dýo pará déka	
a day	**μία μέρα**	*mee-a me-ra*
	mía méra	
a week	**μία εβδομάδα**	*mee-a ev-tho-ma-tha*
	mía evdomáda	
a month	**ένας μήνας**	*e-nas mee-nas*
	énas mínas	
a year	**ένας χρόνος**	*e-nas chro-nos*
	énas chrónos	
Monday	**Δευτέρα**	*thef-te-ra*
	Deftéra	
Tuesday	**Τρίτη**	*tree-tee*
	Tríti	
Wednesday	**Τετάρτη**	*te-tar-tee*
	Tetárti	
Thursday	**Πέμπτη**	*pemp-tee*
	Pémpti	
Friday	**Παρασκευή**	*pa-ras-ke-vee*
	Paraskeví	
Saturday	**Σάββατο**	*sa-va-to*
	Sávvato	
Sunday	**Κυριακή**	*keer-ee-a-kee*
	Kyriakí	
January	**Ιανουάριος**	*ee-a-noo-a-ree-os*
	Ianouários	
February	**Φεβρουάριος**	*fev-roo-a-ree-os*
	Fevrouários	
March	**Μάρτιος**	*mar-tee-os*
	Mártios	
April	**Απρίλιος**	*a-pree-lee-os*
	Aprílios	
May	**Μάιος**	*ma-ee-os*
	Máios	
June	**Ιούνιος**	*ee-oo-nee-os*
	Ioúnios	
July	**Ιούλιος**	*ee-oo-lee-os*
	Ioúlios	
August	**Αύγουστος**	*av-goo-stos*
	Ávgoustos	
September	**Σεπτέμβριος**	*sep-tem-vree-os*
	Septémvrios	
October	**Οκτώβριος**	*ok-to-vree-os*
	Októvrios	
November	**Νοέμβριος**	*no-em-vree-os*
	Noémvrios	
December	**Δεκέμβριος**	*the-kem-vree-os*
	Dekémvrios	

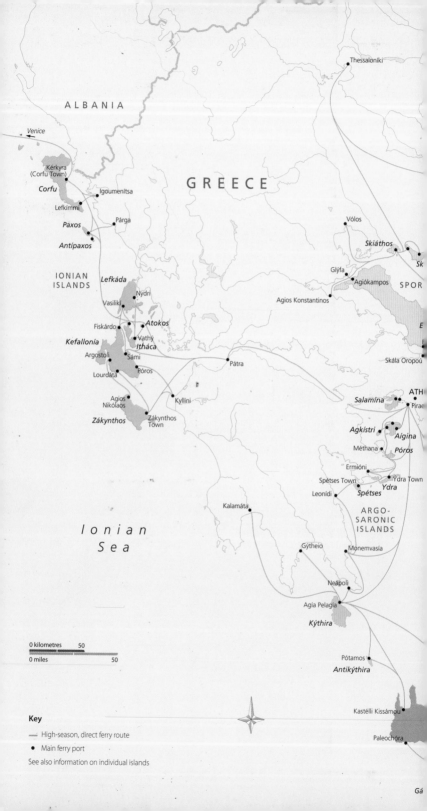